Electrical
Wiring
Commercial

Based on the 2005
NATIONAL ELECTRICAL CODE®

Electrical Wiring
Commercial
Twelfth Edition

Ray C. Mullin   Robert L. Smith

THOMSON
─────────
DELMAR LEARNING™

Australia Canada Mexico Singapore Spain United Kingdom United States

THOMSON

DELMAR LEARNING

Electrical Wiring Commercial, Twelfth Edition
Ray C. Mullin and Robert L. Smith

Vice President, Technology and Trades SBU:
Alar Elken

Editorial Director:
Sandy Clark

Sr. Acquisitions Editor:
Stephen Helba

Developmental Editor:
Jennifer A. Thompson

Marketing Director:
Dave Garza

Channel Manager:
Dennie Williams

Production Director:
Mary Ellen Black

Production Editor:
Barbara L. Diaz

Technology Project Manager:
Kevin Smith

Technology Project Specialist:
Linda Verde

Editorial Assistant:
Dawn Daugherty

Library of Congress Cataloging-in-Publication Data:
Mullin, Ray C.
  Electrical wiring: commercial / Ray C. Mullin and Robert L. Smith.—12th ed.
    p. cm.
  "Based on the 2005 National Electrical Code."
  Includes index.
  ISBN 1-4018-5206-8
  1. Commercial buildings—Electric equipment. 2. Electric wiring—Insurance requirements. I. Smith, Robert L., 1926– II. Title.
  TK3283.M84 2005
  621.319'24—dc22

2004019504

NOTICE TO THE READER

Publisher does not warrant or guarantee any of the products described herein or perform any independent analysis in connection with any of the product information contained herein. Publisher does not assume, and expressly disclaims, any obligation to obtain and include information other than that provided to it by the manufacturer.

The reader is expressly warned to consider and adopt all safety precautions that might be indicated by the activities herein and to avoid all potential hazards. By following the instructions contained herein, the reader willingly assumes all risks in connection with such instructions.

The publisher makes no representation or warranties of any kind, including but not limited to, the warranties of fitness for particular purpose or merchantability, nor are any such representations implied with respect to the material set forth herein, and the publisher takes no responsibility with respect to such material. The publisher shall not be liable for any special, consequential, or exemplary damages resulting, in whole or part, from the readers' use of, or reliance upon, this material.

CONTENTS

Sheet A1   Plot Plan, East Elevation, West Elevation
Sheet A2   Architectural Floor Plan: Basement
Sheet A3   Architectural Floor Plan: First Floor
Sheet A4   Architectural Floor Plan: Second Floor
Sheet A5   Elevations: North and South
Sheet A6   Sections: Longitudinal, Transverse
Sheet E1   Electrical Working Drawing: Basement
Sheet E2   Electrical Working Drawing: Entry Level
Sheet E3   Electrical Working Drawing: Upper Level
Sheet E4   Luminaire-Lamp Schedule, Electrical Symbol Schedule,
            Electrical Power Distribution Diagram, Details of,
            Roof-Type Cooling System Unit, Panelboard Summary

PREFACE

INTENDED USE AND LEVEL

Electrical Wiring: Commercial is intended for use in commercial wiring courses at two-year and four-year colleges, as well as in apprenticeship training programs. The text provides the basics of commercial wiring by offering insight into the planning of a typical commercial installation, carefully demonstrating how the load requirements are converted into branch-circuit, then to feeders, and finally into the building's main electrical service. An accompanying set of plans at the back of the book allows the reader to step through the wiring process by applying concepts learned in each chapter to an actual commercial building, in order to understand and meet *Code* requirements set forth by the *National Electrical Code.*®

SUBJECT AND APPROACH

The twelfth edition of *Electrical Wiring: Commercial* is based on the 2005 *National Electrical Code.*®* The new edition thoroughly and clearly explains the *NEC*® changes that relate to typical commercial wiring.

The *National Electrical Code*® is used as the basic standard for the layout and construction of electrical systems. To gain the greatest benefit from this text, the learner must use the *National Electrical Code*® on a continuing basis.

State and local codes may contain modifications of the *National Electrical Code*® to meet local requirements. The instructor is encouraged to furnish students with any variations from the *NEC*,® as they affect this commercial installation in a specific area.

This book takes the learner through the essential minimum requirements as set forth in the *National Electrical Code*® for commercial installations. In addition to *Code* minimums, the reader will find such information above and beyond the minimum requirements.

The commercial electrician is required to work in three common situations: where the work is planned in advance, where there is no advance planning, and where repairs are needed. The first situation exists when the work is designed by a consulting engineer. In this case, the electrician must know the installation procedures, must be able to read plans, and must be able to understand and interpret specifications. The second situation occurs either during or after construction when changes or remodeling are required. The third situation arises any time after a system is installed. Whenever a problem occurs with an installation, the electrician must understand the operation of all equipment included in the installation in order to solve the problem.

---

* *National Electrical Code*® and *NEC*® are registered trademarks of the National Fire Protection Association, Inc., Quincy, MA 02269. Applicable tables and section references are reprinted with permission from NFPA – 2002, the *National Electrical Code*, Copyright © 2005, National Fire Protection Association, Quincy, Massachusetts 02269. This reprinted material is not the complete and official position of the National Fire Protection Associaion on the referenced subject, which is represented only by the standard in its entirety.

When the electrician is working on the initial installation or is modifying an existing installation, the circuit loads must be determined. Thorough explanations and numerous examples of calculating these loads help prepare the reader for similar problems on the job. The text and assignments make frequent reference to the commercial building drawings at the back of the book.

> Readers should be aware that many of the electrical loads used as *examples in the text were contrived in order to create* Code *problems*. The authors' purpose in putting this building together is to demonstrate, and thus enhance learning, as many Code problems as possible. As an example, there is a single-phase feeder to the Doctor's office. This could have been a three-phase feeder similar to those in the other occupancies. However, using the single-phase feeder allows us to demonstrate many additional Code applications.
>
> *This is not a typical or an ideal design for a commercial building.* It is a composite to demonstrate a range of *Code* applications. The authors also carry many calculations to a higher level of accuracy as compared to the accuracy required in many actual job situations. This is done to demonstrate the correct method. Then, if the reader wants to back off from this level, based upon installation requirements, it can be done intelligently.

FEATURES

- **Safety** is emphasized throughout the book and fully covered in the first chapter. Special considerations in working with electricity, such as how to avoid arc flash, as well as guidelines for safe practices, provide readers with an overview of what dangers are to be expected on the job.

- **Commercial Building Drawings** are included in the back of the book, offering readers the opportunity to apply the concepts that they have learned in each chapter as they step through the wiring process. A description of working drawings and an explanation of symbols can be found in the first chapter.

- *National Electrical Code* references are integrated throughout the chapters, familiarizing readers with the requirements of the *Code* and including explanations of the wiring applications. Revisions to the *NEC* between the 2002 and 2005 editions are carefully highlighted.

- **Review Questions** at the end of each chapter allow readers to test what they have learned in each chapter and to target any sections that require further review.

NEW TO THIS EDITION

Every *Code* reference in the twelfth edition of *Electrical Wiring: Commercial* is the result of comparing each and every past *Code* reference with the 2005 *NEC.* As always, the authors review all comments submitted by instructors across the country, making corrections and additions to the text as suggested. The input from current users of the text ensures that what is covered is what electricians need to know.

1. ►As with all previous editions of this text, all *National Electrical Code* references have been updated to the 2005 *NEC.* Changes between the 2005 and 2002 editions of the *NEC* are marked with these symbols: ◄

2. The section on safety has been expanded to ensure safe practices and optimum working conditions on the job.

3. The *NEC*® use of the terms "compute, computed, and computation" has been replaced with "calculate, calculated, and calculations."

4. The section on arc-flash has been expanded.

5. Added text to clarify the depth of pull boxes.

6. Additional information and illustrations of "isolated ground receptacles."

7. Clarification of where GFCI protection for receptacles in commercial kitchens is required. This now includes employee break areas.

8. Added information on how to size conductors for motor branch-circuits.

9. Information on where to obtain computer programs for calculating motor circuits and arc-flash values.

10. Added information about reducing neutrals where nonlinear loads are involved.

11. More on calculating conductor size when derating, correcting, and adjusting factors are necessary.

12. Explanation of the meaning of emergency systems, legally required standby systems, and optional standby systems.

13. Explanation of why emergency systems and legally required standby systems' over-current protective devices must be selectively coordinated.

14. New requirement that controllers be marked with their short-circuit rating.

15. New requirement that renewable fuses are now only permitted as replacements on existing installations where there is no evidence of overfusing or tampering.

16. Deletion of all reference to Design E motors, as they are no longer manufactured.

17. Added information about K-Rated transformers for application where there is a heavy concentration of nonlinear loads such as computer equipment and electronic ballasts.

18. More information about energy saving lamps and ballasts.

19. Added coverage that circuit breakers for HVAC equipment—referred to as HACR breakers—are no longer subjected to special testing by the testing laboratories.

20. Additional text on series-rated systems, which are becoming more and more popular.

21. The *NEC*® has added technical formulas in Annex H regarding the potential damage to a conductor and the conductor's insulation because of heat generated under short-circuit conditions.

22. The *NEC*® now states that *the earth shall not be considered as an effective ground-fault current path*.

23. Low-voltage remote-control has been updated to reflect current systems.

24. Addition of more Web sites to research for technical information.

SUPPLEMENTS

- The Instructor's Guide contains answers to all review questions included in the book. (Order #: 1-4018-5155-X)

- Visit us at http://www.delmarelectric.com, now LIVE for the 2005 *Code* cycle! The newly designed Web site provides information on other learning materials offered by Delmar, as well as industry links, career profiles, job opportunities, and more!

ABOUT THE AUTHORS

This text was prepared by Ray C. Mullin and Robert L. Smith. Mr. Mullin is a former electrical circuit instructor for the Electrical Trades, Wisconsin Schools of Vocational, Technical and Adult Education. A former member of the International Brotherhood of Electrical Workers, Mr. Mullin is presently an honorary member of the International Association of Electrical Inspectors, an honorary member of the Institute of Electrical and Electronic Engineers, and the National Fire Protection Association, Electrical Section. He served on Code Making Panel 4 for the *National Electrical Code,* NFPA-70 for the National Fire Protection Association.

Mr. Mullin completed his apprenticeship training and has worked as a journeyman and supervisor. He has taught both day and night electrical apprentice and journeyman courses and has conducted engineering seminars. Mr. Mullin has contributed to and assisted other authors in their writing of texts and articles relating to overcurrent protection and conductor withstand ratings. He has had many articles relating to overcurrent protection published in various trade magazines.

Mr. Mullin attended the University of Wisconsin, Colorado State University, and Milwaukee School of Engineering.

He served on the Executive Board of the Western Section, International Association of Electrical Inspectors. He also served on their National Electrical Code Committee and on their Code Clearing Committee. He also serves on the Electrical Commission in his hometown. Mr. Mullin has conducted many technical Code workshops and seminars at state chapter and section meetings of the International Association of Electrical Inspectors and serves on their Code panels.

Mr. Mullin is past Director, Technical Liaison, and Code Coordinator for a large electrical manufacturer and contributed to their technical publications.

Robert L. Smith, P. E., is an Emeritus Professor of Architecture at the University of Illinois where he taught courses in electrical systems, lighting design, acoustics and energy-conscious design. He worked in electrical construction for more than 25 years, and has taught electrical apprentice classes and courses in industrial electricity.

Professor Smith has been affiliated with a number of professional organizations, including the Illinois Society of Professional Engineers, the International Association of Electrical Inspectors, and the National Academy of Code Administration. He is a life-long honorary member of the International Brotherhood of Electrical Workers. He gave frequent public lectures on the application of the *National Electrical Code* and was a regular presenter at the University of Wisconsin Extension.

IMPORTANT NOTE

This edition of *Electrical Wiring: Commercial* was completed after all normal steps of revising the *National Electrical Code* NFPA 70 were taken, and before the actual issuance and publication of the 2005 edition of the *NEC.*

These steps include: NFPA solicits proposals for *2002 NEC,* interested parties submit proposals to NFPA, proposals sent to Code Making Panels (CMPs), proposals reviewed by CMPs and Technical Correlating Committee, Report on Proposals document published, interested parties submit comments on the proposals to NFPA, review of Comments by CMPs and Technical Correlating Committee, Report on Comments document published, review of all Proposals and Comments at NFPA Annual Meeting, new motions permitted to be made at the NFPA Annual Meeting. Finally, the Standard Council meets to review actions made at the NFPA Annual Meeting, and to authorize publication of the *NEC.*

Every effort has been made to be technically correct, but there is always the possibility of typographical errors, or appeals made to the NFPA Board of Directors after the normal review process that could result in reversal of previous decisions by the CMPs.

If changes in the *NEC*® do occur after the printing of this text, these changes will be incorporated in the next printing.

The National Fire Protection Association has a standard procedure to introduce changes between *NEC*® *Code* cycles after the actual *NEC*® is printed. These are called "Tentative Interim Amendments," or TIAs. TIAs and typographical errors can be downloaded from the NFPA website, www.nfpa.org, to make your copy of the *Code* current.

ACKNOWLEDGMENTS

The authors and Publisher wish to thank the following reviewers for their contributions:

Warren DeJardin
Northeast Wisconsin Technical College
Greenbay, WI

Greg Fletcher
Kennebec Valley Technical College
Fairfield, ME

David Gehlauf
Tri-County Vocational School
Glouster, OH

Fred Johnson
Champlain Valley Tech
Plattsburg, NY

Thomas Lockett
Vatterott College
Quincy, IL

Gary Reiman
Dunwoody Institute
Minneapolis, MN

Special thanks to David Williams for his thorough technical review of the content, as well as to Lanny McMahill for his technical review of the *Code* material. Their dedication to ensuring the contents of the twelfth edition of this book are accurate and up to date deserves much praise.

Thanks is owed also to Robin Sterling, whose diligent followup on photo acquisition has allowed the authors to visually portray the latest trends in the industry.

In addition, thanks to Paul Dobrowsky for his tremendous input on the subject of grounding, and to Neil Matthes for his thorough technical review of the *Code* content.

UNIT 1

Commercial Building Plans and Specifications

OBJECTIVES

After studying this unit, the student should be able to

- understand the basic safety rules for working on electrical systems.
- define the project requirements from the contract documents.
- demonstrate the application of building plans and specifications.
- locate specific information on the building plans.
- obtain information from industry-related organizations.
- apply and interchange International System of Units (SI) and English measurements.

SAFETY IN THE WORKPLACE

Before we get started on our venture into the wiring of a typical commercial building, let us talk about safety.

Electricity is dangerous! Working on electrical equipment with the power turned on can result in death or serious injury, either as a direct result of electricity or from an indirect secondary reaction, such as falling off a ladder or falling into the moving parts of equipment. Dropping a metal tool onto live parts or allowing metal shavings from a drilling operation to fall onto live parts of electrical equipment generally results in an *arc blast*, which can cause deadly burns. The heat of an electrical arc has been determined to be hotter than the sun. Pressures developed during an arc blast can blow a person clear across the room. Dirt, debris, and moisture can also set the stage for catastrophic equipment failures and personal injury. Neatness and cleanliness in the workplace are a must.

The Federal regulations in the Occupational Safety and Health Act (OSHA) Number 29, Subpart S, in Part 1910.332 discusses the training needed for those who face the risk of electrical injury. Proper training means "trained in and familiar with the safety-related work practices required by paragraphs 1910.331 through 1910.335." Numerous texts are available that cover the OSHA requirements in great detail.

The *National Electrical Code®* (*NEC®*) defines a *qualified* person as "one familiar with the construction and operation of the equipment and the hazards involved." Merely telling someone or being told "be careful" does not meet the definition of proper training and does not make the person qualified.

Only qualified persons are permitted to work on or near exposed energized equipment. To become qualified, a person must

- have the skill and technique necessary to distinguish exposed live parts from other parts of electrical equipment.
- be able to determine the voltage of exposed live parts and be trained in the use of special precautionary techniques, such as personal protective equipment, insulations, shielding material, and insulated tools.

1

Subpart S, paragraph 1910.333 requires that safety-related work practices be employed to prevent electrical shock or other injuries resulting from either direct or indirect electrical contact. Live parts to which an employee may be exposed shall be de-energized before the employee works on or near them, unless the employer can demonstrate that de-energizing introduces additional or increased hazards.

Working on "live" equipment is acceptable only if there would be a greater hazard if the system were de-energized. Examples of this would be life-support systems, some alarm systems, certain ventilation systems in hazardous locations, and the power for critical illumination circuits. Working on energized equipment requires properly insulated tools, proper nonflammable clothing, rubber gloves, protective shields and goggles, and in some cases, rubber blankets.

OSHA regulations allow only qualified personnel to work on or near electrical circuits or equipment that has not been de-energized. The OSHA regulations provide rules regarding "lockout and tagging" to make sure that the electrical equipment being worked on will not inadvertently be turned on while someone is working on the supposedly dead equipment. As the OSHA regulations state, "a lock and a tag shall be placed on each disconnecting means used to de-energize circuits and equipment. . . . "

Some electricians' contractual agreements require that, as a safety measure, two or more qualified electricians must work together when working on energized circuits. They do not allow untrained apprentices to work on live equipment but do allow apprentices to stand back and observe.

According to *NFPA 70E, Standard for Electrical Safety in the Workplace*, circuits and conductors are not considered to be in an electrically safe condition until all sources of energy are removed, the disconnecting means is under lock out/tag out, and the absence of voltage is verified by an approved voltage tester.

Safety cannot be compromised.

Follow this rule: *turn off* and *lock off* the power and then properly tag the disconnect with a description as to exactly what that particular disconnect serves.

Arc Flash (Arc Blast)

An electrician should not get too complacent when working on electrical equipment. A major short circuit or ground fault at the main service panel, or at the meter cabinet or base, can deliver a lot of energy. On large electrical installations, an arc flash can generate temperatures of 35,000°F (19,427°C). This is hotter than the surface of the sun. This amount of heat will instantly melt copper, aluminum, and steel. The blast will blow hot particles of metal and hot gases all over, resulting in personal injury, fatality, or fire. An arc flash also creates a tremendous air-pressure wave that can cause serious ear damage or memory loss due to the concussion. The blast might blow the victim away from the arc source.

An electrician should not be fooled by the size of the service. Typical residential services are 100, 150, and 200 amperes! Larger services are found on large homes. Commercial installations might have very large services. The commercial building discussed in this text is served by three 500-kcmil (thousand circular mils) copper Type THHN/THWN conductors that total 1140 amperes.

Electricians seem to feel out of harm's way when working on small electrical systems and seem to be more cautious when working on commercial and industrial electrical systems. A fault at a small main service panel, however, can be just as dangerous as a fault on a large service. The available fault current at the main service disconnect for all practical purposes is determined by the kilovolt-ampere rating and impedance of the transformer. Other major limiting factors for fault current are the size, type, and length of the service-entrance conductors. If you want to learn more, visit Bussmann's Web site, http://www.bussmann.com. There you will find an easy-to-use computer program for making arc-flash and fault-current calculations.

Electricians should not be fooled into thinking that if they cause a fault on the load side of the main disconnect that the main breaker will trip off and protect them from an arc flash. An arc flash will release the energy that the system is capable of delivering, for as long as it takes the main circuit breaker to open. How much current (energy) the main breaker will let through depends on the available fault current and the breakers' opening time. A joke in the electrical trade is that a power company will sell power to you a little at a time—or all in one huge arc blast.

▶Although not required for house wiring, *NEC® 110.16* requires that switchboards, panelboards, industrial control panels, meter socket enclosures, and motor control centers in commercial and industrial installations that are likely to be worked on while energized shall be field-marked to warn qualified personnel of potential arc-flash hazard. This marking must be clearly visible to any qualified persons that might have to work on the equipment.◀

Figure 1-1 is an example of a commercially available label.

More information on this subject is found in *NFPA 70E* and in the ANSI Standard Z535.4, *Product Safety Signs and Labels.*

A good suggestion is that when turning a standard disconnect switch On, *don't* stand in front of the switch. Instead, stand to one side. For example, if the handle of the switch is on the right, then stand to the right of the switch, using your left hand to operate the handle of the switch, and turn your head away from the switch. That way, if an arc flash occurs when you turn the disconnect switch On, you will not be standing in front of the switch. You will not have the switch's door fly into your face, and the molten metal particles resulting from the arc flash will fly past you.

Where Do We Go Now?

With safety the utmost concern in our minds, let us begin our venture on the wiring of a typical commercial building.

Figure 1-1 Typical pressure sensitive arc flash and shock hazard label to be affixed to electrical equipment as required by the *National Electrical Code® 110.16.*

COMMERCIAL BUILDING SPECIFICATIONS

When a building project contract is awarded, the electrical contractor is given the plans and specifications for the building. These two contract documents govern the construction of the building. It is very important that the electrical contractor and the electricians employed by the contractor to perform the electrical construction follow the specifications exactly. The electrical contractor will be held responsible for any deviations from the specifications and may be required to correct such deviations or variations at personal expense. Thus, it is important that any changes or deviations be verified — in writing. Avoid verbal change orders.

It is suggested that the electrician assigned to a new project first read the specifications carefully. These documents provide the detailed information that will simplify the task of studying the plans. The specifications are usually prepared in book form and may consist of a few pages to as many as several hundred pages covering all phases of the construction. This text presents in detail only that portion of the specifications that directly affects the electrician; however, summaries of the other specification sections are presented to acquaint the electrician with the full scope of the document.

The specification is a book of rules governing all of the material to be used and the work to be performed on a construction project. The specification is usually divided into several sections.

General Clauses and Conditions

The first section of the specification, *General Clauses and Conditions*, deals with the legal requirements of the project. The index to this section may include the following headings:

Notice to Bidders
Schedule of Drawings
Instructions to Bidders
Proposal
Agreement
General Conditions

Some of these items will affect the electrician on the job and others will be of primary concern to the electrical contractor. The following paragraphs give a brief, general description of each item and how it affects either the electrician on the job or the contractor.

Notice to Bidders. This item is of value to the contractor and his estimator only. The notice describes the project, its location, the time and place of the bid opening, and where and how the plans and specifications can be obtained.

Schedule of Drawings. The schedule is a list, by number and title, of all of the drawings related to the project. The contractor, estimator, and electrician will each use this schedule prior to preparing the bid for the job: the contractor to determine whether all the drawings required are at hand, the estimator to do a take-off and to formulate a bid, and the electrician to determine whether all of the drawings necessary to do the installation are available.

Instructions to Bidders. This section provides the contractor with a brief description of the project, its location, and how the job is to be bid (lump sum, one contract, or separate contracts for the various construction trades, such as plumbing, heating, electrical, and general). In addition, bidders are told where and how the plans and specifications can be obtained prior to the preparation of the bid, how to make out the proposal form, where and when to deliver the proposal, the amount of any bid deposits required, any performance bonds required, and bidders' qualifications. Other specific instructions may be given, depending upon the particular job.

Proposal. The proposal is a form that is filled out by the contractor and submitted at the proper time and place. The proposal is the contractor's bid on a project. The form is the legal instrument that binds the contractor to the owner if: (a) the contractor completes the proposal properly, (b) the contractor does not forfeit the bid bond, (c) the owner accepts the proposal, and (d) the owner signs the agreement. Generally, only the contractor will be using this section.

The proposal may show that alternate bids were requested by the owner. In this case, the electrician on the job should study the proposal and consult with the contractor to learn which of the alternate bids has been accepted in order to determine the extent of the work to be completed.

On occasion, the proposal may include a specified time for the completion of the project. This information is important to the electrician on the job because the work must be scheduled to meet the completion date.

Agreement. The agreement is the legal binding portion of the proposal. The contractor and the owner sign the agreement, and the result is a legal contract. After the agreement is signed, both parties are bound by the terms and conditions given in the specification.

General Conditions. The following items are normally included under the "General Conditions" heading of the *General Clauses and Conditions*. A brief description is presented for each item:

- General Note: Includes the general conditions as part of the contract documents.

- Definition: As used in the contract documents, this item defines the owner, contractor, architect, engineer, and other people and objects involved in the project.

- Contract Documents: Lists the documents involved in the contract, including plans, specifications, and agreement.

- Insurance: Specifies the insurance a contractor must carry on all employees and on the materials involved in the project.

- Workmanship and Materials: Specifies that the work must be done by skilled workers and that the materials must be new and of good quality.

- Substitutions: Specifies that materials used must be as indicated or that equivalent materials must be shown to have the required properties.

- Shop Drawings: Identifies the drawings that must be submitted by the contractor to show how the specific pieces of equipment are to be installed.

- Payments: Specifies the method of paying the contractor during the construction.

- Coordination of Work: Specifies that each contractor on the job must cooperate with every other contractor to ensure that the final product is complete and functional.

- Correction Work: Describes how work must be corrected, at no cost to the owner, if any part of the job is installed improperly by the contractor.

- Guarantee: Guarantees the work for a certain length of time, usually one year.

- Compliance with All Laws and Regulations: Specifies that the contractor will perform all work in accordance with all required laws, ordinances, and codes, such as the *National Electrical Code*® and city codes.

- Others: These sections are added as necessary by the owner, architect, and engineer when the complexity of the job and other circumstances require them. None of the items listed in the "General Conditions" has precedence over another item in terms of its effect on the contractor or the electrician on the job. The electrician must study each of the items before taking a position and assuming responsibilities with respect to the job.

Supplementary General Conditions

The second main section of the specifications is titled *Supplementary General Conditions*. These conditions usually are more specific than the "General Conditions." Although the General Conditions can be applied to any job or project in almost any location with little change, the *Supplementary General Conditions* are rewritten for each project. The following list covers the items normally specified by the *Supplementary General Conditions*:

- The contractor must instruct all crews to exercise caution while digging; any utilities damaged during the digging must be replaced by the contractor responsible.

- The contractor must verify the existing conditions and measurements.

- The contractor must employ qualified individuals to lay out the work site accurately. A registered land surveyor or engineer may be part of the crew responsible for the layout work.

- Job offices are to be maintained as specified on the site by the contractor; this office space may include space for owner representatives.

- The contractor may be required to provide telephones at the project site for use by the architect, engineer, subcontractor, or owner.

- Temporary toilet facilities and water are to be provided by the contractor for the construction personnel.

- The contractor must supply an electrical service of a specified capacity to provide temporary light and power at the site.

- The contractor may have to supply a specified type of temporary heating to keep the temperature at the level specified for the structure.

- According to the terms of the guarantee, the contractor agrees to replace faulty equipment and correct construction errors for a period of one year.

The previous listing is by no means a complete catalog of all of the items that can be included in the section *Supplementary General Conditions*.

Other names may be applied to the *Supplementary General Conditions* section, including *Special Conditions* and *Special Requirements*. Regardless of the name used, these sections contain the same types of information. All sections of the specifications must be read and studied by all of the construction trades involved. In other words, the electrician must study the heating, plumbing, ventilating, air-conditioning, and general construction specifications to determine whether there is any equipment furnished by the other trades, where the contract specifies that such equipment is to be installed and wired by the electrical contractor. The electrician must also study the general construction specifications because the roughing in of the electrical system will depend on the types of construction that will be encountered in the building.

This overview of the *General Conditions* and *Supplementary General Conditions* of a specification is intended to show the student that the construction worker on the job is affected by parts of the specification other than the part designated for each particular trade.

Contractor Specification

In addition to the sections of the specification that apply to all contractors, separate sections exist for each of the contractors, such as the general contractor who constructs the building proper, the plumbing contractor who installs the water and sewage systems, the heating and air-conditioning contractor, and the electrical contractor. The contract documents usually do not make one contractor responsible for work specified in another section of the specifications. However, it is always considered good practice for each contractor to be aware of how they are involved in each of the other contracts in the total job.

WORKING DRAWINGS

The construction plans for a building are often called **blueprints**. This term is a carryover from the days when the plans were blue with white lines. Today, a majority of the plans used have black lines on white because this combination is considered easier to read. The terms *plans* and *working drawings* will be commonly used in this text.

A set of ten plan sheets is included at the back of the text showing the general and electrical portions of the work specified:

- *Sheet A1—Plot Plan, East Elevation, West Elevation, Index to Drawings:* The plot plan shows the location of the commercial building and gives needed elevations. The east elevation is the street view of the building, and the west elevation is the back of the building. The index lists the content of all the plan sheets.

- *Sheet A2—Architectural Floor Plan; Basement*

- *Sheet A3—Architectural Floor Plan; First Floor*

- *Sheet A4—Architectural Floor Plan; Second Floor*

  The architectural floor plans give the wall and partition details for the building. These sheets are dimensioned; the electrician can find exact locations by referring to these sheets. The electrician should also check the plans for the materials used in the general construction as these will affect when and how the system will be installed.

- *Sheet A5—Elevations; North and South:* The electrician must study the elevation dimensions, which are given in feet and hundredths of a foot above sea level. For example, the finished second floor, which is shown at 218.33', is 218 ft 4 in. above sea level.

- *Sheet A6—Sections; Longitudinal, Transverse:* This sheet gives detail drawings of the more important sections of the building. The location of the section is indicated on the floor plans. When looking at a section, imagine that you are looking in the direction of the arrows at a building that is cut in two at the place indicated. You should see the section exactly as you view the imaginary building from this point.

- *Sheet E1—Electrical Working Drawing; Basement, Owner—Emergency Panelboard Directory, Telephone Riser Diagram*

- *Sheet E2—Electrical Working Drawing; First Floor, Drug Store Panelboard Directory, Bakery Panelboard Directory*

- *Sheet E3—Electrical Working Drawing; Second Floor, Insurance Office Panelboard Directory, Beauty Salon Panelboard Directory, Doctor's Office Panelboard Directory*

  These sheets show the detailed electrical work on an outline of the building. Because dimensions usually are not shown on the electrical plans, the electrician must consult the other sheets for this information. It is recommended that the electrician refer frequently to the other plan sheets to ensure that the electrical installation does not conflict with the work of the other construction trades.

- *Sheet E4—Luminaire—Lamp Schedule, Electrical Symbol Schedule, Electrical Power Distribution Diagram, Detail of Typical Roof-Type Cooling System Unit, Panelboard Summary*

To assist the electrician in recognizing components used by other construction trades, the following illustrations are included: Figure 1-2A and Figure 1-2B, Architectural drafting symbols; Figure 1-3, Standard symbols for plumbing, piping, and valves; Figure 1-4, Sheet metal ductwork symbols; and Figure 1-5, Generic symbols for electrical plans. However, the electrician should be aware that variations of these symbols may be used, and the specification and/or plans for a specific project must always be consulted.

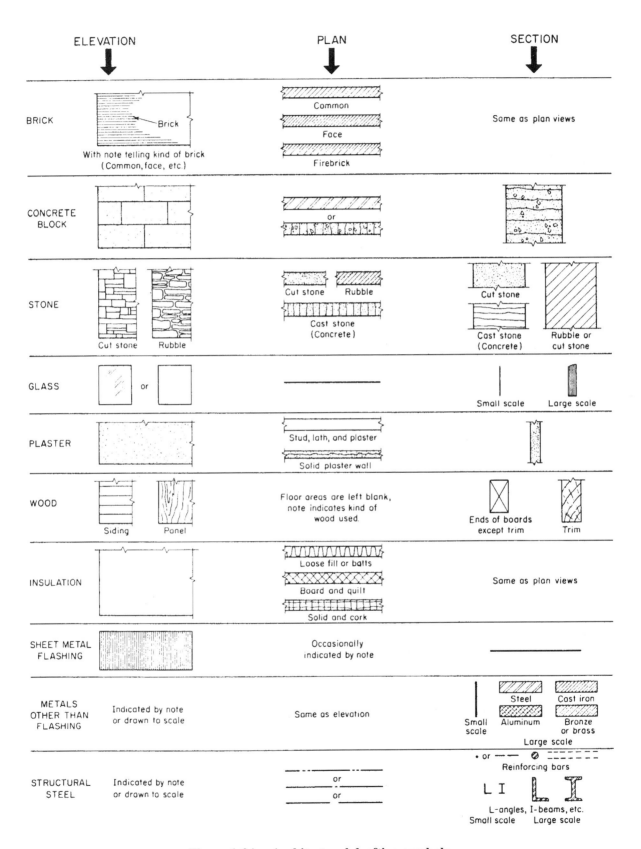

Figure 1-2A   Architectural drafting symbols.

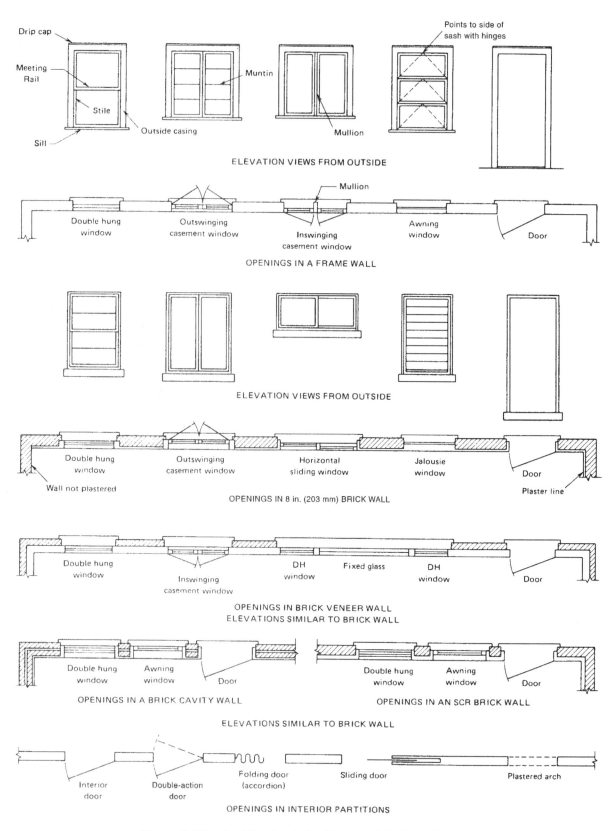

Figure 1-2B   Architectural drafting symbols (continued).

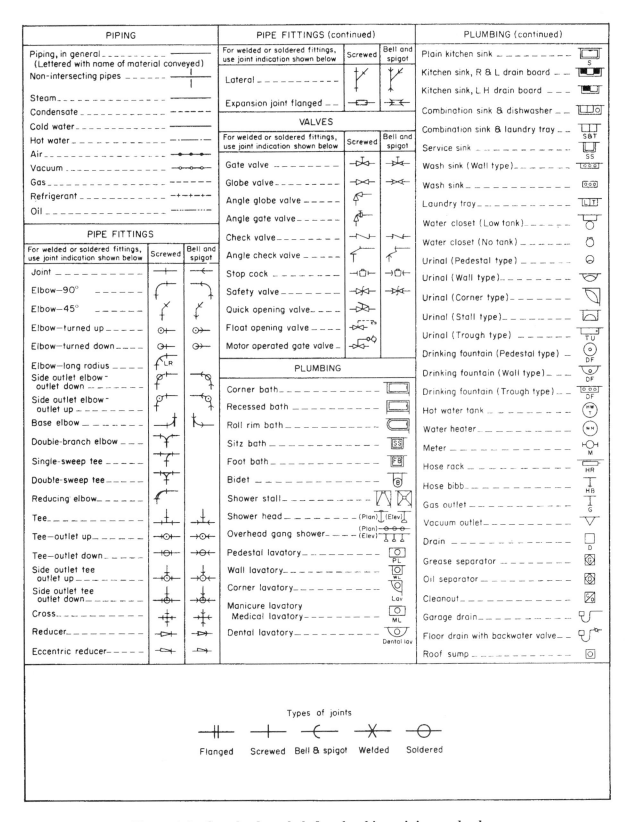

Figure 1-3 Standard symbols for plumbing, piping, and valves.

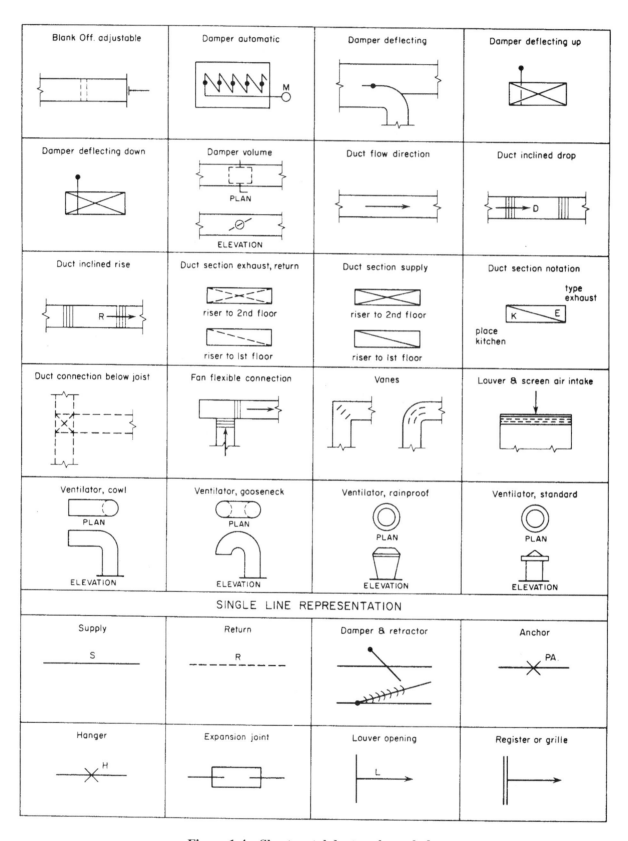

Figure 1-4   Sheet metal ductwork symbols.

GENERAL OUTLETS

CEILING   WALL

		Outlet.
⊕ⓑ	⊸ⓑ	Blanked Outlet.
ⓓ		Drop Cord.
ⓔ	⊸ⓔ	Electrical Outlet—for use only when circle used alone might be confused with columns, plumbing symbols, etc.
ⓕ	⊸ⓕ	Fan Outlet.
ⓙ	⊸ⓙ	Junction Box.
ⓛ	⊸ⓛ	Lampholder.
ⓛPS	⊸ⓛPS	Lampholder with Pull Switch.
ⓢ	⊸ⓢ	Pull Switch.
ⓥ	⊸ⓥ	Outlet for Vapor Discharge Lamp.
ⓧ	⊸ⓧ	Exit Light Outlet.
ⓒ	⊸ⓒ	Clock Outlet. (Specify Voltage).

RECEPTACLE OUTLETS

⊸⊖	Single Receptacle Outlet.
⊸⊜	Duplex Receptacle Outlet.
⊸⊜	Triple Receptacle Outlet.
⊸⊜	Duplex Receptacle Outlet, Split Circuit.
⊸⊜	Duplex Receptacle Outlet with NEMA 5-20R Receptacle.
⊸⊜w	Weatherproof Receptacle Outlet.
⊸⊜R	Range Receptacle Outlet.
⊸⊜s	Switch and Receptacle Outlet.
⊸⊖Ⓡ	Radio and Receptacle Outlet.
▲	Special Purpose Receptacle Outlet.
⊙	Floor Receptacle Outlet.

SWITCH SYMBOLS

S	Single-pole Switch.
S2	Double-pole Switch.
S3	Three-way Switch.
S4	Four-way Switch.
SD	Automatic Door Switch.
SE	Electrolier Switch.
SK	Key Operated Switch.
SP	Switch and Pilot Lamp.
SCB	Circuit Breaker.
SWCB	Weatherproof Circuit Breaker.
SMC	Momentary Contact Switch.
SRC	Remote Control Switch.
SWP	Weatherproof Switch.
SF	Fused Switch.
SWF	Weatherproof Fused Switch.

SPECIAL OUTLETS

○ a,b,c - etc.	Any Standard Symbol as given above with the addition of a lower case subscript letter may be used to designate some special variation of Standard Equipment of particular interest in a specific set of architectural plans.
⊸⊖ a,b,c - etc.	
S a,b,c - etc.	

When used they must be listed in the Key of Symbols on each drawing and if necessary further described in the specifications.

PANELS, CIRCUITS AND MISCELLANEOUS

▬	Lighting Panel.
▨	Power Panel.
——	Branch Circuit; Concealed in Ceiling or Wall.
– – –	Branch Circuit; Concealed in Floor.
- - - -	Branch Circuit; Exposed.
→→	Home Run to Panelboard. Indicate number of Circuits by number of arrows.

Note: Any circuit without further designation indicates a two-wire circuit. For a greater number of wires indicate as follows: ⫻ (3 wires) ⫻⫻ (4 wires), etc.

▬	Feeders. Note: Use heavy lines. It is recommended that feeders and homeruns with more than one set of circuits be labeled to identify the count and size of the conductors and the size of the raceway, such as "six 10 AWG, trade size ¾ EMT."
⊒⊏	Underfloor Duct and Junction Box. Triple System. Note: For double or single systems, eliminate one or two lines. This symbol is equally adaptable to auxiliary system layouts.
Ⓖ	Generator.
Ⓜ	Motor.
Ⓘ	Instrument.
Ⓣ	Power Transformer. (Or draw to scale.)
⊠	Controller.
⊏	Isolating Switch.
⏦	Overcurrent device, (fuse, breaker, thermal overload)
⏛	Switch and fuse

AUXILIARY SYSTEMS

▣	Push Button.
⊏⟋	Buzzer.
⊏⊃	Bell.
⊸◇	Annunciator.
◀	Outside Telephone.
◁	Interconnecting Telephone.
◁	Telephone Switchboard.
Ⓣ	Bell Ringing Transformer.
Ⓓ	Electric Door Opener.
Ⓕ⊃	Fire Alarm Bell.
Ⓕ	Fire Alarm Station.
⊠	City Fire Alarm Station.
⟦FA⟧	Fire Alarm Central Station.
⟦FS⟧	Automatic Fire Alarm Device.
Ⓦ	Watchman's Station.
⟦W⟧	Watchman's Central Station.
Ⓗ	Horn.
Ⓝ	Nurse's Signal Plug.
Ⓜ	Maid's Signal Plug.
Ⓡ	Radio Outlet.
⟦SC⟧	Signal Central Station.
☐	Interconnection Box.
⫼⫼	Battery.
- - - -	Auxiliary System Circuits.
☐ a,b,c	Special Auxiliary Outlets. Subscript letters refer to notes on plans or detailed description in specifications.

Figure 1-5   Generic symbols for electrical plans.

CODES AND ORGANIZATIONS

Many organizations, such as cities and power companies, develop electrical codes that they enforce within their areas of influence. These codes generally are concerned with the design and installation of electrical systems. In all cases, the latest *National Electrical Code*® is used as the basis for the local code. Consult these organizations before starting work on any project. The local codes may contain special requirements that apply to specific and particular installations. Additionally, the contractor may be required to obtain special permits and/or licenses before construction work can begin.

National Fire Protection Association

Organized in 1896, the National Fire Protection Association (NFPA) is an international, nonprofit organization dedicated to the twin goals of promoting the science of fire protection and improving fire protection methods. The NFPA publishes an eleven-volume series covering the national fire codes. The *NEC*®[1] is a part of volume 3 of this series. The purpose and scope of this code are set forth in *NEC*® *Article 90*.

Although the NFPA is an advisory organization, the recommended practices contained in its published codes are widely used as a basis for local codes. Additional information concerning the publications of the NFPA and membership in the organization can be obtained by writing to

National Fire Protection Association
1 Batterymarch Park
PO Box 9101
Quincy, Massachusetts 02269-9101

National Electrical Code®

The original *NEC*® was developed in 1897. Sponsorship of the *Code* was assumed by the NFPA in 1911.

The *National Electrical Code*® generally is the bible for the electrician. However, the *NEC*® does not have a legal status until the appropriate authorities adopt it as a legal standard. In May 1971, the Department of Labor through OSHA adopted the *NEC*® as a national consensus standard. Therefore,

---

[1]*National Electrical Code*® and *NEC*® are Registered Trademarks of the National Fire Protection Association, Inc., Quincy, MA.

in the areas where OSHA is enforced, the *NEC*® is the law.

Throughout this text, references are made to chapters, articles, sections, and tables of the *National Electrical Code*.® The use of the term *section* has been attenuated in the *Code*. In this text, it is used extensively to ensure proper identification of the *Code* references.

The student, and any other person interested in electrical construction, should obtain and use a copy of the latest edition of the *NEC*.® To help the user of this text, relevant *Code* sections are paraphrased where appropriate. However, the *NEC*® must be consulted before any decision related to electrical installation is made.

The *NEC*® is revised and updated every three years.

Code **Terms.** The following terms are used throughout the *Code*. It is important to understand the meanings of these terms.

APPROVED: Acceptable to the authority having jurisdiction.

AUTHORITY HAVING JURISDICTION: The organization, office, or individual responsible for approving equipment, materials, an installation, or a procedure.

IDENTIFIED (as applied to equipment): Recognizable as suitable for the specific purpose, functions, use, environment, application, and so forth, where described in a particular *Code* requirement.

LABELED: Equipment or materials to which has been attached a label, symbol, or other identifying mark of an organization that is acceptable to the authority having jurisdiction and concerned with product evaluation, that maintains periodic inspection of production of labeled equipment or materials, and by whose labeling the manufacturer indicates compliance with appropriate standards or performance in a specified manner.

LISTED: Equipment, materials, or services included in a list published by an organization that is acceptable to the authority having jurisdiction and concerned with the evaluation of products or services, that maintains periodic inspection of the production of listed equipment or materials or periodic inspection of services, and whose listing states that either the equipment, material, or services meets appropriate designated standards or has been tested and found suitable for a specified purpose.

RULES:

- Mandatory rules are required or prohibited, and use the term *shall* or *shall not.*

- Permissive rules are actions that are allowed, but not required. Permissive rules use the term *shall be permitted* or *shall not be required.*

FPN: *Fine Print Notes* (FPNs) are found throughout the *National Electrical Code.*® FPNs are explanatory in nature, in that they make reference to other sections of the *Code. FPNs* also define things where further description is necessary. *FPNs,* by themselves, are NOT *Code* rules.

Copies of the *NEC*® are available from the NFPA, the International Association of Electrical Inspectors, and from many bookstores.

Citing Code References.   Every time that an electrician makes a decision concerning the electrical wiring, the decision should be checked by reference to the *Code.* Usually this is done from memory without actually using the *Code* book. If there is any doubt in the electrician's mind, then the *Code* should be referenced directly—just to make sure. When the *Code* is referenced, it is a good idea to record the location of the information in the *Code* book—this is referred to as citing the *Code* reference. There is a very exact way that the location of a *Code* item is to be cited. The various levels of *Code* referencing are shown in Table 1-1. Starting at the top of the table, each step becomes a more specific reference. If a person references Chapter 1, this reference includes all the information and requirements that are set forth in several pages. When citing a specific *Section* or an *Exception,* only a few words may be included in the citation. The electrician wants to be as specific as possible when citing the *Code.* The word *section* does not precede the section numbers in the *Code.*

Underwriters Laboratories, Inc.

Founded in 1894, Underwriters Laboratories (UL) is a nonprofit organization that operates laboratories to investigate materials, devices, products, equipment, and construction methods and systems to define any hazards that may affect life and property. The organization provides a listing service to manufacturers, Figure 1-6. Any product authorized to carry an Underwriters listing has been evaluated with respect to all reasonable foreseeable hazards to life and property, and it has been determined that the product provides safeguards to these hazards to an acceptable degree. A listing by the Underwriters Laboratories does not mean that a product is

Citing the *NEC*®		
Division	Designation	Example
Chapter	1 through 9	*Chapter 1*
Article	90 through 820	*Article 250*
Part	Roman numeral	*Article 250, Part II*
Section	Article number, a dot (period), plus one, two, or three digits	*Section 250.20*
Paragraph	Section designation, plus uppercase letter in ( ), followed by digit in ( ), followed by a lowercase letter in ( ) as if required	*Section 250.119(A)(1)*
Exception to	Precedes or follows a section designation	*Exception No. 1 to Section 250.24(B)* or *Section 250.61(B) Exception 3a*
Annex	A, B, C, D, E, F, G (Formerly Appendix. The Annex is not part of the *NEC*® and is not enforceable.)	*Annex A*

Table 1-1

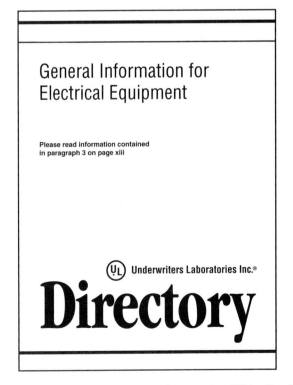

Figure 1-6   Underwriters Laboratories *White Book.*

approved by the *National Electrical Code.*® However, many local agencies do make this distinction.

The *National Electrical Code*® defines *approved* as *acceptable to the authority having jurisdiction.* *NEC*® *110.3(B)* stipulates that, *Listed or labeled equipment shall be installed and used in accordance with any instructions included in the listing or labeling.*

Useful UL publications are

- *Electrical Construction Equipment Directory (Green Book)*

- *Electrical Appliances and Utilization Equipment Directory (Orange Book)*

- *Hazardous Locations Equipment Directory (Red Book)*

- *General Information for Electrical Equipment (White Book)*

- *Recognized Component Directories (Yellow Books)* (a 3-volume set)

Many inspection authorities continually refer to these books in addition to the *National Electrical Code.*® If the answer to a question cannot be found readily in the *National Electrical Code,*® it generally can be found in the UL publications listed. An index of publications and information concerning the UL can be obtained by writing to

Public Information
Underwriters Laboratories, Inc.
333 Pfingsten Road
Northbrook, IL 60062
847-272-8800
http://www.ul.com

Recent harmonizing of standards has resulted in Underwriters Laboratories now evaluating equipment for a manufacturer under both a UL Standard as well as a Canadian Standard. This can save a manufacturer time and money because the manufacturer can have the testing and evaluation done in one location, acceptable to both U.S. and Canadian authorities enforcing the *Code.*

Nationally Recognized Testing Laboratories (NRTL)

When legal issues arise regarding electrical equipment, the courts must rely on the testing, evaluation, and product certification by qualified testing organizations. OSHA has recognized a number of organizations that meet the legal requirements found in OSHA 29 CFT 1910.7. These organizations are referred to in the industry as a NRTL (pronounced *nurtle,* as in *turtle*). The letters stand for *Nationally Recognized Testing Laboratory.* Visit the OSHA Web site at http://www.osha.gov for more details.

Intertek Testing Services (ITS)

Formerly known as Electrical Testing Laboratories, ITS is a nationally recognized testing laboratory. They provide testing, evaluation, labeling, listing, and follow-up service for the safety testing of electrical products. This is done in conformance to nationally recognized safety standards, or to specifically designated requirements of jurisdictional authorities. Information can be obtained by writing to

Intertek Testing Services, NA, Inc.
ETL SEMKO
3933 U.S. Route 11
Cortland, NY 13045
Phone: 607-753-6711
http://www.etlsemko.com

National Electrical Manufacturers Association (NEMA)

NEMA is a nonprofit organization supported by the manufacturers of electrical equipment and supplies. NEMA develops standards that are designed to assist the purchaser in selecting and obtaining the correct product for specific applications. A typical standard is illustrated in Figure 1-7. Information concerning NEMA standards may be obtained by writing to

National Electrical Manufacturers Association
1300 North 17th Street, Suite 1847
Rosslyn, VA 22209
703-841-3200
http://www.nema.org

American National Standards Institute (ANSI)

Various working groups in the organization study the numerous codes and standards. An American National Standard implies "a consensus of those concerned with its scope and provisions." The *National*

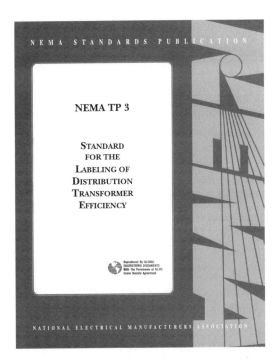

Figure 1-7   A typical NEMA standard.

Electrical Code® is approved by ANSI and is numbered ANSI/NFPA 70-1999.

ANSI
25 West 43rd Street, 4th Floor
New York, NY 10036
212-642-4900
http://www.ansi.org

Canadian Standards Association (CSA)

The Canadian Electrical Code is significantly different than the *National Electrical Code.®* Those using this text in Canada must follow the Canadian Electrical Code. *Electrical Wiring—Commercial* (© Thomas Nelson Holdings) is available based on the Canadian Electrical Code.

The Canadian Electrical Code is a voluntary code suitable for adoption and enforcement by electrical inspection authorities. The Canadian Electrical Code is published by and available from

CSA International
178 Rexdale Boulevard
Toronto, Ontario, Canada
M9W 1R3
416-747-4044, 800-463-6727
Fax: 416-747-2510
http://www.csa-international.org

International Association of Electrical Inspectors (IAEI)

IAEI is a nonprofit organization whose membership consists of electrical inspectors, electricians, contractors, and manufacturers throughout the United States and Canada. One goal of the IAEI is to improve the understanding of the *NEC.®* Representatives of this organization serve as members of the various panels of the *National Electrical Code®* Committee and share equally with other members in the task of reviewing and revising the *NEC.®* The IAEI publishes a bimonthly magazine, the *IAEI News.* Additional information concerning the organization may be obtained by writing to

International Association of
    Electrical Inspectors
901 Waterfall Way, Suite 602
Richardson, TX 75080-7702
800-786-4234, 972-235-1455
http://www.iaei.org

Illuminating Engineering Society of North America (IESNA)

IESNA was formed more than sixty-five years ago. The objective of this group is to communicate information about all facets of good lighting practice to its members and to consumers. The IESNA produces numerous publications that are concerned with illumination.

The *IESNA Lighting Handbooks* are regarded as the standard for the illumination industry and contain essential information about light, lighting, and luminaires (fixtures). Information about publications or membership may be obtained by writing to

Illuminating Engineering Society
    of North America
120 Wall St., 17th Floor
New York, NY 10005-4001
212-248-5000
http://www.iesna.org

National Electrical Installation Standards (NEIS)

NEIS, an ongoing project of the National Electrical Contractors Association (NECA), cover installation standards for the electrical trade. These installation standards are not about the *National Electrical Code*,® but rather, they cover those issues not covered in the *NEC*,® such as housekeeping; how to properly handle, receive, and store electrical equipment; checking out the equipment before energizing; and so on. These are actual on-the-job issues that are not *Code*-related, but issues electricians need to know. In the past, this was "hands-on" on-the-job training. Now, there are installation standards that can be followed by everyone in the electrical industry. These installation standards are also recognized by ANSI.

Installation Standards. At the moment, installation standards are available on the following subjects:

- *Standard Practices for Good Workmanship in Electrical Contracting*

- *Standard for Installing Steel Conduit*

- *Standard for Installing Nonmetallic Raceways*

- *Standard for Installing and Testing Fiber Optic Cables*

- *Standard for Installing Commercial Building Telecommunications Systems*

- *Symbols for Electrical Construction Drawings*

- *Standard for Fire Alarm System Job Practices*

- *Recommended Practice for Installing Aluminum Building Wire and Cable*

- *Recommended Practice for Installing Metal Cable Trays*

- *Recommended Practice for Installing and Maintaining Temporary Electric Power at Construction Sites*

- *Recommended Practice for Installing and Maintaining Industrial Heat Tracing Systems*

- *Recommended Practice for Installing and Maintaining Switchboards*

- *Recommended Practice for Installing and Maintaining Motor Control Centers*

- *Recommended Practice for Installing Generator Sets*

- *Recommended Practice for Installing Residential Generator Sets*

- *Recommended Practice for Installing and Commissioning Interconnected Generation Systems*

- *Recommended Practice for Installing and Maintaining Panelboards*

- *Recommended Practice for Installing and Maintaining Busways*

- *Recommended Practice for Installing and Maintaining Dry-Type Transformers*

- *Recommended Practice for Installing Indoor Commercial Lighting Systems*

- *Recommended Practice for Installing Exterior Lighting Systems*

- *Recommended Practice for Installing Industrial Lighting Systems*

- *Recommended Practice for Installing and Maintaining Medium-Voltage Cable*

Check price and availability by contacting:

National Electrical Contractors Association
3 Bethesda Metro Center, Suite 1100
Bethesda, MD 20814
301-215-4500
Fax: 301-215-4500
http://www.necanet.org

Registered Professional Engineer (PE)

Although the requirements may vary slightly from state to state, the general statement can be made that a registered professional engineer has demonstrated his or her competence by graduating from college and passing a difficult licensing examination. Following the successful completion of the examination, the engineer is authorized to practice engineering under the laws of the state. A requirement is usually made that a registered professional engineer must supervise the design of any building that is to be used by the public. The engineer must indicate approval of the design by affixing a seal to the plans.

Information concerning the procedure for becoming a registered professional engineer and a definition of the duties of the professional engineer can be obtained by writing the state government department that supervises licensing and registration.

METRICS (SI) AND THE *NEC*®

The United States is the last major country in the world not using the metric system as the primary system. We have been very comfortable using English (United States Customary) values, but this is changing. Manufacturers are now showing both inch-pound and metric dimensions in their catalogs. Plans and specifications for governmental new construction and renovation projects started after January 1, 1994, have been using the metric system. You may not feel comfortable with metrics, but metrics are here to stay. You might just as well get familiar with the metric system.

Some common measurements of length in the English (Customary) system are shown with their metric (SI) equivalents in Table 1-2.

The *NEC*® and other NFPA Standards are becoming international standards. All measurements

in the 2005 *NEC*® are shown with metrics first, followed by the inch-pound value in parentheses. For example, 600 mm (24 in.).

In *Electrical Wiring—Commercial*, ease in understanding is of utmost importance. Therefore, inch-pound values are shown first, followed by metric values in parentheses. For example, 24 in. (600 mm).

A *soft metric conversion* is when the dimensions of a product already designed and manufactured to the inch-pound system have their dimensions converted to metric dimensions. The product does not change in size.

A *hard metric measurement* is where a product has been designed to SI metric dimensions. No conversion from inch-pound measurement units is involved. A *hard conversion* is where an existing product is redesigned into a new size.

In the 2005 edition of the *NEC*,® existing inch-pound dimensions did not change. Metric conversions were made, then rounded off. Please note that when comparing calculations made by both English and metric systems, slight differences will occur due to the conversion method used. These differences are not significant, and calculations for both systems are therefore valid. Where rounding off would create a safety hazard, the metric conversions are mathematically identical.

For example, if a dimension is required to be 6 ft, it is shown in the *NEC*® as 1.8 m (6 ft). Note that the 6 ft remains the same, and the metric value of 1.83 m has been rounded off to 1.8 m. This edition of *Electrical Wiring—Commercial* reflects these rounded-off changes. In this text, the inch-pound measurement is shown first, i.e., 6 ft (1.8 m).

Trade Sizes

A unique situation exists. Strange as it may seem, what electricians have been referring to for years has not been correct!

Raceway sizes have always been an approximation. For example, there has never been a ½-in. raceway! Measurements taken from the *NEC*® for a few types of raceways are shown in Table 1-3.

You can readily see that the cross-sectional areas, critical when determining conductor fill, are different. It makes sense to refer to conduit, raceway, and tubing sizes as *trade sizes*. The *NEC*® in

Customary and Metric Comparisons		
Customary	*NEC*® SI Units	SI Units
0.25 in.	6 mm	6.3500 mm
0.5 in.	12.7 mm	12.7000 mm
0.62 in.	15.87 mm	15.8750 mm
1.0 in.	25 mm	25.4000 mm
1.25 in.	32 mm	31.7500 mm
2 in.	50 mm	50.8000 mm
3 in.	75 mm	76.2000 mm
4 in.	100 mm	101.6000 mm
6 in.	150 mm	152.4000 mm
8 in.	200 mm	203.2000 mm
9 in.	225 mm	228.6000 mm
1 ft	300 mm	304.8000 mm
1.5 ft	450 mm	457.2000 mm
2 ft	600 mm	609.6000 mm
2.5 ft	750 mm	762.0000 mm
3 ft	900 mm	914.4000 mm
4 ft	1.2 m	1.2192 m
5 ft	1.5 m	1.5240 m
6 ft	1.8 m	1.8288 m
6.5 ft	2.0 m	1.9182 m
8 ft	2.5 m	2.4384 m
9 ft	2.7 m	2.7432 m
10 ft	3.0 m	3.0480 m
12 ft	3.7 m	3.6576 m
15 ft	4.5 m	4.5720 m
18 ft	5.5 m	5.4864 m
20 ft	6.0 m	6.0960 m
22 ft	6.7 m	6.7056 m
25 ft	7.5 m	7.6200 m
30 ft	9.0 m	9.1440 m
35 ft	11.0 m	10.6680 m
40 ft	12.0 m	12.1920 m
50 ft	15.0 m	15.2400 m
75 ft	23.0 m	22.8600 m
100 ft	30.0 m	30.4800 m

Table 1-2

Trade Size	Inside Diameter (I.D.)
½ Electrical Metallic Tubing	0.622 in.
½ Electrical Nonmetallic Tubing	0.560 in.
½ Flexible Metal Conduit	0.635 in.
½ Rigid Metal Conduit	0.632 in.
½ Intermediate Metal Conduit	0.660 in.

Table 1-3   Table showing trade sizes of raceways vs. actual inside diameters.

90.9(C)(1) states that *where the actual measured size of a product is not the same as the nominal size, trade size designators shall be used rather than dimensions. Trade practices shall be followed in all cases.* This edition of *Electrical Wiring—Commercial* uses the term *trade size* when referring to conduits, raceways, and tubing. For example, instead of referring to a ½-in. EMT, it is referred to as trade size ½ EMT.

The *NEC®* also uses the term *metric designator.* A ½-in. EMT is shown as *metric designator 16 (½).* A 1-in. EMT is shown as *metric designator 27 (1).* The numbers 16 and 27 are the *metric designator* values. The (½) and (1) are the *trade sizes.* The metric designator is the raceway's inside diameter—in rounded-off millimeters (mm). Table 1-4 shows some of the more common sizes of conduit, raceways, and tubing. A complete table is found in the *NEC®, Table 300.1(C).* Because of possible confu-

Metric Designator & Trade Size	
Metric Designator	Trade Size
12	⅜
16	½
21	¾
27	1
35	1¼
41	1½
53	2
63	2½
78	3

Table 1-4   This table shows the metric designators for raceways through trade size 3.

Trade Size Knockout	Actual Measurement
½	⅞ in.
¾	1³⁄₃₂ in.
1	1⅜ in.

Table 1-5   This table compares the trade size of a knockout to the actual measurement of the knockout.

sion, this text uses only the term *trade size* when referring to conduit and raceway sizes.

Conduit knockouts in boxes do not measure up to what we call them. Table 1-5 shows trade size knockouts and their actual measurements.

Outlet boxes and device boxes use their nominal measurement as their *trade size.* For example, a 4 in. × 4 in. × 1½ in. does not have an internal cubic-inch area of 4 in. × 4 in. × 1½ in. = 24 cubic inches. *Table 314.16(A)* shows this size box as having a 21 cubic-in. area. This table shows *trade sizes* in two columns—millimeters and inches.

Table 1-6 provides the detailed dimensions of some typical sizes of outlet and device boxes in both metric and English units.

In this text, a square outlet box is referred to as 4 × 4 × 1½-inch square box, 4" × 4" × 1½" square box, or trade size 4 × 4 × 1½ square box. Similarly, a single-gang device box might be referred to as a 3 × 2 × 3-inch device box, a 3" × 2" × 3"-deep device box, or a trade size 3 × 2 × 3 device box. The box type should always follow the trade size numbers.

Trade sizes for construction material will not change. A 2 × 4 is really a *name,* not an actual dimension. A 2 × 4 stud will still be referred to as a 2 × 4 stud. This is its *trade size.*

In this text, measurements directly related to the *NEC®* are given in both inch-pound and metric units. In many instances, only the inch-pound units are shown. This is particularly true for the examples of raceway calculations, box fill calculations, and load calculations for square foot areas, and on the plans (drawings). To show both English and metric measurements on a plan would certainly be confusing and would really clutter up the plans, making them difficult to read.

Because the *NEC®* rounded off most metric conversion values, a calculation using metrics results in a different answer when compared to the same calcula-

Box Dimensions

Box Dimensions		Box Type	Minimum Capacity	
mm	in.		cm³	in.³
100 × 32	4 × 1¼	round/octagonal	205	12.5
100 × 38	4 × 1½	round/octagonal	254	15.5
100 × 54	4 × 2⅛	round/octagonal	353	21.5
100 × 32	4 × 1¼	square	295	18.0
100 × 38	4 × 1½	square	344	21.0
100 × 54	4 × 2⅛	square	497	30.3
75 × 50 × 38	3 × 2 × 1½	device	123	7.5
75 × 50 × 50	3 × 2 × 2	device	164	10.0
75 × 50 × 57	3 × 2 × 2¼	device	172	10.5

See *NEC® Table 314.16(A)* for complete listing.

Table 1-6

tion done using inch-pounds. For example, load calculations for a residence are based on 3 volt-amperes per square foot or 33 volt-amperes per square meter.

For a 40 ft × 50 ft dwelling:

3 VA × 40 ft × 50 ft = 6000 volt-amperes.

In metrics, using the rounded-off values in the *NEC®*:

33 VA × 12 m × 15 m = 5940 volt-amperes.

The difference is small, but nevertheless, there is a difference.

To show calculations in both units throughout this text would be very difficult to understand and would take up too much space. Calculations in either metrics or inch-pounds are in compliance with *NEC® 90.9(D)*. In *90.9(C)(3)* we find that metric units are not required if the industry practice is to use inch-pound units.

It is interesting to note that the examples in *Annex D* of the *NEC®* use inch-pound units, not metrics.

Guide to Metric Usage

The metric system is a "base-10" or "decimal" system in that values can be easily multiplied or divided by "ten" or "powers of ten." The metric system as we know it today is known as the International System of Units (SI) derived from the French term "*le Système International d'Unités.*"

In the United States, it is the practice to use a period as the decimal marker and a comma to separate a string of numbers into groups of three for easier reading. In many countries, the comma has been used in lieu of the decimal marker and spaces are left to separate a string of numbers into groups of three. The SI system, taking something from both, uses the period as the decimal marker and the space to separate a string of numbers into groups of three, starting from the decimal point and counting in either direction. For example, 12 345.789 99. An exception to this is when there are four numbers on either side of the decimal point. In this case, the third and fourth numbers from the decimal point are not separated. For example, 2015.1415.

In the metric system, the units increase or decrease or decrease in multiples of 10, 100, 1000, and so on. For instance, one megawatt (1 000 000 watts) is 1000 times greater than one kilowatt (1000 watts).

By assigning a name to a measurement, such as a watt, the name becomes the unit. Adding a prefix to the unit, such as *kilo-* forms the new name *kilowatt*, meaning 1000 watts. Refer to Table 1-7 for prefixes used in the numeric systems.

Certain prefixes shown in Table 1-7 have a preference in usage. These prefixes are *mega-*, *kilo-*, the unit itself, *centi-*, *milli-*, *micro-*, and *nano-*. Consider that the basic metric unit is a meter (one). Therefore, a kilometer is 1000 meters, a centimeter is 0.01 meter, and a millimeter is 0.001 meter.

The advantage of the SI metric system is that recognizing the meaning of the proper prefix lessens the possibility of confusion.

In this text, when writing numbers, the names are often spelled in full, but when used in calculations, they are abbreviated. For example: *m* for

Numeric Presentations				
Name	Exponential	Metric (SI)	Script	Customary
mega	(10^6)	1 000 000	one million	1,000,000
kilo	(10^3)	1 000	one thousand	1000
hecto	(10^2)	100	one hundred	100
deka		10	ten	10
unit		1	one	1
deci	(10^{-1})	0.1	one-tenth	1/10 or 0.1
centi	(10^{-2})	0.01	one-hundredth	1/100 or 0.01
milli	(10^{-3})	0.001	one-thousandth	1/1000 or 0.001
micro	(10^{-6})	0.000 001	one-millionth	1/1,000,000 or 0.000,001
nano	(10^{-9})	0.000 000 001	one-billionth	1/1,000,000,000 or 0.000,000,001

Table 1-7

meter, *mm* for *millimeter*, *in.* for *inch*, and *ft* for *foot*. It is interesting to note that the abbreviation for inch is followed by a period (12 in.), but the abbreviation for foot is not followed by a period (6 ft). Why? Because *ft.* is the abbreviation for *fort*.

Summary

As time passes, there is no doubt that metrics will be commonly used in this country. In the meantime, we need to take it slow and easy. The transition will take time. Table 1-8 shows useful conversion factors for converting English units to metric units. The Appendix of this text contains a comprehensive table showing the conversion values for inch-pound and metric units.

Useful Conversions and Their Abbreviations
inches (in.) × 0.0254 = meter (m)
inches (in.) × 0.254 = decimeters (dm)
inches (in.) × 2.54 = centimeters (cm)
centimeters (cm) × 0.3937 = inches (in.)
millimeters (mm) = inches (in.) × 25.4
millimeters (mm) × 0.039 37 = inches (in.)
feet (ft) × 0.3048 = meters (m)
meters (m) × 3.2802 = feet (ft)
square inches (in.²) × 6.452 = square centimeters (cm²)
square centimeters (cm²) × 0.155 = square inches (in.²)
square feet (ft²) × 0.093 = square meter (m²)
square meters (m²) × 10.764 = square feet (ft²)
square yards (yd.²) × 0.8361 = square meters (m²)
square meters (m²) × 1.196 = square yards (yd.²)
kilometers (km) × 1 000 = meters (m)
kilometers (km) × 0.621 = miles (mi)
miles (mi) × 1.609 = kilometers (km)

Table 1-8

REVIEW QUESTIONS

Refer to the *National Electrical Code®* or the working drawings when necessary. Where applicable, responses should be written in complete sentences.

1. What section of the specification contains a list of contract documents?
   General clauses + conditions

2. The requirement for temporary light and power at the job site will be found in what portion of the specification?
   Supplementary General conditions

3. The electrician uses the Schedule of Working Drawings for what purpose?
   To see what is needed (to be sure) not interfear with other trades

Complete the following items by indicating the letter(s) designating the correct source(s) of information for

4. ___F___ Ceiling height A 6		A. Architectural floor plan
5. ___B___ Electrical receptacle style		B. Details
6. ___C___ Electrical outlet location		C. Electrical layout drawings
7. _F / E_ Exterior wall finishes		D. Electrical symbol schedule
8. ___G___ Grading elevations		E. Elevations
9. ___C___ Panelboard schedules		F. Sections
10. ___A___ Room width		G. Site plan
11. ___A___ Swing of door		H. Specification
12. ___F___ View of interior wall		

Match the items on the left with those on the right by writing the letter designation of the appropriate organization from the list on the right.

13. ___C___ Electrical *Code*		A. IAEI
14. ___A___ Electrical inspectors		B. IESNA
15. ___E___ Fire codes		C. *NEC*®
16. ___B___ Lighting information		D. NEMA
17. ___G___ Listing service		E. NFPA
18. ___D___ Manufacturers' standards		F. PE
19. ___F___ Seal		G. UL

Match the items on the left with those on the right by writing the letter designation of the proper level of *NEC*® interpretation from the list on the right.

20. ___B___ Allowed by the *Code*		A. never
21. ___C.___ May be done		B. shall
22. ___B___ Must be done		C. with special permission
23. ___B___ Required by the *Code*		
24. ___A___ Up to the electrician		

25. Find *NEC*® 250.52(A)(5) and record the first four words.
    ___Rod and Pipe Electrodes___

Show required conversion calculations for Question 26 and Question 27.

26. Luminaire (fixture) style F is four feet long. The length in SI units, as specified by the *NEC*® is ___1.2 m   4 × .3048 =___

27. The gross area of the drugstore basement is 1395 square feet. The area in square meters is ___1395 × .093 =   129.735 M²___

Determine the following dimensions. Write the dimensions using unit names, not symbols (for example, 1 foot, not 1'), and indicate the source of the information.

28. What is the inside clear distance of the interior stairway to the drugstore basement?
    ___3 Feet   A2 Basement Floor Plan___

29. What is the gross square footage of each floor of the building?

    2580 feet² 1st Floor A3 _2580 Feet² 2nd Floor A 4_

30. What is the distance in the drugstore from the exterior block wall to the block wall separating the drugstore from the bakery?

    22 Feet 11 inches   1st Floor Plan   A 3

31. What is the finished floor to finished ceiling height on the second floor?

    8 Feet 6 inches   A 1

For Questions 32 through 34, cite the *NEC*® source.

32. The standard ampere ratings for fuses and fixed trip circuit breakers. _____

    _240.6 A_

33. The minimum bending radius for metal clad cable with a smooth sheath with an external diameter of 1 inch. _330. 24A² not less than 12 times the external diameter._
    (3/4") mm, not more than 38 mm (1 1/2 in) in external diameter.

34. The permission to use splices in busbars as grounding electrode conductor. _____

    250.64 Grounding & Electrode

Perform the following.

35. In examining the layout of the boiler room, you become concerned about the working clearance between the engine generator and the control panel. Write a memo to your superior supporting your concern. Cite the applicable section(s) of the *NEC*® (Propose a solution, if you wish.) _Arthur upon examination of Boiler_
    Room the clearance between generator And control
    Panel does not meet code 110.26 condition 1.
    Looks like there is only 2 ft 6 inches can we
    Have the generator turned as to meet the
    code.

36. Write a letter to one of the organizations listed in this unit requesting information about the organization and the services they provide.
    To whom it May concern, I was interested
    in Purchassing your electrical construction
    equipment directory (Greenbook) and also
    electrical Appliances and Utilization Equipment
    directory (Orangebook). Thank you very
    Much
    Sincerly
    Mark D. Turner

UNIT 2

Reading Electrical Working Drawings – Entry Level

OBJECTIVES

After studying this unit, the student should be able to

- read and interpret electrical symbols used in construction drawings.
- identify the electrical installation requirements for the drugstore.
- identify the electrical installation requirements for the bakery.

Electrical and architectural working drawings are the maps that the electrician must read and understand. Having this skill is essential to being able to install a complete electrical system and to coordinating related activities with those of workers in the other crafts. This unit will provide the first step in developing the ability to read symbols that appear on the drawings and applying them to the electrical work.

The units in this text that address electrical working drawings will apply information presented in preceding units and will introduce special features of the building area being discussed. The questions at the end of each of these units will require that the student use the specifications, the drawings, and the 2005 *National Electrical Code.*®

The user of this text is encouraged to peruse not only the electrical working drawings but the architectural working drawings, the loading schedules, the panelboard worksheets, and the panelboard directories. (The first two items are in the Appendix; the third is on the working drawings.) Most of the information given on and in these items is yet to be discussed in detail, but much is self-evident. See Figure 2-1.

ELECTRICAL SYMBOLS

On Sheet E4 of the Commercial Building working drawings is an Electrical Symbol Schedule that lists all of the electrical symbols that are used in this set of drawings. Knowing the special characteristics of these symbols will improve your ability to remember them and to interpret other symbols that are not used in these drawings.

Surface Raceway

This is, as the name implies, a raceway that is installed on a surface. The requirements for construction and installation are given in *NEC*® *Article 386, Metallic* and *Article 388, Nonmetallic.* The symbol may be used to indicate a variety of raceway types, and the electrician must always check on the specific installation requirements. This raceway use is common in remodeling applications and in situations where electrical power, communications, and electronic signals need to be available at a series of locations. The installation of surface raceways is discussed in Unit 9, "Special Systems."

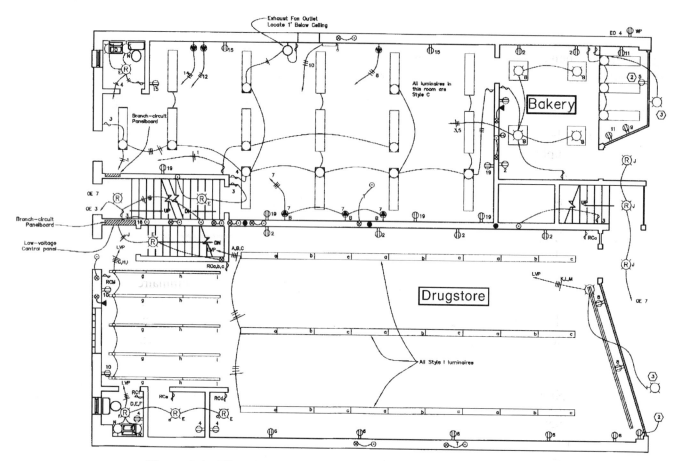

Figure 2-1   Electrical working drawing for drugstore and bakery.

Panelboard

Panelboards are distribution points for electrical circuits. They contain circuit protective devices. See *NEC® Article 100* for the official definition.

The two basic classes are power and lighting and appliance. A power panelboard is often installed as a distribution point serving lighting and appliance panelboards and other loads. The requirements for construction and application of panelboards are set forth in *NEC® Article 408*. In this building, a panelboard is located in each occupancy so the tenant will have ready access to the overcurrent devices. The symbol is not drawn to scale, and the installer must consult the shop drawings for specific dimensions.

Lighting Outlets

The next three symbols are for different applications of the same basic symbol, which is a circle with four ticks, at right angles, on the circumference as is shown for the ceiling outlet. In this case, the lumi-

naire (fixture) would be surface mounted or stem hung with the outlet box installed so that the opening is flush with the ceiling surface. According to *314.20*, the front edge of a box:

- shall not be set back more than ¼ in. (6mm) from the finished surface of a noncombustible material.

- shall be flush with the finished surface of a combustible material.

The second symbol has a stem protruding that, when drawn on the plans, will attach to a wall. This stem indicates that the outlet box is to be installed in the wall. The box must be installed so that its opening is flush with the wall surface.

This third lighting outlet symbol is used when the luminaire (fixture) is recessed. The outlet box will be attached to the luminaire (fixture) when it is installed, often with a short section of flexible raceway. Usually a rough-in kit will be available as a part of the luminaire (fixture) package and will be set in place before the ceiling material is installed.

Receptacle Outlets

The next nine symbols are for receptacle outlets. The first four are for duplex receptacle outlets, indicated by two parallel lines drawn within a circle.

 For the first of these symbols the receptacle symbol is drawn within a box, which indicates an outlet box is to be installed in the floor and that a duplex receptacle is installed.

In the second of this first set of four, the two parallel lines are extended beyond the box as stems. This indicates that the outlet box is to be installed flush mounted in the wall. The receptacle that should be installed is a NEMA Type 5, 15-ampere.

 The next symbol is identical to the previous one, except the space between the lines is darkened to indicate that a NEMA Type 5, 20-ampere receptacle should be installed.

 This fourth symbol is the same as the previous one except that it is a very special hospital-grade receptacle. It has transient voltage suppression and a grounding connection that is isolated from the raceway, and the metal box should be installed.

The next five symbols are the same except for the uppercase letter. The stem indicates that the outlet box should be installed flush in the wall. The uppercase letter is used to indicate the type of receptacle that should be installed. This symbol is used when there is a limited number of that receptacle type to be installed.

Switches

The final three symbols in the left side of the schedule are for switches. The actual switch symbol is the S and the line drawn through the S and extended indicates that the switch box should be installed in the wall. For the first symbol, a single-pole switch should be installed.

 This symbol indicates a three-way switch.

 The last symbol on this side of the schedule indicates that a four-way switch is to be installed.

Other Symbols

 A small number located adjacent to the symbol for an outlet, not a switch, indicates the branch-circuit number. This receptacle is connected to branch-circuit number 2.

 A large rectangular or square box represents a luminaire (fixture). These are drawn to scale and may be individual or several in a row. A lowercase letter within the symbol indicates which switch is used to operate the luminaire(s) (fixture[s]). An uppercase letter in the symbol indicates the style of luminaire (fixture) to be installed. The style designation is defined in the luminaire (fixture) schedule. This luminaire (fixture) is a style "A" and is operated by a switch marked with an "a."

 This symbol indicates the installation of a manual disconnecting means that may, or may not, be equipped with an overcurrent device. The construction specifications would need to be consulted to determine the exact requirements of this device.

 This symbol indicates that both a motor controller and a disconnect switch are to be installed. The symbol for a disconnect switch is not usually drawn to scale.

 When concealed raceways are drawn on plans, they usually are shown as curved lines. A straight line generally indicates a surface-mounted raceway. Hash marks drawn across raceway or cable lines indicate the *number* and *use* of the installed conductors. There are different acceptable ways of using hash marks. One way is to use full slashes to indicate "hot" (or switch leg) conductor(s), and half slashes to indicate grounded neutral conductor(s). No slashes indicate one "hot" and one grounded conductor. A dot(s) indicates an equipment grounding conductor(s). The letters "IG" are added near the dot to indicate an isolated-insulated grounding conductor(s). Another way is to use long hash marks to indicate a neutral (white) conductor and short hash marks to indicate "hot" ungrounded conductors. Check the plans and/or specifications for a Symbol Schedule to be sure you fully understand the meaning of the hash marks indicated on the plans you are working with.

An arrowhead at the end of a branch-circuit symbol indicates that the raceway goes from this point to the panelboard but will no longer be drawn on the plans. This symbol is used to avoid the graphic congestion created if all the lines came into a single point on the plans. The small numbers indicate which branch-circuits are to be installed in the raceway. As the overcurrent devices in a panelboard are usually numbered with odd numbers on the left and even on the right, it is common to see groups of odd or even numbers.

When raceways are installed vertically in the building from one floor to another, the vertical direction may be represented by the arrow symbol inside a circle. A dot represents the head of the arrow and indicates that the raceway is headed upward. A cross represents the tail of the arrow and that the raceway is headed downward. When both are shown, it means the raceway passes through. This symbol should be shown in the exact same location on the next, or previous, floor plan but should indicate the opposite direction.

 If the raceway is for use of the telephone system, the line will be broken and an uppercase *T* will be inserted.

The three parallel lines are used to indicate a lighting track that will accept special luminaires (fixtures) that the occupant can exchange or adjust.

 The last four symbols indicate different ways of labeling switches. The first indicates that the switch is a part of a low-voltage wiring system.

 An RCM indicates that it is a master switch on a low-voltage control that will override the action of individual

switches or perform the same action that several of the single switches would perform.

 The lowercase letter identifies that this is switch "a." This letter will appear at all the luminaires (fixtures) that this switch controls.

 The final symbol indicates that this is a switch that can also be used to dim the connected lighting.

THE DRUGSTORE

A special feature of the drugstore wiring is the low-voltage remote-control system. See Unit 20, "Low-Voltage Remote-Control," for a complete discussion. This system offers control flexibility that is not available in the traditional control system. The switches used in this system operate on 24 volts, and the power wiring, at 120 volts, goes directly to the electrical load. This reduces branch-circuit length and voltage drop. A switching schedule gives details on the system operation, and a wiring diagram provides valuable information to the installer.

One of the reasons for the low popularity of this system is the scarcity of electricians who are prepared to install a low-voltage control system. The system specified for the drugstore has been around a long time and is still being used. It is the most basic type of low-voltage control available today and is discussed in Unit 20.

The student is encouraged to request manufacturer's literature from any electrical distributor. Check out the Web sites listed in the back of this text and browse the Web. An amazing number of companies manufacture low-voltage control systems.

Different types of illumination systems have been selected for many of the spaces in the building. The student should observe the differences in the wiring requirements.

In the merchandise area, nine luminaires (fixtures) are installed in a continuous row. It is necessary to install electrical power to only one point of a continuous row of luminaires (fixtures). From this point, the conductors are installed in the wiring channel of the luminaire. In the pharmacy area, a luminous ceiling is shown. This illumination system consists of rows of strip fluorescents and a ceiling that will transmit light. The installation of the ceiling, in many jurisdictions, is the work of the electrician. For this system to be efficient, the surfaces above the luminous ceiling must be highly reflective (white).

THE BAKERY

For the production area of the bakery, a special luminaire (fixture) is selected to prevent contamination of the bakery materials. These units are totally enclosed individuals requiring a separate electrical connection to each luminaire (fixture). They may be supplied by installing a conduit in the upper-level slab or on the ceiling surface. In the sales area, more attractive luminaires (fixtures) have been selected.

A conventional control system is to be installed for the lighting in the bakery. The goal of the system is to provide control at every entry point so that

a person is never required to walk through an unlighted space. Often this requires long switching circuits, such as the three-point control of the main lighting in the work area.

The electrician may be responsible for making changes in conductor size to compensate for excessive voltage drop. This requires the electrician to be alert for high loads on long circuits, such as the control circuit on the bakery work area lighting.

The loading schedule calculations and panelboard details for the drugstore, bakery, insurance office, doctor's office, beauty salon, and the owners loads are found in the Appendix of this text in Tables A-4 through A-27.

Table A-28 in the Appendix is a table of useful electrical formulas.

You will want to refer to these tables often as you study *Electrical Wiring—Commercial*.

REVIEW QUESTIONS

Refer to the *National Electrical Code*® or the working drawings when necessary. Where applicable, responses should be written in complete sentences.

Answer Questions 1–7 by identifying the symbol and the type of installation (wall, floor, or ceiling) for the boxes.

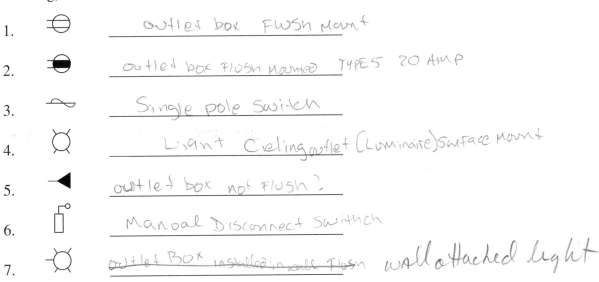

1. outlet box Flush Mount
2. outlet box Flush Mounted TYPES 20 AMP
3. Single pole Switch
4. Light Cieling outlet (Luminaire) surface Mount
5. outlet box not Flush?
6. Manual Disconnect Switch
7. outlet Box installed in wall Flush wall attached light

Note: The drugstore basement does not include the service equipment area or the boiler room.

8. How many duplex receptacle outlets are to be installed in the drugstore basement?

9

9. The duplex receptacle outlets in the drugstore basement are supplied from which panelboard? _____Drug stor_____

10. The duplex receptacle outlets in the drugstore basement are to be connected to which branch-circuit(s)? _____ relay s_____

11. How many lighting outlets are to be installed in the drugstore basement? _____11_____

12. What style(s) of luminaires (fixtures) is/are to be installed in the drugstore basement? _____Style L_____

13. How may the luminaires (fixtures) for the drugstore basement be installed? _____recessed_____

14. List the luminaires (fixtures) required for the drugstore. Give the style, count, and mounting method for each style.

Style __L__   Count __11__   Mounting method __Recessed__
Style __I__   Count __27__   Mounting method __Surface__
Style __D__   Count __18__   Mounting method __Surface__
Style __E__   Count __3__   Mounting method __recessed__
Style __N__   Count __2__   Mounting method __Surface__

15. From the information you have studied about the various electrical symbols, identify the following symbols, some of which are not shown in the symbol schedule.

_____Floor Switch_____

_____Raceway through Floor_____

_____Nema type 5 20 Amp_____

_____Lumanair Fixture_____

16. Where is the source of supply located for the disconnect switch installed in the drugstore basement? _____Comming from The Roof wall facing west_____

Comming from the Roof

UNIT 3

Calculating the Electrical Load

OBJECTIVES

After studying this unit, the student should be able to

- determine the minimum lighting loading for a given area.
- determine the minimum receptacle loading for a given area.
- determine the minimum equipment loading.
- determine a reasonable connected load.
- tabulate the unbalanced load.

THE ELECTRICAL LOAD

To plan any electrical wiring project, the first step is to determine the load that the electrical system is to serve. Only with this information can components for the branch-circuit, the feeders, and the service be properly selected. The *NEC®* provides considerable guidance in determining the minimum loading that is appropriate for a given occupant. Often the electrician is asked to generate this information. This unit will provide a foundation for proper selection of electrical circuit components.

NEC® Article 220 establishes the procedure that is to be used to calculate electrical loads. Using the drugstore as an example case, the application of this procedure will be illustrated.

- The phrase "calculated load" will be used to designate when the value is in compliance with the requirements of *NEC® Article 220*.
- The phrase "connected load" will be used to designate the value of the load as it actually exists.

The values discussed and calculated are shown in Table 3-1.

LIGHTING LOADING CALCULATIONS

The following guidelines should be followed when calculating the lighting load.

- Use *NEC® Table 220.12*, shown in this text as Table 3-1, to find the value of volt-amperes per square foot (or per square meter) for general lighting loads, or use the actual volt-amperes if that value is higher.

- For show window lighting, allow 200 volt-amperes per linear foot (660 volt-ampere per linear meter) or use the connected load if that value is higher. See *NEC® 220.14(G)*.

- Where lighting track is installed, 150 volt-amperes is to be allowed for every two feet (600 mm) of track or any fraction thereof. See *NEC® 220.43(B)*.

- At least one receptacle outlet shall be installed above each 12-ft (3.7-m) section of show window. See *NEC® 210.62*.

It is not a requirement that the connected load be equal to or greater than these allowances. For example, some energy codes may limit the connected lighting load to less than the values given in *NEC® Table 220.12*. It is a requirement that the electrical system have sufficient capacity for these allowances.

The application of these guidelines is illustrated in Table 3-2.

Table 220.12 General Lighting Loads by Occupancy

| Type of Occupancy | Unit Load | |
	Volt-Amperes per Square Meter	Volt-Amperes per Square Foot
Armories and auditoriums	11	1
Banks	39[b]	3½[b]
Barber shops and beauty parlors	33	3
Churches	11	1
Clubs	22	2
Court rooms	22	2
Dwelling units[a]	33	3
Garages — commercial (storage)	6	½
Hospitals	22	2
Hotels and motels, including apartment houses without provision for cooking by tenants[a]	22	2
Industrial commercial (loft) buildings	22	2
Lodge rooms	17	1½
Office buildings	39[b]	3½[b]
Restaurants	22	2
Schools	33	3
Stores	33	3
Warehouses (storage)	3	¼
In any of the preceding occupancies except one-family dwellings and individual dwelling units of two-family and multifamily dwellings:		
Assembly halls and auditoriums	11	1
Halls, corridors, closets, stairways	6	½
Storage spaces	3	¼

[a]See 220.14(J).
[b]See 220.14(K).

Table 3-1

Reprinted with permission from NFPA 70-2005

General Lighting

The minimum lighting load to be included in the calculations for a given type of occupancy is determined from *NEC* Table 220.12, which is shown in Table 3-1. For a "store" occupancy, the table indicates that the unit load per square foot is 3 volt-amperes and that the unit load per square meter is 33 volt-amperes. Therefore, for the first floor of the drugstore, the lighting load allowance is

60 ft × 23.25 ft × 3 VA per sq ft = 4185 VA

(18 m × 7 m × 33 VA per sq m = 4158 VA)

Because this value is a minimum, it is also necessary to determine the connected load. The greater of the two values becomes the calculated lighting load in accordance with the requirements of *NEC* Article 220.

As presented in detail in a later unit, the illumination in the sales area of the drugstore is provided by fluorescent luminaires (lighting fixtures) equipped with two lamps and a ballast. *This ballast can have a significant effect on the volt-ampere requirement for a luminaire (fixture).* According to the luminaire schedule shown on Sheet E4 of the working drawings, this lamp-ballast combination consumes 75 watts of power but has a load rating of 87 volt-amperes. (Watts can be used to determine the cost of operation; volt-amperes are used to determine the size of the conductors and overcurrent devices.)

The connected lighting load for the first floor of the drugstore is

27 Style I luminaires @ 87 volt-amperes	2349 volt-amperes
4 Style E luminaires @ 144 volt-amperes	576 volt-amperes
15 Style D luminaires @ 74 volt-amperes	1110 volt-amperes
2 Style N luminaires @ 60 volt-amperes	120 volt-amperes
Total connected load	4155 volt-amperes

It is important that the connected load be tabulated as accurately as possible. The values will not only be used to select the proper electrical components, but they may also be used to predict the cost of energy for operating the building.

Storage Area Lighting

According to *NEC* Table 220.12, the minimum load for basement storage space is 0.25 volt-ampere per square foot (3 volt-amperes per square meter).

The load allowance for this space is

1012 sq. ft × 0.25 VA per sq. ft = 253 VA

(94 sq. m × 3 VA per sq. m = 282 VA)

The connected load for the storage space is

9 Style L luminaires @ 87 VA	783 VA
1 Style E luminaire @ 144 VA	144 VA
Total connected load	927 VA

Drugstore—Loading Schedule							
General Lighting:	Count	VA/Unit	NEC®	Connected	Calculated	Balanced	Nonlinear
NEC® 220.12	1395	3	4185				
Style I luminaire	27	87		2349			2349
Style E luminaire	4	144		576			576
Style D luminaire	15	74		1110			1110
Style N luminaire	2	60		120			
Totals:			4185	4155	4185		
Storage:	Count	VA/Unit	NEC®	Connected			
NEC® 220.12	1012	0.25	253				
Style L luminaire	9	87		783			783
Style E luminaire	1	144		144			144
Totals:			253	927	927		
Show Window:	Count	VA/Unit	NEC®	Connected			
NEC® 220.14(G)	16	200	3200				
Receptacle outlets	3	500		1500			
Lighting track	8	150		1200			
Totals:			3200	2700	3200		
Other Loads:	Count	VA/Unit	NEC®	Connected			
Receptacle outlets	21	180		3780			
NEC® 220.14(L)							
Roof receptacle	1	1500		1500			
Sign outlet	1	1200		1200			
Totals:				6480	6480		
Motors & Appliances:	Amperes	VA/Unit		Connected			
Air conditioning:							
Compressor	20.2	360		7272		7272	
Condenser	3.2	208		666		666	
Evaporator	3.2	208		666		666	
Totals:				8604	8604		
TOTAL LOADS:					23,396	8604	4962

Table 3-2

Show Window Lighting

The load allowance for a show window is given in *NEC® 220.14(G)* and *220.43(A)* as 200 volt-amperes per linear foot (660 volt-amperes per linear meter), or major fraction of the show window. The drugstore window is 16 ft (4.85 m) long. The load allowance is

$$16 \text{ ft} \times 200 \text{ VA per ft} = 3200 \text{ VA}$$

$$(4.85 \text{ m} \times 660 \text{ VA per m} = 3200 \text{ VA})$$

The allowance for lighting track is set forth in *NEC® 220.43(B)*. That *Section* stipulates that 150 VA be allotted for each 2 ft (600 mm), or fraction of track. The track in the drugstore is 15 ft (5.2 m), which is 8 units of track. The allowance is

$$8 \text{ units @ } 150 \text{ VA/unit} = 1200 \text{ VA}$$

As the actual number of luminaires (fixtures) supplied by the track is unknown and will vary during usage, the allowance is considered as the connected load. The case with receptacles in the show window is that the exact load is unknown and will vary, but it is expected that as much as 500 volt-amperes will be installed at each receptacle. Given this information, the total connected load is

3 outlets @ 500 VA per outlet	1500 VA
8 unit of track @ 150 VA per unit	1200 VA
Total connected load	2700 VA

OTHER LOADS

The remaining loads consist of miscellaneous receptacle outlets, a receptacle on the roof, and a sign outlet.

Receptacle Outlets

- Each single or multiple receptacle on one strap shall be considered at not less than 180 volt-amperes, *NEC® 220.14(I)*. See Figure 3-1.

- Allow actual rating for specific loads, *NEC® 220.14(A)*.

The allowance for the receptacle outlets is

21 outlets @ 180 VA = 3780 VA

Roof Receptacle

The roof receptacle outlet is a requirement of *NEC® 210.63*. This weatherproof receptacle outlet must be located within 25 ft of the equipment: *NEC® 210.8(B)(3)* requires that the receptacle be GFCI protected. No load allowance is stipulated. The arbitrary allowance is

1 roof receptacle outlet @ 1500 VA per outlet = 1500 VA

Sign Outlet

The sign outlet is a requirement of *NEC®* 220.14(F) and is assigned a minimum allowance of 1200 volt-amperes.

1 sign outlet @ 1200 VA per outlet = 1200 VA

MOTORS AND APPLIANCES

- *NEC® Article 422* sets forth the installation requirement for appliances.
- *NEC® Article 430*, unless "specifically amended" by *NEC® Article 422*, should be referred to for motor-operated appliances.
- *NEC® Article 440* addresses air-conditioning and refrigeration equipment that incorporates hermetic refrigeration motor-compressor(s).

Hermetic Motor-Compressor

Included as a part of the specification, in the Appendix, are the electrical characteristics of the air-conditioning equipment to be installed in the commercial building. The load requirements for the drugstore are copied here for convenience. The connection scheme is shown in Figure 3-2.

Supply voltage:
208-volt, three-phase, three-wire, 60 hertz

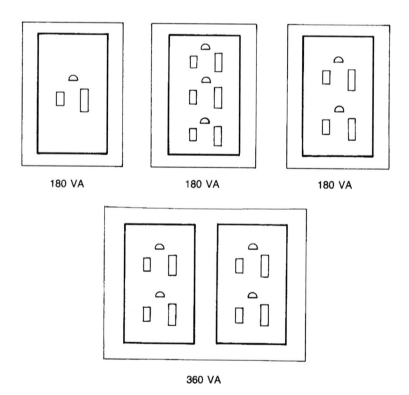

180 VA   180 VA   180 VA

360 VA

Figure 3-1   Minimum receptacle outlet allowance.

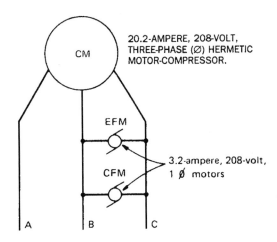

Figure 3-2   Typical air-conditioner load connections.

Hermetic refrigeration compressor-motor:

Rated load current 20.2 amperes @ 208-volt, three-phase

Evaporator motor:

Full-load current 3.2 amperes @ 208-volt, single-phase

Condenser motor:

Full-load current 3.2 amperes @ 208-volt, single-phase

It is necessary to convert these loads to volt-ampere values before they can be added to the other loads. For a 208-volt, three-phase supply, the rated load current is to be multiplied by 360. For a 208-volt load, single-phase supply, the rated load current is to be multiplied by 208.

The air-conditioning load is

Compressor: 3 ph., 208 V, 20.2 A	7272 VA
Evaporator: 1 ph., 208 V, 3.2 A	666 VA
Condenser: 1 ph., 208 V, 3.2 A	666 VA
Total air-conditioning load	8604 VA

This is a load calculation and does not indicate the current that will occur in the conductors supplying the equipment. A typical connection scheme is shown in Figure 3-2. It can be seen that the current in the three phases would not be equal. This will be discussed in detail in later units.

SUMMARY OF DRUGSTORE LOADS
General Lighting

The connected load (4155 volt-amperes) is less than the load allowance (4185 volt-amperes), so the allowance is used.

Storage

The connected load (927 volt-amperes) is greater than the load allowance (253 volt-amperes), so the actual load is used.

Show Window

The connected load (2700 volt-amperes) is less than the load allowance (3200 volt-amperes), so the allowance is used.

Other Loads

Specific allowances are used.

Motor Loads

The connected motor loads are used.

REVIEW QUESTIONS

Refer to the *National Electrical Code®* or the working drawings when necessary. Where applicable, responses should be written in complete sentences.

For Questions 1–4, indicate the unit load that would be included in the branch-circuit calculations as set forth in *NEC® Table 220.12*.

1. A restaurant   _____

2. A schoolroom   _____

3. A corridor in a school   _____

4. A corridor in a dwelling   _____

A different Style N luminaire (fixture) is selected for use in the drugstore. The new luminaire (fixture) is rated for 150 watts, but only 100 watts will be installed.

5. The *NEC®* load for the two Style N luminaires (fixtures) is _____.

6. After this revision, the calculated load for the general lighting will be _____.

Indicate the effect, if any, a change in loads can have on the feeders serving the area.

7. An increase in the calculated load.

   _____

8. An increase in the balanced load.

   _____

9. An increase in the nonlinear load.

   _____

Answer Questions 10, 11, and 12 using the following information.

An addition is being planned to a rural school building. You have been asked to determine the load that will be added to the panelboard that will serve this addition.

The addition will be a building 80 ft × 50 ft. It will consist of four classrooms, each 40 ft × 20 ft and a corridor that is 10-ft wide.

The following loads will be installed:

Each classroom:
    12 fluorescent luminaires (fixtures), 2 ft × 4 ft @ 85 VA each
    20 duplex receptacles
    1 AC unit 208 volt 1 phase @ 5000 VA

Corridor:
    5 fluorescent luminaires (fixtures) 1 ft × 8 ft @ 85 VA each
    8 duplex receptacles

Exterior:
   4 wall-mounted luminaires (fixtures) @ 125 VA each
   4 duplex receptacles

Load	Count	VA/Unit	NEC®	Connected	Calculated	Balanced	Nonlinear

10. The calculated load is   _____ VA.

11. The balanced load is   _____ VA.

12. The nonlinear load is   _____ VA.

UNIT 4

Branch-Circuits

OBJECTIVES

After studying this unit, the student should be able to

- determine the required number of branch-circuits for a set of loads.
- determine the correct rating for branch-circuit protective devices.
- determine the preferred type of wire for a branch-circuit.
- determine the required minimum size conductor for a branch-circuit.

CONDUCTOR SELECTION

When called upon to connect an electrical load such as lighting, motors, heating, or air-conditioning equipment, the electrician must have a working knowledge of how to select the proper type and size of conductors to be installed. Installing a conductor of the proper type and size will ensure that the voltage at the terminals of the equipment is within the minimums as set forth by the *NEC®* and that the circuit will have a long uninterrupted life.

One of the first steps in understanding conductors is to refer to *NEC® Article 310*. This article contains such topics as insulation types, conductors in parallel, wet and dry locations, marking, maximum operating temperatures, permitted use, trade names, direct burial, ampacity tables, adjustment factors, and correction factors.

Conductor Type Selection

An important step in selecting a conductor is the selection of an insulation type appropriate for the installation. An examination of *NEC® Table 310.16* will reveal that although several different types are listed, there are only three different temperature ratings: 140°F (60°C), 167°F (75°C), and 194°F (90°C). It should also be noted that the higher the temperature rating, the higher the ampacity for a given conductor size.

Should the operational temperature of a conductor become excessive, the insulation may soften, or melt, causing grounds or shorts and possible equipment damage and personal injury. This heat comes from two sources, the surrounding room temperature, which is referred to as ambient heat, and the current in the wire. Heat generated in the wire is expressed by the formula

$$H = I^2Rt$$

In this formula, H is the heat in watt hours; I is the current in amperes; R is the resistance in ohms and t is the time in seconds. Many of the rules in the *NEC®* concerning the selection of conductors are related to this formula.

The resistance of a conductor is dependent on the size, material, and length. Resistance values are listed in *NEC® Chapter 9, Table 8*. Conductor resistance may be lowered by selecting copper instead of aluminum or by increasing the size. (The circuit length is usually established by other factors.) The resistance of a conductor does not change appreciably in response to current or ambient conditions.

The current has a dramatic effect on the heat. A 25 percent load change, from 16 to 20 amperes, will result in a 56 percent change in heat. The *NEC®* incorporates many features to address this issue. The most notable of these is in the selection of the conductor type. Certain types of conductor insulation can tolerate higher heat levels, thus permitting a higher current with no change in wire size. The *NEC®* also restricts the connected load, thus limiting the current. In addition, it recognizes the heating effect of bundling conductors and requires the application of an adjustment factor.

Many factors should be considered when selecting a conductor, *NEC® Table 310.13*:

- Column 1 provides the technical names for the conductors, which are seldom used in the trade.

- Column 2 gives the type designation. These are used extensively in the trade, and there is some logic to their origin. In general, R indicates a rubber-based covering; H indicates a high temperature rating; W is used when the conductor is usable in wet locations; T indicates a thermoplastic covering; and N is used when there is a nylon outer covering.

- Column 3 lists the temperature ratings. The 60, 75, and 90°C (140, 167, and 194°F) ratings are used extensively in building wiring.

- Column 4 is highly used as it lists the applications where the conductor is approved for use. Turning to *NEC® Article 100*, Definitions, and finding "Location" will provide complete definitions for dry, damp, and wet. For example, if a circuit is run underground in a conduit, conductors rated for wet locations must be used.

- Column 5 gives a brief, technical description of the insulation.

- Columns 6, 7, and 8 give the insulation thickness for various sizes and types of conductors. These values become very important when calculating conduit fill. For example, for electrical metallic tubing (EMT) trade size ¾, *NEC® Annex C, Table C1* shows that sixteen Type THHN, fifteen Type TW, or eleven Type THW, size 12 AWG conductors may be installed.

- Column 9 provides a description of the outer coating. Popular Types THHN and THWN are listed as having a nylon jacket. This jacket tends to make installation easier but is not critical to the insulation value.

Wire Selection

The wires available for building construction are placed in two categories by *NEC® Table 310.16*: copper and aluminum or copper-clad aluminum.

Copper wire is usually the wire of choice partly because the allowable ampacity of a given size of conductor is less for the aluminum wire than it is for the copper wire. For example, the allowable ampacity of a Type THHN trade size 12 AWG is 30 amperes for copper and 25 amperes for the aluminum.

In addition, serious connection problems exist with the aluminum conductors:

- If moisture is present, there may be corrosive action when dissimilar metals come in contact, such as an aluminum conductor in a copper connector.

- The surface of aluminum oxidizes when exposed to air. If the oxidation is not broken, a poor connection will result. Thus, when installing larger aluminum conductors, an inhibitor is brushed onto the conductor and then it is scraped with a stiff brush. The scraping breaks through the oxidation, thus allowing the inhibitor to cover the wire surface. This prevents air from coming in contact with the aluminum surface. Pressure-type aluminum connectors usually have a factory-installed inhibitor in the connector.

- Aluminum wire expands and contracts to a greater degree than copper wire for a given current. This action can loosen connections that were not correctly made. Thus, it is extra important that an aluminum connection be tightened properly.

Table 4-1 gives terminal markings that shall be followed to ensure proper connection.

How Equipment Affects Conductor Selection

NEC® 110.14(C) indicates that for equipment rated 100 amperes or less (or marked for 14 AWG through 1 AWG), the allowable ampacity values found in the 140°F (60°C) column of *NEC® Table 310.16* are to be used. For example, when installing Type THHN conductors that have 194°F (90°C) insulation, the allowable ampacity values found in the 140°F (60°C) column are used. The reasoning is rather simple: a wire has two ends! A receptacle connected to one end is listed for 140°F (60°C). The other end of the conductor might be connected to a

Devices and Conductors		
Type of Device	**Markings**	**Conductors Permitted**
15- or 20-ampere receptacles and switches	CO/ALR	Copper, aluminum, copper-clad aluminum
15- and 20-ampere receptacles and switches	NONE	Copper, copper-clad aluminum
30-ampere and greater receptacles and switches	AL/CU	Copper, aluminum, copper-clad aluminum
30-ampere and greater receptacles and switches	NONE	Copper only
Screwless pressure terminal connectors of the push-in type	NONE	Copper or copper-clad aluminum
Wire connectors	AL	Aluminum
Wire connectors	AL/CU	Copper, aluminum, copper-clad aluminum
Wire connectors	CC	Copper-clad aluminum only
Wire connectors	CC/CU	Copper or copper-clad aluminum
Wire connectors	CU	Copper only

Table 4-1

circuit breaker listed for 167°F (75°C). But the panelboard is listed as a complete assembly, and the marking on the panelboard governs the allowable ampacity of the conductors.

NEC® 110.14(C) says that for equipment rated over 100 amperes (or marked for conductors larger than 1 AWG), the allowable ampacity values found in the 167°F (75°C) column of *NEC® Table 310.16* are to be used. For example, when installing Type THHN conductors that have 194°F (90°C) insulation, the allowable ampacity values found in the 167°F (75°C) column are used. The reasoning is the same as that given previously.

Why even have 194°F (90°C) insulation? Boiler rooms, attics, and outside installations in the hot sun are examples of where the high-temperature rating might be needed. The big advantage of high-stemperature insulations is that they permit the use of the allowable ampacity found in the 194°F (90°C) column for applying derating, adjusting, and correction factors.

Many conductors have a dual rating. For example, a Type THWN is rated 194°F (90°C) in a dry location and 167°F (75°C) in a wet location. A THWN-2 is rated 194°F (90°C) in a dry or wet location.

As always, check the markings on the conductor and refer to the appropriate *NEC®* tables.

Size Selection

NEC® Table 310.13 (Table 4-2) and *NEC® Table 310.16* (Table 4-3 and Table 4-4) are referred to regularly by electricians, engineers, and electrical inspectors when information about wire sizing is needed. *Table 310.16* shows the allowable ampacities

Conductor Specification and Application								
Trade name	**Type**	**Maximum Operating Temperature**	**Application provisions**	**Insulation**	**Insulation Thickness**			**Outer covering**
					AWG-kcmil	**mm**	**mils**	
Heat-resistant thermoplastic	THHN	194°F (90°C)	Dry and damp locations	Flame-retardant, heat resistant thermoplastic	14-12	0.38	15	Nylon jacket or equivalent
					10	0.51	20	
					8-6	0.76	30	
					4-2	1.02	40	
					1-4/0	1.27	50	
					250-500	1.52	60	
					501-1000	1.78	70	
Moisture- and heat-resistant thermoplastic	THHW	167°F (75°C)	Wet location	Flame-retardant, moisture- and heat- resistant thermoplastic	14-10	0.76	30	None
					8	1.14	45	
					6-2	1.52	60	
		194°F (90°C)	Dry location		1-4/0	2.03	80	
					213-500	2.41	95	
					501-1000	2.79	110	

See *NEC® Table 310.13* for complete listing.

Table 4-2

Allowable Ampacities—Copper Conductors			
Size	Temperature ratings of conductors		
AWG or kcmil	140°F (60°C)	167°F (75°C)	194°F (90°C)
	Types: TW, UF	Types: RHW, THHW, THW, THWN, XHHW, USE, ZW	Types: TBS, SA, SIS, FEP, FEPB, MI, RHH, RHW-2, THHN, THHW, THW-2, THWN-2, USE-2, XHH, XHHW-2, ZW-2
18			14
16			18
14*	20	20	25
12*	25	25	30
10*	30	35	40
8	40	50	55
6	55	65	75
4	70	85	95
3	85	100	110
2	95	115	130
1	110	130	150
1/0	125	150	170
2/0	145	175	195
3/0	165	200	225
4/0	195	230	260
250	215	255	290
300	240	285	320
350	260	310	350
400	280	335	380
500	320	380	430
600	355	420	475
700	385	460	520
750	400	475	535
800	410	490	555
900	435	520	585
1000	455	545	615
1250	495	590	665
1500	520	625	705
1750	545	650	735
2000	560	665	750
For aluminum or copper-clad aluminum values, consult the NEC® Table 310.16.			

Table 4-3

Correction Factors			
Ambient Temperature	Multiply allowable ampacities by the appropriate factors shown below		
	140°F (60°C)	167°F (75°C)	194°F (60°C)
70–77°F, 21–25°C	1.08	1.05	1.04
78–86°F, 26–30°C	1.00	1.00	1.00
87–95°F, 31–35°C	0.91	0.94	0.96
96–104°F, 36–40°C	0.82	0.88	0.91
105–113°F, 41–45°C	0.71	0.82	0.87
114–122°F, 46–50°C	0.58	0.75	0.82
123–131°F, 51–55°C	0.41	0.67	0.76
132–140°F, 56–60°C		0.58	0.71
141–158°F, 61–70°C		0.33	0.58
159–176°F, 71–80°C			0.41
For other values consult the NEC® Tables 310.16 and 310.17.			

Table 4-4

Correction Factors

The correction factors are given in Table 4-4. This set of factors applies to both copper and aluminum or copper-clad aluminum and is given in both NEC® Tables 310.16 and 310.17. They appear at the bottom of these tables.

If the environmental temperature anywhere along the length of a raceway is less than 78°F (26°C) or higher than 86°F (30°C), the ampacity of the conductors in the raceway, or cable, must be modified by a correction factor. A discussion of the conditions that require this action are set forth in NEC® 310.10. The selection is dependent on the conductor's temperature rating. As Type THHN is the conductor of choice for the commercial building, the correction factor will be taken from the 194°F (90°C) column.

As an example, the air-conditioning equipment, circuit 16, is located on the roof where the temperature would reach 100°F (38°C). A correction factor of 0.91 is applied to the Type THHN conductors serving this equipment.

Of particular concern is when raceways are run across rooftops. In direct sunlight, the temperature on raceways and the conductors within raceways reaches a dangerously high level. See the *Fine Print Notes* to *Table 310.10*.

Adjustment Factors

The use of adjustment factors is another response to excessive heat affecting the current-carrying

of insulated conductors. The temperature limitations, the types of insulation, the material the wire is made of, the conductor size in American Wire Gauge (AWG) or kcmil, the ampacity of the conductor, and the correction factors for a range of ambient temperatures are also shown.

NEC® Table 310.16 is used most often because it is always referred to when conductors are to be installed in raceways. It is important to read and understand all the notes and footnotes to these tables.

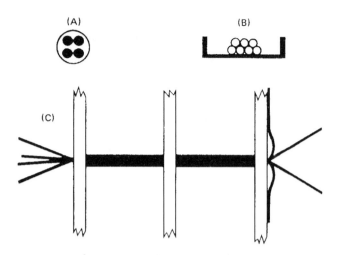

Figure 4-1   Bundled and stacked conductors.

Adjustment Factors	
Number of Current-Carrying Conductors	**Adjustment Factor**
4 through 6	80%
7 through 9	70%
10 through 20	50%
21 through 30	45%
31 through 40	40%
41 and above	35%
See *NEC®* Table 310.15(B)(2)(a).	

Table 4-5

Reprinted with Permission from NFPA 70-2005

capacity of a conductor. This is the collective heat generated by a group of conductors, such as shown in Figure 4-1. Those conductors are placed in a conduit of cable or otherwise bundled in a way that prevents airflow from dissipating the heat. Whenever a grouping of more than three conductors exists, a factor from *NEC® Table 310.15(B)(2)(a)* must be applied. Drugstore circuits 16 and 17 are examples of the application of adjustment factors. Five current-carrying conductors are installed in a raceway serving the air-conditioning equipment and the roof receptacle. The adjustment factor for five conductors is 80 percent or 0.8. See Table 4-5.

Conductors must be derated according to *NEC® 310.15(B):*

- when more than three current-carrying conductors are installed in a raceway or cable. See Figure 4-1(A) and *NEC® 310.15(B)(2)*.

- when single conductors or cable assemblies are stacked or bundled without spacing, as in

a cable tray, for lengths greater than 24 in. (600 mm). See Figure 4-1(B) and *NEC® 310.15(B)(2)*.

- when cable assemblies (i.e., nonmetallic sheathed cable) are run together for distances more than 24 in. (600 mm). See Figure 4-1(C) and *NEC® 310.15(B)(2)*.

It is important to note that the *NEC®* refers to current-carrying conductors for the purpose of derating when more than three conductors are installed in a raceway or cable. Following are the basic rules:

- DO count all current-carrying wires.

- DO count the neutral of a three-wire, single-phase circuit when the circuit is taken from a four-wire, three-phase, wye-connected system. See Figure 4-2 and *NEC® 310.15(B)(4)(b)*.

- DO count the neutral of a four-wire, three-phase, wye-connected circuit when the major portion of the load is electric discharge lighting (fluorescent, mercury vapor, high-pressure sodium, etc.), data processing, and other loads where the

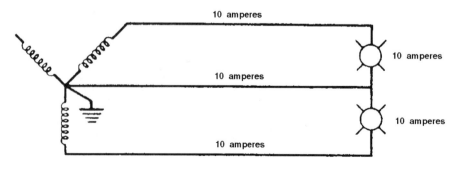

Figure 4-2   Four-wire wye system, three-wire circuit.

neutral carries third harmonic current. See Figure 4-3 and *NEC® 310.15(B)(4)(c)*. In branch-circuits and feeders that supply major nonlinear loads, the current in the neutral can be as high as two times that of the current in the phase conductors. In these situations, it is recommended that the neutral be double sized. Special K-rated transformers are recommended for these applications.

- DO NOT count the neutral of a three-wire, single-phase circuit where the neutral carries only the unbalanced current. See Figure 4-4 and *NEC® 310.15(B)(4)(a)*.

- DO NOT count equipment-grounding conductors that are run in the same raceway with cir-

cuit conductors. See Figure 4-5. Grounding conductors must be included when calculating raceway fill. See *NEC® 310.15(B)(5)*.

- DO NOT apply an adjustment factor to conductors in sections of raceway 24 in. (600 mm) or less in length. See Figure 4-6 and *NEC® 310.15(B)(2)(a)*.

- DO NOT count the noncurrent-carrying "dummy" or "traveler" conductors on three-way or four-way switching arrangements. Only one of the travelers is carrying current at any one time, so only one conductor need be counted when determining ampacity adjustment factor. See Figure 4-7. Both travelers are counted when computing raceway fill.

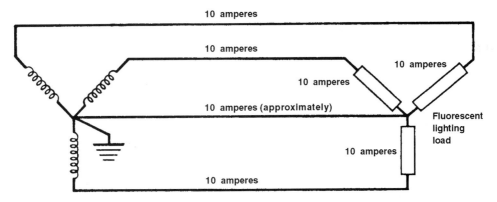

Figure 4-3   Four-wire wye system, nonlinear load.

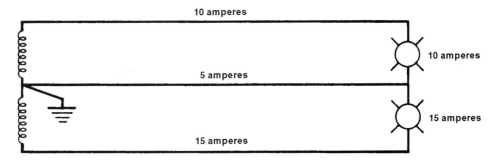

Figure 4-4   Single-phase, three-wire circuit.

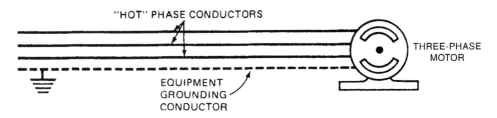

Figure 4-5   Counting equipment-grounding conductors.

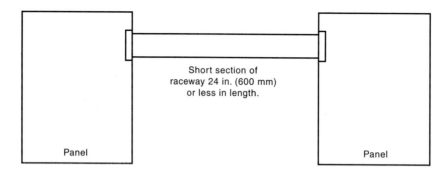

Figure 4-6 Ampacity adjustment not required.

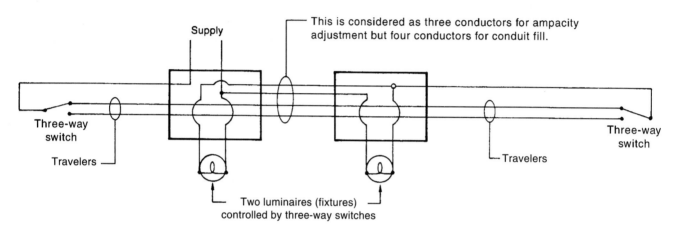

Figure 4-7 Counting three-way switch travelers.

DETERMINING CIRCUIT COMPONENTS

Conductors and overcurrent protection are key components of an electrical system.

Conductors

- The conductor type must be selected to meet the criteria of *NEC® 110.14(C)*. If the circuit (i.e., the overcurrent protective device [OCPD]) rating is 100-ampere or less, the conductor may be selected from any of the three columns in *NEC® Table 310.16*. If the circuit rating is greater than 100-ampere, it must either be a 167°F (75°C) or 194°F (90°C) rated conductor.

- The conductor type must comply with the requirements of the location, for example, dry, damp, or wet.

- Branch-circuit conductors shall have an ampacity not less than the load to be served, *NEC® 210.19(A)*.

- In compliance with *NEC® 210.19(A)*, the allowable ampacity (the value given in *NEC® Table 310.16*) of the conductor must be equal to or greater than the noncontinuous load plus 125 percent of the continuous load.

- For circuits rated 100-ampere, or if any of the terminations are marked for size 1 AWG conductors or less, enter the 140°F (60°C) column of *NEC® Table 310.16* and identify the allowable ampacity that is equal to or next greater than the OCPD ampere rating. This conductor is the minimum size allowed for the load. See *NEC® 110.14* for details on electrical connections.

- For circuits with a rating greater than 100 amperes, or if all the terminations are rated for 167°F (75°C) or higher, enter the 167°F (75°C) column of *NEC® Table 310.16* and identify the allowable ampacity that is equal to or next greater than the OCPD ampere rating. The size of this conductor is the minimum size allowed for the load.

- The allowable ampacity of the conductor must be sufficient to allow the use of the required OCPD and not less than the calculated load (noncontinuous load plus 125 percent of the continuous load).

- When serving a motor, the conductors must have an ampacity of not less than 125 percent of the nameplate current rating, *NEC® 430.6(A)(1)* and *430.22*. The current ratings for single-phase motors is found in *NEC® Table 430.248* and for three-phase motors in *NEC® Table 430.250*.

- When serving an air-conditioning unit, the conductors must have an ampacity not less than the *Minimum Circuit Ampacity* as marked on the nameplate of the unit. This marking on the nameplate makes it easy to size the conductors. UL Standards require that this data be provided by the manufacturer. *Article 440* also provides the basis for this requirement.

Overcurrent Protection

The purpose of an OCPD is to protect the circuit wiring and devices and to some extent the equipment served by the circuit. The following *NEC®* references must be consulted when selecting an OCPD. The student should read these references with great care because only a project-specific synopsis is presented.

- A continuous load will operate at 100 percent of the maximum current for a period of three hours or more. Office and store lighting are common examples of continuous loads. See *NEC® Article 100*.

- *NEC® 210.20(A)* states that the OCPD shall have a rating not less than the noncontinuous load plus 125 percent of the continuous load.

- The rating of the OCPD becomes the rating of the circuit regardless of the conductor size or load type. See Figure 4-8 and *NEC® 210.3*.

- In general, branch-circuit conductors shall be protected in accordance with their ampacities. See *NEC® 240.4*.

- A footnote to *NEC® Table 310.16* must be observed when selecting 14, 12, or 10 AWG conductors. This footnote refers to *NEC® 240.4(D)*. In this *Section*, the OCPD rating for copper conductors of trade sizes 14, 12, and 10 AWG are limited to 15-, 20-, and 30-ampere. For aluminum and copper-clad aluminum conductors of trade sizes 12 and 10 AWG, the ratings are limited to 15- and 25-ampere. *NEC® 240.4(E)*, *(F)*, and *(G)* should be consulted for exceptions.

- The ampacity of the circuit conductors may be less than the circuit rating if the circuit is rated at 800 amperes or less and it is not a multioutlet branch-circuit supplying cord- and plug-connected portable loads. See *NEC® 240.4(B)*.

- Multioutlet branch-circuit ratings are limited to 15-, 20-, 30-, 40-, and 50-ampere except on industrial premises where they may have a higher rating. See *NEC® 210.3*.

- OCPDs are available in the standard ampere ratings listed in *NEC® 240.6*.

- Branch-circuits serving storage-type water heaters with a capacity of 120 gallon (450 L) or less shall have a rating of 125 percent of the nameplate load. See *NEC® 422.13*.

- For motor loads of less than 100 amperes, the rating of an inverse-time circuit breaker (as used in the commercial building) shall not exceed 400 percent of the load. A rating of 250 percent is considered as appropriate. See *NEC® 430.52(C)* and *Table 430.52*.

- When a branch-circuit supplies an air-conditioning system that consists of a hermetic motor compressor and other loads (as occurs in the commercial building), the circuit rating shall not exceed 225 percent of the hermetic motor load plus the sum of the additional loads. See *NEC® 440.22(B)(1)*. The nameplate on an AC unit will be marked with the *maximum overcurrent protection size and type*.

Selection Criteria

An abbreviated circuit component selection criteria for motors, motor-compressor units, and other circuits is shown in Table 4-6. For motors, the OCPDs are inverse-time circuit breakers.

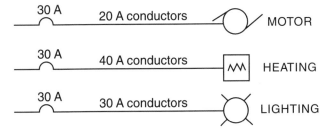

Figure 4-8   Three circuits rated 30-ampere.

Circuit Component Selection			
Circuit Size/Rating Load Type	Minimum Allowable Ampacity	OCPD Rating	Maximum OCPD Rating
C – Continuous	Not less than 125% of load	Not less than 125% load	Next higher over ampacity
H – Hermetic Motor Compressor	125% compressor load plus other loads	175% compressor load plus other loads	225% compressor load plus other loads[1]
N – Noncontinuous	Not less than load	Not less than load	Next higher over ampacity
M – Motor (100-ampere maximum)	125% load plus other loads	225% load plus other loads	400% load[2]
R – Cord and Plug Branch-Circuit	Not less than OCPD rating	Not less than load	Not higher than ampacity

1. See *NEC*® 430.52 before applying this value.
2. See *NEC*® 440.22(C) before applying this value.

Table 4-6

DEFINING THE BRANCH-CIRCUITS

In Unit 3, the electrical loads to be installed in the drugstore were identified and quantified. The next step usually would be to draw, or sketch, a layout of the space. For this exercise review Working Drawing E2, observing the location of the various electrical outlets.

The information in this unit concentrates on the branch-circuits, including the selection of the overcurrent devices and the conductor type and sizes. It will be useful to refer to the panelboard worksheet shown in this unit. The first five columns essentially define the branch-circuit. Much of this information is taken from the loading schedule. The next four columns result in the determining of the circuit rating. The remaining columns are related to the sizing of conductors.

Laying out an electrical system usually starts with a sketch or drawing of the space(s). Symbols, as discussed in Unit 2, are placed on the plans representing the location of the equipment. Then lines are drawn showing the circuiting of the equipment. Following this, branch-circuits can be assigned. The actual number of branch-circuits to be installed in the drugstore is determined by referring to the plans or specifications. The following discussion illustrates the process employed to arrive at the final number.

The specifications (see Appendix) for this commercial building direct the electrician to install copper conductors of a size not smaller than 12 AWG. The maximum load on a circuit is limited by the rating of the overcurrent protective device. For a 12 AWG copper conductor, the maximum overcurrent protection is 20 amperes, per *NEC*® 240.4(D). On a 120-volt circuit, this limits the load to

$$120 \text{ V} \times 20 \text{ A} = 2400 \text{ VA}$$

Minimum Number Branch-Circuits

Referring to the loading schedule developed in Unit 3, the calculated lighting load is 8312 volt-amperes. Add the 3780-volt-ampere allowance for the 21 receptacles for a total calculated load of 12,092 volt-amperes. Divide this total load by the maximum load per branch-circuit, rounding up to the next integer, to determine the minimum number of branch-circuits for general lighting and power.

12,092 volt-amperes at 2400 volt-amperes per branch-circuit requires six branch-circuits, as any fractional part must be rounded up.

Add the three special circuits, one for the roof receptacle, one for the sign, and one for the cooling system to arrive at nine circuits, the absolute minimum number of circuits for the drugstore.

Actual Number Branch-Circuits

This simplified procedure does not account for the continuous loads that limit branch-circuit loads to 80 percent of the rating of the overcurrent device.

Other factors, such as switching arrangements and the convenience of installation, are additional important considerations in determining the number of circuits to be used. This is illustrated in the drugstore, as seventeen circuits are scheduled.

At this point, it is necessary to refer to the loading schedule developed in the previous unit. An examination of the first five columns of the panelboard worksheet will reveal the first steps in arriving at a final number of branch-circuits.

Example 1

Ten luminaires (fixtures) are to be installed in a 1000 sq. ft (92.9 sq. m) industrial loft where the summer temperature will be as high as 97°F (36.11°C). Each luminaire (fixture) is designed for one 300-watt, 120-volt incandescent lamp, although a 100-watt lamp will be installed. The activities in the room will be continuous from 7 AM to 6 PM. The power system is 120/240-volt, single-phase. In compliance with *NEC® 210.23(A)*, branch-circuits with a 20-ampere rating shall be used. The conductors shall be no smaller than 12 AWG and shall be Type THHN.

For this problem, the circuit component selection procedure is divided into three steps.

The first step is to determine the number of circuits and the load per circuit.

1a. First, calculate the *minimum load* by multiplying the area by the volt-ampere value given in *NEC® Table 220.12* for the occupancy.

1b. Then the *connected load* is calculated by multiplying the number of luminaires (fixtures) by the luminaire (fixture) rating in volt-amperes.

1c. The greater of the *minimum load* and *connected load* becomes the calculated load in volt-amperes.

1d. The volt-amperes of the calculated load are divided by the circuit voltage to determine the calculated load in amperes.

1e. The calculated load in amperes is then divided by the branch-circuit rating. That quotient,

rounded up if required to a whole number, indicates the number of circuits.

1f. The calculated load per circuit is determined by dividing the calculated load by the number of circuits.

In the second step, the rating of the OCPD is determined. As this is a branch-circuit, *NEC® Article 210* is consulted.

2a. First, establish the load that it is expected to operate continuously for three hours or more. The remainder of the load is classified as noncontinuous.

2b. The *minimum circuit ampacity* is determined by adding together 125 percent of the continuous load to 100 percent of the noncontinuous load.

2c. The rating of the OCPD is selected from *NEC® 240.6 Standard Ampere Ratings*. It is the standard rating equal to, or next higher than, the OCPD selection ampacity.

2d. The conductor size and type is selected. It shall have an ampacity not less than the *minimum circuit ampacity* and of a type consistent with the problem statement.

The third step is to verify the conductor selection.

3a. Read the allowable ampacity for the selected size and type of conductor from *NEC® Table 310.16*.

3b. Record the *maximum ambient temperature*.

3c. Read the correction factor for the selected conductor type and the *maximum ambient temperature* from *NEC® Table 310.16*.

3d. Count the number of current-carrying conductors that are to be installed in the raceway.

3e. Read the adjustment factor, for the number of *current-carrying conductors*, from *NEC® Table 310.15(B)(2)(a)*.

3f. Calculate the ampacity by multiplying the allowable ampacity by the adjustment and correction factors. If it is not less than the minimum circuit ampacity (2c), the conductor size is verified. If it fails, increase the conductor size and reverify.

Application of Procedure

Step 1: Determine the number of circuits and load per circuit.

1a. Minimum load = 91 square meters × 22 volt-amperes per square meter = 2000 volt-amperes (1000 square feet × 2 volt-amperes per square foot = 2000 volt-amperes)

1b. Connected load = 300 volt-amperes per luminaire × 10 luminaires (fixtures) = 3000 volt-amperes

1c. Calculated load = 3000 volt-amperes

1d. Calculated load amperes = 3000 volt-amperes/ 120 volts = 25 amperes

1e. Number of circuits = 25 amperes/20 amperes per circuit = 2 circuits

1f. Calculated load per circuit = 25 amperes/ 2 circuits = 12.5 amperes per circuit

Step 2: Determine the OCPD (branch-circuit) rating.

2a. Continuous load = 12.5 amperes

   Noncontinuous load = 0 ampere

2b. OCPD selection ampacity = (12.5 amperes × 1.25) + 0 = 15.6 amperes

2c. OCPD rating = 20-ampere

2d. Conductor size and type = 12 AWG, Type THHN

Step 3: Verify the conductor size.

3a. Conductor allowable ampacity = 30 amperes

3b. 22°C

3c. Correction factor = 0.88 (for 22°C)

3d. Current-carrying conductors = 3

3e. Adjustment factor = 1.0

3f. Conductor ampacity = 30 × 0.88 × 1.0 = 26.4 amperes. The conductor ampacity is not less than the minimum circuit ampacity (2c), thus the conductor size is verified.

USING THE PANELBOARD WORKSHEET, COLUMNS A–E

The information studied in this unit is now applied to the drugstore and is illustrated in a panelboard worksheet. This worksheet is presented in three parts. In the first part, the information necessary to do the calculations is recorded in columns A–E, Table 4-7. In the second part, columns F–K, Table 4-8, the rating of the OCPD is determined along with the minimum conductor size and ampacity. In the final part, columns L–S, Table 4–9, the wire size is selected and listed in column Q. The final two columns verify the selection by demonstrating that the ampacity of the selected conductor is not less than the minimum ampacity given in column K.

Following each part is a discussion of the action performed by each of the columns. If there is a question, students are encouraged to review the information presented at the beginning of this unit.

Panelboard Worksheet – Columns A–E				
A	**B**	**C**	**D**	**E**
Phase Connection	Circuit Number	Load/Area Served	Calculated Volt-ampere	Calculated Ampere
A	1	9 Style I, Merchandise Area	783	6.5
A	2	4 Rec., Merchandise North	720	6
B	3	18 Style I, Merchandise Area	1566	13
B	4	3 Rec., Toilet Area	540	4.5
C	5	3 Style E, 2 N, Toilet Area	552	4.6
C	6	5 Rec., Merchandise S. Wall	900	7.5
A	7	2 Rec., Pharmacy	360	3
A	8	3 Rec., Show Window	1500	13
B	9	9 Style L, 1 E, Basement	927	7.7
B	10	15 Style D, Pharmacy	1110	9.3
C	11	Track Show Window	600	5
C	12	4 Rec., S. Basement	720	6
A	13	Track Show Window	600	5
A	14	3 Rec., N. Basement	540	4.5
B	15	Sign	1200	10
B		Evaporator		3.2
C	16	Compressor	8604	20
A		Condenser		3.2
C	17	1 Receptacle, Roof	1500	13

Table 4-7

	Panelboard Worksheet – Columns F–K				
F	G	H	I	J	K
Load Type	Load Modifier	OCPD Selection Ampere	OCPD Rating	Minimum Conductor Size (AWG)	Minimum Ampacity
C	1.25	9	20	12	16
R	1	6	20	12	21
C	1.25	17	20	12	17
R	1	5	20	12	21
C	1.25	6	20	12	16
R	1	8	20	12	21
R	1	3	20	12	21
C	1.25	16	20	12	16
C	1.25	10	20	12	16
C	1.25	12	20	12	16
C	1.25	7	20	12	16
R	1	6	20	12	21
C	1.25	7	20	12	16
R	1	5	20	12	21
C	1.25	13	20	12	16
H	1 / 2.25 / 1	52	50	*(handwritten: 8 / Breaker)*	32
N	1	13	20	12	16

Table 4-8

(handwritten) 23.2 × 2.25% = 52
(handwritten) 310% / Toget Size
(handwritten) 20 +3.02 3.02
(handwritten) 26.4 × 125%

	Panelboard Worksheet – Columns L–S						
L	M	N	O	P	Q	R	S
Ambient Temp. (°C)	Correction Factor	Current Carrying Conductors	Adjustment Factor	Minimum Allowable Ampacity	Conductor Size (AWG)	Allowable Ampacity	Ampacity
30	1		1	16	12	30	30
30	1		1	21	12	30	30
30	1		1	17	12	30	30
30	1		1	21	12	30	30
30	1		1	16	12	30	30
30	1		1	21	12	30	30
30	1		1	21	12	30	30
30	1		1	16	12	30	30
30	1		1	16	12	30	30
30	1		1	16	12	30	30
30	1		1	16	12	30	30
30	1		1	21	12	30	30
30	1		1	16	12	30	30
30	1		1	21	12	30	30
30	1		1	16	12	30	30
38	0.9	5	0.8	44	8	55	40
38	0.9	5	0.8	22	12	30	22

Table 4-9

(handwritten) 310% / 90° column / IF conductor allows Thhn
(handwritten) 55 A ×.9 × 80% = 40
(handwritten) 55 × .9 × .8 = 40

Phase Connection (Column A)

The letter in this column indicates which phase(s) is/are being used by the circuits. The letters A, B, and C are used to designate the different phases.

Circuit Number (Column B)

If the viewer is facing the panelboard, the circuits connected on the left side are given odd numbers with the even numbers being assigned to the circuits on the right side. Thus, Circuits 1, 3, and 5 would be the top three circuits on the left side, 2, 4, and 6 on the right side. If it is a three-phase panelboard, two or three circuits can be grouped together with a single neutral and the conductors installed in a raceway, or cable, to serve two or three single-phase 120-volt loads.

A typical case is Circuits 1 and 3. These are connected to Phases A and B and installed in a conduit serving the luminaires (fixtures) in the main sales area. The lighting load served by these circuits is 2349 volt-amperes. This load will continue for more than three hours, thus it is a continuous load and is limited to

$$120 \text{ V} \times 20 \text{ A} \times 0.8 = 1920 \text{ VA}$$

Three switches for this lighting could be served by two or three circuits. To place all on one circuit would overload the circuit:

$$783 \text{ VA} + 1566 \text{ VA} = 2349 \text{ VA}$$

A circuit could have been provided for each switch, but that strategy would have resulted in an excessive use of circuits. They would have to be loaded to

$$2349 \text{ VA} / 3 = 783 \text{ VA}$$

The choice was to connect two switches to one circuit and the other switch to a second circuit; thus one circuit has a load of 1566 volt-amperes, and the other a load of 783 volt-amperes.

Load/Area Served (Column C)

In this column is a listing of the loads and the areas of the occupancy. The information in this column comes from the drugstore loading schedule prepared in the preceding unit.

Calculated Volt-Ampere (Column D)

The loads in VA (volt-ampere) were also taken from the information provided by the drugstore loading schedule.

Calculated Ampere (Column E)

The value given in Column D is converted to ampere. For a single-pole circuit (Circuits 1–15 and 17), the volt-ampere is divided by 120; if the circuit were two-pole, the divisor is 208; and if the circuit is three-pole (Circuit 16), the divisor is 360.

USING THE PANELBOARD WORKSHEET, COLUMNS F–K

Columns F (Load Type), G (Load Modifier), H (OCPD Selection Ampere), I (OCPD Rating), J (Minimum Conductor Size AWG), and K (Minimum Ampacity) are used to record the critical first steps in the circuit component determination process. The content of these columns is to a great extent dictated by compliance with *NEC® 110.14(C)*.

How these conditions impact the circuit values will be demonstrated in this and later examples. In the commercial building, the lighting loads are in general considered continuous loads. Receptacles may be placed in C, N, or R class depending on how they are to be used. Circuit 8 serves receptacles in the show window, which will supply lighting, thus the circuit is considered a continuous load. Circuit 17 is used only when the AC unit is being repaired

or maintained, so it is considered a noncontinuous load. Circuit 16 is a hermetic unit consisting of three motors: the evaporator, the compressor, and the condenser.

Load Types (Column F)

The following load types will appear in this column:

- **C (Continuous Loads)**
  When the load is continuous: the OCPD ampere rating must not be less than 125 percent of the load or 20-ampere as required by the contract specifications. The minimum ampacity shall not be less than 125 percent of the load and shall permit the use of the OCPD. This would be an ampacity greater than the next lower rating taken from *NEC® 240.6*. For example in this Section, 15- is the next lower rating to 20-ampere. Thus, an ampacity of 16-ampere would permit the use of a 20-ampere OCPD. Actually, the ampacity could be as low as 15.51-ampere because rounding up is permitted.

- **N (Noncontinuous Loads)**
  When the load is noncontinuous: the OCPD ampere rating shall be not less than the load, and the ampacity of the conductor shall not be less than the load.

- **H (Hermetic Units)**
  When the circuit serves a hermetic refrigeration unit: the OCPD selection ampere is determined by adding the condenser and evaporator loads to 175 percent of the motor-compressor rated-load current (RLC); the OCPD rating shall not be greater than the sum of the condenser and evaporator loads plus 225 percent of the motor-compressor RLC; the circuit conductors shall have an ampacity not less than 125 percent of the motor-compressor RLC plus the sum of the other loads; and the motor leads and controller terminals have 167°F (75°C) temperature ratings, thus that column of *NEC® Table 310.16* shall be used when determining the minimum size conductor regardless of the load or the conductor size. See *NEC® 110.14*.

- **M (Motors)**
  When the circuit supplies a motor: if the OCPD is an inverse-time circuit breaker and the motor

load is 100-ampere or less, the preferred maximum rating is 250 percent, but it may be increased to 400 percent if required for starting the motor. See *NEC® 430.52(C)(1)* and *Table 430.52*.

The minimum ampacity of the circuit conductors shall not be less than 125 percent of the motor load. See *NEC® 430.22(A)*.

The leads and controller terminals have a 167°F (75°C) temperature rating, thus that column in *NEC® Table 310.16* shall be used when determining the minimum size conductor regardless of the load or the conductor size. See *NEC® 110.14*.

- **R (Receptacles Supplying Cord- and Plug-Connected Portable Loads)**
  ▶If a branch circuit supplies cord- and plug-connected equipment that is not fastened in place, the equipment shall not exceed 80% of the branch-circuit ampere rating, *210.23(A)(1)*. If the cord- and plug-connected equipment is fastened in place, and the circuit also supplies cord- and plug-connected equipment not fastened in place, or lighting, or both, the load of the fastened in place equipment shall not exceed 50% of the branch-circuit ampere rating, *210.23(A)(2)*.◀

Load Modifier (Column G)

If the load type is C, the load is continuous and is to be increased by 25 percent, in accordance with *NEC® 210.19*.

If the load type is M, the load is to be increased by 25 percent in accordance with *NEC® 430.22(A)*.

If the load is an H, it shall comply with *NEC® 440.22*.

If the load is a water heater, it shall be included as a continuous load. Other loads remain unchanged.

OCPD Selection Ampere (Column H)

The value in this column is the product of the connected load in amperes and the load modifier. This value is used to select the OCPD.

OCPD Rating (Column I)

The OCPD ampere rating is selected from *NEC® 240.6*. It is the rating next larger than the OCPD

selection ampere. On this project, the minimum rating for a circuit is 20-ampere.

Minimum Conductor Size (AWG) (Column J)

For loads 100-ampere or less, where one or more of the terminations are rated at 140°F (60°C), using size 1 AWG or smaller conductors, the minimum conductor size is determined by

- locating the ampacity in the 140°F (60°C) column of *NEC® Table 310.16*, which is equal to or next greater than the OCPD selection ampere and recording the conductor size of that ampacity.

If the load is greater than 100 amperes or if conductors are larger than size 1 AWG or if all the terminations have a rating of 167°F (75°C), the minimum conductor size is determined by

- locating the ampacity in the 167°F (75°C) column of *NEC® Table 310.16*, which is equal to or greater than the OCPD selection ampere and recording the conductor size of that ampacity.

Minimum Ampacity (Column K)

The values in two of the previous columns are used to determine the minimum ampacity value.

- If the load type is C or N, the minimum ampacity must be greater than the ampere value in column H and the rating from *NEC® 240.6* that is not less than the ampere rating in column I.
- If the load type is R, the minimum ampacity must be greater than the OCPD ampere rating.
- If the load type is H, the minimum ampacity must not be less than the sum of all other loads plus 125 percent of the motor-compressor load.
- If the load type is M, the minimum ampacity must not be less than 125 percent of the largest motor load plus the sum of other loads.

USING THE PANELBOARD WORKSHEET, COLUMNS L–S

Beginning with the column titled Ambient Temperature, the remaining columns in Table 4-9 are used to develop the information necessary to select the minimum conductor size and applicable data.

Ambient Temperature (°C) (Column L)

The ambient temperature in degrees Celsius, as discussed previously and in *NEC® 310.10*, is recorded in this column. A default value of 30°C is prerecorded.

Correction Factor (Column M)

This factor is taken from the table located at the bottom of *NEC® Table 310.16*. The factor is based on the ambient temperature of the environment where the circuit is installed. *NEC® 310.10* sets forth the requirements for the use of correction factors.

Current-Carrying Conductors (Column N)

The number of current-carrying conductors in a specific raceway is recorded in this column. The column is blank unless there are four or more.

Adjustment Factor (Column O)

This factor is taken from *NEC® Table 310.15(B)(2)(a)* and is used to compensate for the increase in temperature caused by grouping current-carrying conductors in a raceway or cable.

Minimum Allowable Ampacity (Column P)

The minimum allowable ampacity is the minimum ampacity divided by the product of the adjustment factor and correction factor.

Conductor Size (AWG) (Column Q)

The conductor size is determined by entering *NEC® Table 310.16* with the minimum allowable ampacity for the required Type (THHN) and material (copper). A size is selected that has an ampacity equal to or greater than the minimum allowable ampacity (Column P) and a size that is as large or larger than the minimum conductor size (Column J).

Allowable Ampacity (Column R)

The conductor ampacity taken from *NEC® Table 310.16* for the specified Type (THHN) and the calculated conductor size.

Ampacity (Column S)

The ampacity is the allowable ampacity from Column R multiplied by the correction and adjustment factors. This value should be equal to or greater than the minimum ampacity of Column K.

REVIEW QUESTIONS

Refer to the *National Electrical Code®* or the working drawings when necessary. When applicable, responses should be written in complete sentences.

Complete the following table using the following circuits:

1. Eleven 120-volt duplex receptacles for general use.

2. Ten 120-volt fluorescent luminaires (fixtures), rated at 150 volt-amperes each.

3. Eight 120-volt incandescent luminaires (fixtures), rated at 200 watts each.

4. One 240-volt exhaust fan, rated at 5000 watts.

5. One 120/240-volt single-phase, three-wire feeder, 40-kilovolt-amperes continuous load, originating in a room with an ambient temperature of 100°F.

- Circuits 1, 2, 3, and 4 are installed in the same raceway for a distance of 10 ft in an environment where the maximum ambient temperature will be 86°F.

- Type THHN conductors shall be used for all circuits.

Circuit number	1	2	3	4	5
Calculated load volt-ampere					
Calculated load ampere					
Load type					
Load modifier					
OCPD selection ampere					
OCPD rating					
Minimum conductor size					
Minimum ampacity					
Ambient temperature					
Correction factor					
Current-carrying conductors					
Adjustment factor					
Minimum allowable ampacity					
Conductor size					
Allowable ampacity					
Ampacity					

UNIT 5

Switches and Receptacles

OBJECTIVES

After studying this unit, the student should be able to

- select switches and receptacles with the proper rating for a particular application.
- install various types of receptacles correctly.
- connect single-pole, three-way, four-way, and double-pole switches into control circuits.

Electricians select and install numerous receptacles and switches. Therefore, it is essential that electricians know the important characteristics of these devices and how they are to be connected into the electrical system.

RECEPTACLES

The National Electrical Manufacturers Association (NEMA) has developed standards for the physical appearance of locking and nonlocking plugs and receptacles. The differences in the plugs and receptacles are based on the ampacity and voltage rating of the device. For example, the two most commonly used receptacles are the NEMA 5-15R, Figure 5-1, and the NEMA 5-20R, Figure 5-2. The NEMA 5-15R receptacle has a 15-ampere, 125-volt

Figure 5-1   NEMA 5-15R.

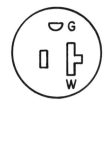

Figure 5-2   NEMA 5-20R.

52

Figure 5-3   NEMA 5-15P.   **Figure 5-4   NEMA 5-20P.**   **Figure 5-6   NEMA 6-15P or 6-20P.**

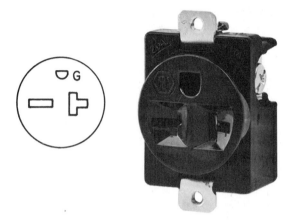

Figure 5-5   NEMA 6-20R.

Figure 5-7   NEMA 6-30R.

rating and has two parallel slots and a ground pin-hole. This receptacle will accept the NEMA 5-15P plug only, Figure 5-3. The NEMA 5-20R receptacle has two parallel slots and a T slot. This receptacle is rated at 20 amperes, 125 volts. The NEMA 5-20R will accept either a NEMA 5-15P or 5-20P plug, Figure 5-4. A NEMA 6-20R receptacle is shown in Figure 5-5. This receptacle has a rating of 20 amperes at 250 volts and will accept either the NEMA 6-15P or 6-20P plug, Figure 5-6. The acceptance of 15- and 20-ampere plugs complies with *NEC® Table 210.21(B)(3)*, which permits the installation of either the NEMA 5-15R or 5-20R receptacle on a 20-ampere branch-circuit. Another receptacle, which is specified for the commercial building, is the NEMA 6-30R, Figure 5-7. This receptacle is rated at 30 amperes, 250 volts.

The NEMA standards for general-purpose non-locking and locking plugs and receptacles are shown in Table 5-1, Table 5-2, and Table 5-3. A special note should be made of the differences between the 125/250 (NEMA 14) devices and the three-phase, 250-volt (NEMA 15) devices. The connection of the 125/250-volt receptacle requires a neutral, a ground-ing wire, and two-phase connections. For the three-phase, 250-volt receptacle, a grounding wire and three-phase connections are required.

NEMA Terminal Identification for Receptacles and Plugs	
Green-Colored Terminal (marked G, GR, GN, or GRND)	This terminal has a hexagon shape. Connect equipment grounding conductor ONLY to this terminal. (Green, bare, or green with yellow stripe.)
White (Silver)-Colored Terminal (marked W)	Connect white or gray grounded circuit conductor ONLY to this terminal.
Brass-Colored Terminal (marked X, Y, Z)	Connect HOT conductor to this terminal (black, red, blue, etc.). There is no NEMA standard for a specific color conductor to be connected to a specific letter. It is good practice to establish a color code and stick to it throughout the installation.

Table 5-1

NEMA Configurations for General-Purpose Nonlocking Plugs and Receptacles

(handwritten in left margin: Single phase loAD)

		#	15 AMPERE RECEPTACLE	15 AMPERE PLUG	20 AMPERE RECEPTACLE	20 AMPERE PLUG	30 AMPERE RECEPTACLE	30 AMPERE PLUG	50 AMPERE RECEPTACLE	50 AMPERE PLUG	60 AMPERE RECEPTACLE	60 AMPERE PLUG
2-POLE 2-WIRE	125 V	1	1-15R	1-15P		1-20P		1-30P				
	250 V	2		2-15P	2-20R	2-20P	2-30R	2-30P				
	277 V AC	3	(RESERVED FOR FUTURE CONFIGURATIONS)									
	600 V	4	(RESERVED FOR FUTURE CONFIGURATIONS)									
2-POLE 3-WIRE GROUNDING	125 V	5	5-15R	5-15P	5-20R	5-20P	5-30R	5-30P	5-50R	5-50P		
	250 V	6	6-15R	6-15P	6-20R	6-20P	6-30R	6-30P	6-50R	6-50P		
	277 V AC	7	7-15R	7-15P	7-20R	7-20P	7-30R	7-30P	7-50R	7-50P		
	347 V AC	24	24-15R	24-15P	24-20R	24-20P	24-30R	24-30P	24-50R	24-50P		
	480 V AC	8	(RESERVED FOR FUTURE CONFIGURATIONS)									
	600 V AC	9	(RESERVED FOR FUTURE CONFIGURATIONS)									
3-POLE 3-WIRE	125/250 V	10			10-20R	10-20P	10-30R	10-30P	10-50R	10-50P		
	3 Ø 250 V	11	11-15R	11-15P	11-20R	11-20P	11-30R	11-30P	11-50R	11-50P		
	3 Ø 480 V	12	(RESERVED FOR FUTURE CONFIGURATIONS)									
	3 Ø 600 V	13	(RESERVED FOR FUTURE CONFIGURATIONS)									
3-POLE 4-WIRE GROUNDING	125/250 V	14	14-15R	14-15P	14-20R	14-20P	14-30R	14-30P	14-50R	14-50P	14-60R	14-60P
	3 Ø 250 V	15	15-15R	15-15P	15-20R	15-20P	15-30R	15-30P	15-50R	15-50P	15-60R	15-60P
	3 Ø 480 V	16	(RESERVED FOR FUTURE CONFIGURATIONS)									
	3 Ø 600 V	17	(RESERVED FOR FUTURE CONFIGURATIONS)									
4-POLE 4-WIRE	3 Ø 208Y/120 V	18	18-15R	18-15P	18-20R	18-20P	18-30R	18-30P	18-50R	18-50P	18-60R	18-60P
	3 Ø 480Y/277 V	19	(RESERVED FOR FUTURE CONFIGURATIONS)									
	3 Ø 600Y/347 V	20	(RESERVED FOR FUTURE CONFIGURATIONS)									
4-POLE 5-WIRE GROUNDING	3 Ø 208Y/120 V	21	(RESERVED FOR FUTURE CONFIGURATIONS)									
	3 Ø 480Y/277 V	22	(RESERVED FOR FUTURE CONFIGURATIONS)									
	3 Ø 600Y/347 V	23	(RESERVED FOR FUTURE CONFIGURATIONS)									

Table 5-2

NEMA Configurations for General-Purpose Locking Plugs and Receptacles

Table 5-3

Hospital-Grade Receptacles

In locations where severe abuse or heavy use is expected, hospital-grade receptacles are recommended. These are a high-quality product and meet UL requirements. These receptacles are marked with a small green dot; see Figure 5-8.

Electronic Equipment Receptacles

Circuits with receptacle outlets that serve microcomputers, solid-state cash registers, or other sensitive electronic equipment should be served by receptacles that are specially designed. These receptacles may be constructed for an isolated ground, Figure 5-9, they may have transient voltage surge protection, they may be hospital grade, or they may be any combination of these three. Where the isolated grounding is desired, an insulated grounding conductor is to be run from the neutral terminal at the service entrance to the isolated terminal on the receptacle. Isolated ground receptacles are required by *NEC® 406.2(D)* to have an orange triangle on their faces as illustrated in Figure 5-9. Prior to the 1996 edition of the *NEC®* the entire face of the receptacle was permitted to be orange in color.

Isolated Ground Receptacle

In a standard conventional receptacle outlet, Figure 5-10(A), the green hexagonal grounding screw, the grounding contacts, the yoke (strap), and the metal wall box are all "tied" together to the building's equipment grounding system. Thus, the many receptacles in a building create a multiple-ground situation.

In an isolated ground receptacle, Figure 5-10(B), the green hexagonal grounding screw and the grounding contacts of the receptacle are isolated from the metal yoke (strap) of the receptacle and also from the building's equipment grounding system. Permission is given in *NEC® 408.40* to install a separate green insulated grounding conductor from the green hexagonal screw on the receptacle all the way back to the main service disconnect (see Figure 5-13). This separate green grounding conductor does not connect to any panels, load centers, or other ground reference points in between. The system grounding is now "clean," and the result is less transient noise (disturbances) transmitted to the connected load.

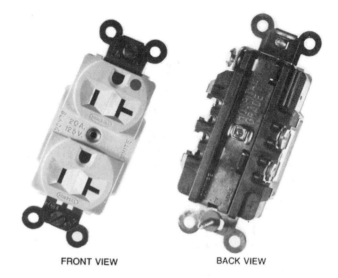

FRONT VIEW   BACK VIEW

Figure 5-8   Hospital-grade receptacle.

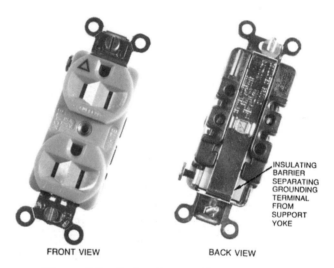

INSULATING BARRIER SEPARATING GROUNDING TERMINAL FROM SUPPORT YOKE

FRONT VIEW   BACK VIEW

Figure 5-9   Isolated grounding receptacle.

Equipment grounding conductors and isolated equipment grounding conductors are two different items. The FPN to *NEC® 250-146(D)* calls attention to the fact that metal raceways and outlet boxes may still need to be grounded with an equipment grounding conductor. An additional isolated equipment grounding conductor may then be installed to minimize noise as discussed earlier.

When nonmetallic boxes are used in conjunction with an isolated ground receptacle, nonmetallic faceplates should be used because the metal yoke on the receptacle is not connected to the receptacle's equipment grounding conductor terminal. A metallic faceplate would not be properly grounded.

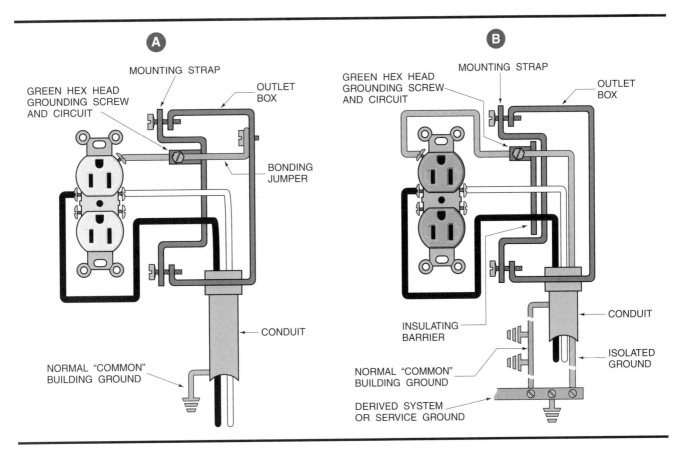

Figure 5-10   (A) Standard conventional receptacle and (B) isolated grounded receptable.

Electronic Equipment Grounding

Most commercial buildings today contain electronic equipment such as computers, copy machines, fax machines, data-processing equipment, telephone systems, security systems, medical diagnostic instruments, HVAC and similar electronic controls, electronic cash registers.

Electronic equipment is extremely sensitive and susceptible to line disturbances caused by such things as

voltage sags, spikes, and surges
static charges
lightning strikes and surges on the system
electromagnetic interference (EMI)
radio frequency interference (RFI)
improper grounding

Metal raceways in a building act like a large antenna and can pick up electromagnetic interference (electrical noise) that changes data in computer/data-processing equipment.

Electric utilities are responsible for providing electricity within certain voltage limits. Switching surges or lightning strikes on their lines can cause problems with electronic equipment. Static, lightning energy, radio interference, and electromagnetic interference can cause voltages that exceed the tolerance limits of the electronic equipment. These can cause equipment computing problems. Electronic data-processing equipment will put out garbage that is unacceptable in the business world. Electronic equipment must be protected from these disturbances.

Quality voltage provides a clean sine wave that is free from distortion, Figure 5-11. Poor voltage (dirty power) might show up in the form of sags, surges, spikes or impulses, notches or dropouts, or total loss of power, Figure 5-12. Disturbances that

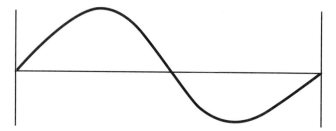

Figure 5-11   Quality voltage.

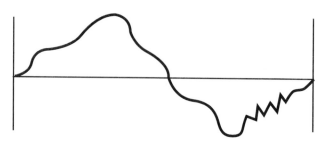

Figure 5-12   Poor voltage.

can change the sine wave may be caused by other electronic equipment operating in the same building.

Proper grounding of the ac distribution system and the means by which electronic equipment is grounded are critical, as these can affect the operation of electronic equipment. The ungrounded (HOT) conductor, the grounded (NEUTRAL) conductor, and the equipment grounding (GROUND) conductor must be properly sized, tightly connected, and correctly terminated. The equipment grounding conductor serves a vital function for a computer in that the computer's dc logic has one side directly connected to the metal frame of the computer, thus using ground as a reference point for the processing of information. An acceptable impedance for the grounding path associated with normal equipment grounding for branch-circuit wiring is 1 to 2 ohms. Applying Ohm's law:

$$\text{Amperes} = \frac{\text{Volts}}{\text{Ohms}} = \frac{120}{1} = 120 \text{ amperes}$$

A ground-fault value of 120 amperes will easily cause a 15- or 20-ampere branch-circuit breaker to trip off. *NEC® 250.118* lists the different items that are considered acceptable equipment grounding conductors.

The acceptable impedance of the ground path for branch-circuit wiring is *not* acceptable for grounding electronic equipment such as computer/data-processing equipment. The acceptable imped-ance of a ground path for grounding electronic equipment is much more critical than the safety equipment ground impedance for branch-circuit wiring. The maximum impedance for grounding electronic equipment is 0.25 ohms. That is why a separate insulated isolated equipment grounding conductor is needed for the grounding of computer/data-processing equipment to ensure a low impedance path to ground. Furthermore, stray currents on the equipment grounding conductor can damage the electronic equipment or destroy or alter the accuracy of the information being processed. Proper equipment grounding minimizes these potential problems. A separate insulated isolated equipment grounding conductor is sized according to *NEC® Table 250.122.* For longer runs, a larger size equipment grounding conductor is recommended to keep the impedance to 0.25 ohms or less. The electrical engineer designing the circuits and feeders for computer rooms and data-processing equipment considers all of this. *NEC® Article 645* covers information technology equipment and systems. The National Fire Protection Association publication *NFPA-75* titled *Protection of Electronic Computer/Data Processing Equipment* contains additional information.

For circuits supplying computer/data-processing equipment, *never* use the grounded metal raceway as the *only* equipment ground for the electronic data-processing equipment. Instead of allowing the grounded metal raceway system to serve as the equipment ground for sensitive electronic equipment, use a separate green *insulated* equipment grounding conductor that:

- connects to the isolated equipment grounding terminal of an isolated type receptacle.

- does not connect to the metal yoke of the receptacle.

- does not connect to the metal outlet or switch box.

- must be isolated from the metal raceway system.

- does not touch anywhere except at the connection at its source and at the grounding terminal of the isolated receptacle.

- must be run in the same raceway as the HOT and NEUTRAL wires.

- provides a ground path having an impedance of 0.25 ohms or less.

This separate insulated equipment grounding conductor is permitted by the *NEC®* to pass through a panelboard (see Figure 5-13) and be carried all the way back to the main service or to the transformer such as a 480-volt primary, 208/120-volt wye connected secondary step-down transformer. See *NEC® 250.146(D)* and *408.20, Exception.*

To summarize grounding when electronic computer/data-processing equipment is involved, generally there are two equipment grounds: (1) the metal raceway system that serves as the safety equipment grounding means for the metal outlet boxes, switch boxes, junction boxes, and similar metal equipment that is part of the premise wiring; and (2) a second insulated isolated equipment ground that serves the electronic equipment. This is sometimes referred to as a dedicated equipment grounding conductor.

NEC® 250.96(B), FPN and *250.146(D), FPN* state that the "use of an isolated equipment grounding conductor does not relieve the requirement for grounding the raceway system and outlet box."

Figure 5-13 illustrates one way that receptacles serving computer-type equipment can be connected. The panel might have been provided with an equipment ground bus insulated from the panel. This would provide a means of splicing one or more separate insulated isolated equipment grounding conductors running from the ground pin of the isolating-type receptacles. One larger size insulated isolated equipment grounding conductor would then be run from the source (in this figure, the ground bus in the main switch) to the ground bus in the panel.

Another problem that occurs where there is a heavy concentration of electronic equipment is that

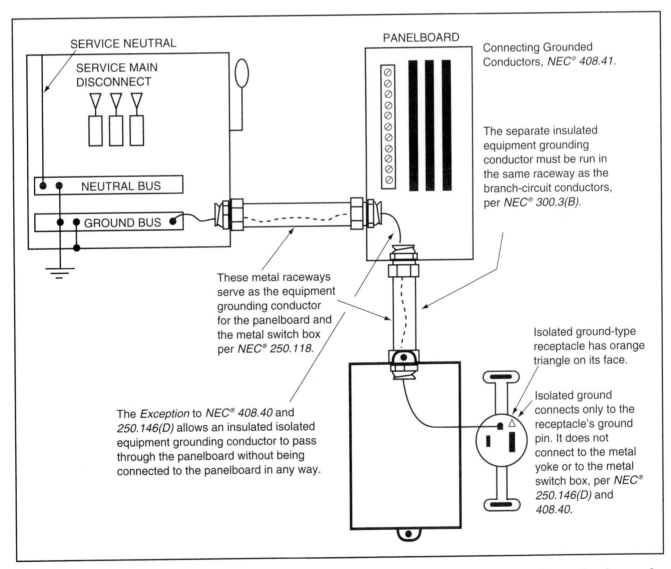

Figure 5-13   Diagram showing the installation of an insulated isolated ground to reduce noise on circuits supplying computers.

of overheated neutral conductors. To minimize this problem on branch-circuits that serve electronic loads such as computers, it is highly recommended that separate neutrals be run for each phase conductor of the branch-circuit wiring, rather than running a common neutral for multiwire branch-circuits. This is done to eliminate the problem of harmonic currents overheating the neutral conductor.

If surge protection is desired, this will absorb high-voltage surges on the line and further protect the equipment, Figure 5-14. These are highly recommended in areas of the country where lightning strikes are common. When a comparison is made between the cost of the equipment or the cost of re-creating lost data and the cost of the receptacle, probably under $50, this is inexpensive protection.

Ground-Fault Circuit Interrupter Receptacles

▶*NEC® 210.8* sets forth the requirements for the installation of ground-fault circuit interrupters (GFCI) in all occupancies.◀

▶In commercial buildings, GFCI protection is required for all 125-volt, single-phase, 15- and 20-ampere receptacles installed in bathrooms, in kitchens, on rooftops, outdoors where accessible to the public, and outdoors for receptacles required for servicing HVAC equipment. Receptacles on rooftops and outdoors in public places are exempt if they are not readily accessible and are supplied by a dedicated branch circuit.◀

▶The definition in *NEC® 210.8(B)(2)* of a kitchen is *an area with a sink and permanent facilities for food preparation and cooking.* This includes employee break areas.◀

GFCI protection can be provided by GFCI circuit breakers or GFCI receptacles. For complete information on how GFCIs operate, refer to *Electrical Wiring—Residential.*

Electrocutions and personal injury have resulted because of electrical shock from appliances, such as radios, shavers, and electric heaters. This shock hazard exists whenever a person touches both the defective appliance and a conducting surface, such as a water pipe, metal sink, or any conducting material that is grounded. To protect against this possibility of shock, 15- and 20-ampere branch-circuits can be protected with GFCI receptacles or GFCI circuit breakers. The receptacles are the most commonly used.

Figure 5-14   Surge suppressor receptacle.

The Underwriters Laboratories require that Class A GFCIs trip on ground-fault currents of 4 to 6 milliamperes (0.004 to 0.006 ampere). Figure 5-15 illustrates the principle of how a GFCI operates.

▶**Cord- and Plug-Connected Vending Machines.** Cord- and plug-connected vending machines (new and remanufactured) are required to have integral GFCI protection either in the attachment plug cap or within 12 in. (300 mm) of the attachment plug cap. If the vending machine does not have integral GFCI protection, *NEC® 422.51* requires that the outlet into which the vending machine will be plugged into be GFCI protected.◀

Receptacles in Electric Baseboard Heaters

Electric baseboard heaters are available with or without receptacle outlets. Figure 5-16 shows the relationship of an electric baseboard heater to a receptacle outlet. *NEC® Article 424* of the *Code* states the requirement for fixed electric space heating equipment. This receptacle may be counted as one of the required outlets for the wall space utilized by the heater. See *NEC® 210.52.* This receptacle shall not be connected to the heater's branch-circuit.

SNAP SWITCHES

The term *snap switch* is rarely used today except in the *National Electrical Code®* and the Underwriters Laboratories standards. Most electricians refer to snap switches as toggle switches or wall switches. Refer to Figure 5-17. These switches are divided into two categories.

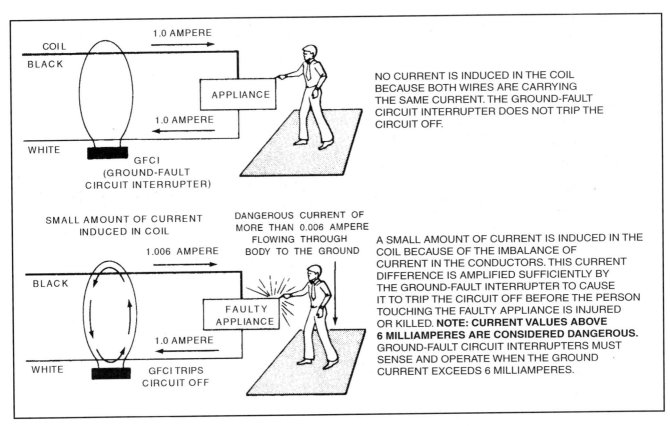

Figure 5-15   Basic principle of ground-fault circuit interrupter operation.

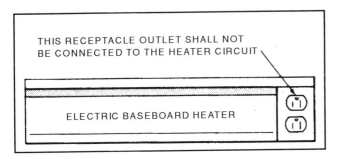

Figure 5-16   Factory-mounted receptacles on permanently installed electric baseboard heater.

NEC® 404-14(B) contains those ac–dc general-use snap switches that are used to control:

- alternating-current or direct-current circuits.
- resistive loads not to exceed the ampere rating of the switch at rated voltage.
- inductive loads not to exceed one-half the ampere rating of the switch at rated voltage.
- tungsten filament lamp loads not to exceed the ampere rating of the switch at 125 volts when marked with the letter T. (A tungsten filament lamp draws a very high current at the instant the

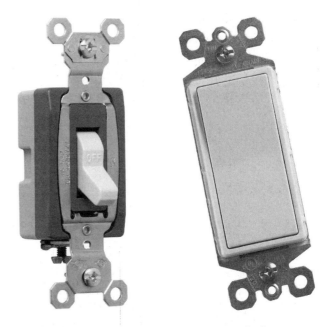

Figure 5-17   General-use snap switches.

circuit is closed. As a result, the switch is subjected to a severe current surge.)

The ac–dc general-use snap switch normally is not marked ac–dc. However, it is always marked with the current and voltage rating, such as 10A-125V or 5A-250V-T.

NEC® 404-14(A) contains those ac general-use snap switches that are used to control

- alternating-current circuits only.

- resistive, inductive, and tungsten-filament lamp loads not to exceed the ampere rating of the switch at 120 volts.

- motor loads not to exceed 80 percent of the ampere rating of the switch at rated voltage.

Ac general-use snap switches may be marked ac only, or they may be marked with the current and voltage rating markings. A typical switch marking is *15A, 120-277V ac*. The 277-rating is required on 277/480-volt systems. See *NEC® 210.6* for additional information pertaining to maximum voltage limitations.

Terminals of switches rated at 20 amperes or less, when marked CO/ALR, are suitable for use with aluminum, copper, and copper-clad aluminum conductors, *NEC® 404.14*. Switches not marked CO/ALR are suitable for use with copper and copper-clad aluminum conductors only.

Screwless pressure terminals of the conductor push-in type may be used with copper and copper-clad aluminum conductors only. These push-in-type terminals are not suitable for use with ordinary aluminum conductors.

Further information on switch ratings is given in *NEC® 404.14(C)* and in the Underwriters Laboratories *Electrical Construction Equipment Directory*.

Snap Switch Types and Connections

Snap switches are readily available in four basic types: single-pole, three-way, four-way, and double-pole.

Single-Pole Switch. A single-pole switch is used where it is desired to control a light or group of lights, or other load, from one switching point. This type of switch is used in series with the ungrounded (hot) wire feeding the load. Figure 5-18 shows typical applications of a single-pole switch controlling a light from one switching point, either at the switch or at the light. Note that red has been used for the switch return (switch leg/switch loop) in the raceway (conduit) diagrams. Any color other than white, gray, or green could have been used for the switch return. When using cable for the wiring of switches,

NEC® 200.7(C)(2) permits the use of a conductor with insulation that is white, gray, or has three continuous white stripes as a switch supply *only* if the conductor is permanently reidentified by painting or other effective means at each location where the conductor is visible and accessible.

Three diagrams are shown for each of the switching connections discussed. The first diagram is a schematic drawing and is valuable when visualizing the current path. The second diagram represents the situation where a raceway will be available for installing the conductors. The third diagram illustrates the connection necessary when using a cable such as armored cable (AC), metal-clad cable (MC), or nonmetallic-sheathed cable (NMC). Grounding conductors and connections are not shown in order to keep the diagrams as simple as possible.

Three-Way Switch. A three-way switch, shown in Figure 5-19, has a *common terminal* to which the switch blade is always connected. The other two terminals are called the *traveler terminals*. In one position, the switch blade is connected between the common terminal and one of the traveler terminals. In the other position, the switch blade is connected between the common terminal and the second traveler terminal. The three-way switch can be identified readily because it has no on or off position. Note that on and off positions are not marked on the switch handle in Figure 5-19. The three-way switch is also identified by its three terminals. The common terminal is darker in color than the two traveler terminals, which have a brass color. Figure 5-20 shows the application of three-way switches to provide control at the light or at the switch. Travelers are also referred to as kickers or dummies.

In Figure 5-20A and Figure 5-20B, red has been used for the switch return (switch leg/switch loop) in the raceway (conduit) diagrams. Any color other than white, gray, or green could have been used for the switch return. For the *travelers*, select a color that is different than the switch return, such as a pair of yellows, a pair of blues, a pair of browns, etc. When using cable for the wiring of three-way switches, *NEC® 200.7(C)(2)* permits the use of a conductor with insulation that is white, gray, or has three continuous white stripes as a switch supply *only* if the conductor is permanently reidentified by painting or other effective means at each location where the conductor is visible and accessible.

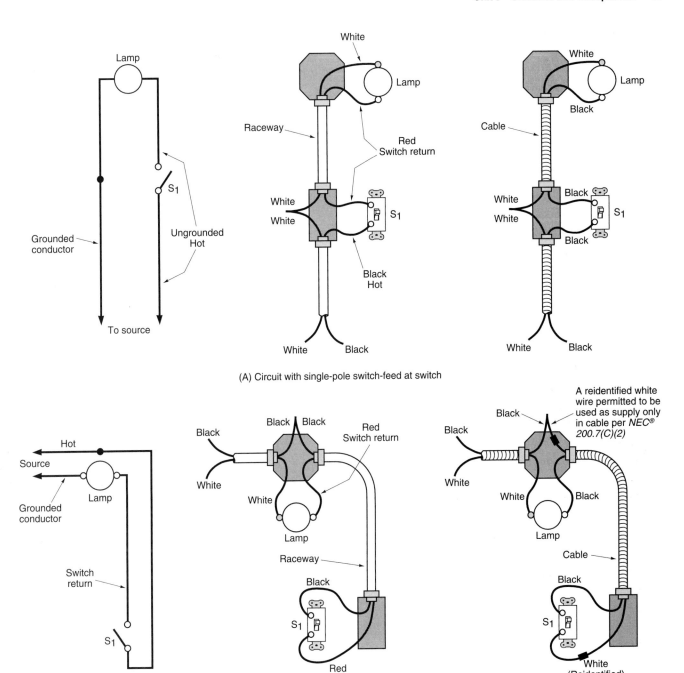

(A) Circuit with single-pole switch-feed at switch

(B) Circuit with single-pole switch-feed at light

Figure 5-18   Single-pole switch connections.

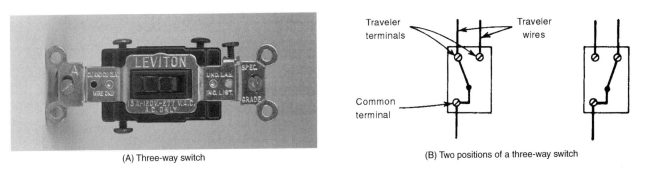

(A) Three-way switch

(B) Two positions of a three-way switch

Figure 5-19   (A) Three-way switch and (B) positions.

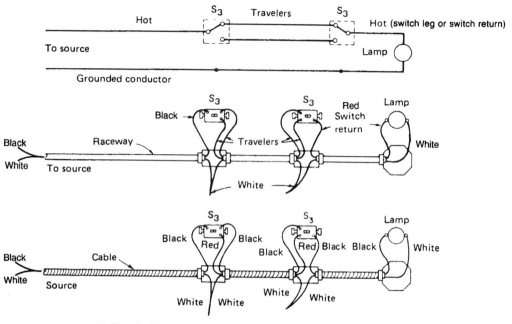

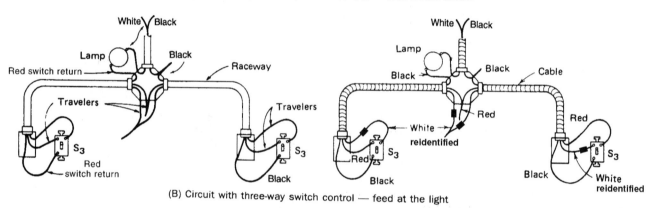

(A) Circuit with three-way switch control — feed at the switch

(B) Circuit with three-way switch control — feed at the light

Figure 5-20   Three-way switch connections.

Four-Way Switch. A four-way switch is similar to the three-way switch in that it does not have on and off positions. However, the four-way switch has four terminals. Two of these terminals are connected to traveler wires from one three-way switch, and the other two terminals are connected to traveler wires from another three-way switch, Figure 5-21. In Figure 5-21, terminals A1 and A2 are connected to one three-way switch and terminals B1 and B2 are connected to the other three-way switch. In position 1, the switch connects A1 to B2 and A2 to B1. In position 2, the switch connects A1 to B1 and A2 to B2.

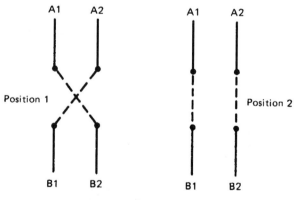

Two positions of a four-way switch

Figure 5-21   Four-way switch operation.

The four-way switch is used when a light, a group of lights, or other load must be controlled from more than two switching points. The switches that are connected to the source and the load are three-way switches. At all other control points, however, four-way switches are used. Figure 5-22 illustrates a typical circuit in which a lamp is controlled from any one of three switching points. Note that red has been used for the switch return (switch leg/switch loop) in the raceway (conduit) diagrams. Any color other than white, gray, or green could have been used for the switch return. For the *travelers*, select a color that is different than the switch return, such as a pair of yellows, a pair of blues, a pair of browns, etc. For cable wiring, the white conductor is used as the supply to a switch, and shall be permanently reidentified as being an ungrounded conductor. The white conductor shall not be used as the switch return, *200.7(C)(2)*. In Figure 5-22, the arrangement is such that white conductors were spliced together. Care must be used to ensure that the traveler wires are connected to the proper terminals of the four-way switch. That is, the two traveler wires from one three-way switch must be connected to the two terminals on one end of the four-way switch. Similarly, the two traveler wires from the other three-way switch must be connected to the two terminals on the other end of the four-way switch.

Double-Pole Switch. A double-pole switch is rarely used on lighting circuits. As shown in Figure 5-23, a double-pole switch can be used for those installations where two separate circuits are to be controlled with one switch. All conductors of circuits supplying gasoline-dispensing pumps, or running through such pumps, must have a disconnecting means. Thus, the lighting mounted on gasoline-dispensing islands may require two-pole switches, *NEC® 514.11.*

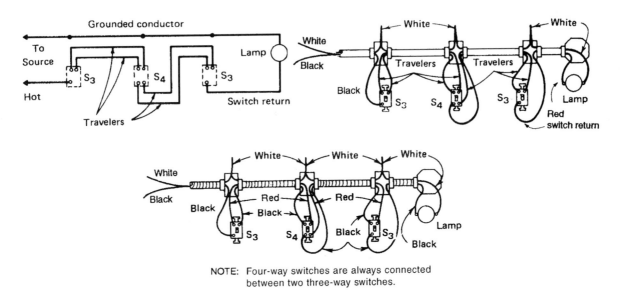

NOTE: Four-way switches are always connected between two three-way switches.

Figure 5-22 Four-way switch connections.

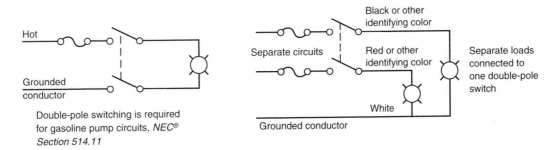

Figure 5-23 Double-pole switch connections.

Grounding of Snap Switches

Snap switches are required to be grounded, *NEC® 404.9(B)*. To meet this *NEC®* requirement, you will see a green hexagon shaped grounding screw on all snap switches and receptacles. This is one of the requirements of *NEC® 404.9(B)*. Where the wiring method is a grounded metallic wiring method such as EMT, AC, or MC, this serves as the equipment grounding conductor, *NEC® 250.118*.

When there is direct metal-to-metal contact of the metal yoke between the switch and the metal wall box, or if the switch or receptacle has a self-grounding clip on the yoke, there is no need to connect a separate equipment grounding conductor to this grounding screw.

▶ If proper equipment grounding cannot be provided, such as a system that does not have an equipment grounding conductor as defined in *NEC® 250.118*, we find in *NEC® 404.9(B)* that snap switches are considered to be effectively grounded if protected by a ground-fault circuit-interrupter. ◀

After proper grounding of a snap switch or receptacle has been accomplished, we must then think about how to ground the face plate. A metal face plate becomes grounded when the No. 6-32 faceplate screws are in place, provided the snap switch or receptacle is properly grounded.

CONDUCTOR COLOR CODING

Some color coding is mandated by the *National Electrical Code.®* These are the grounded conductors and equipment grounding conductors. Other color coding choices are left up to the electrician. Reference to color coding is found in *NEC® 110.15, 200.6, 200.7, 210.5, 215.8, 230.56,* and *310.12*.

Ungrounded Conductor

An ungrounded hot conductor must have an outer finish that is other than green, white, gray, or with three continuous white stripes. This conductor is commonly called the hot conductor. You will get a shock if this hot conductor and a grounded conductor or grounded surface such as a metal water pipe are touched at the same time.

Popular colors are black, white, red, blue, green, yellow, brown, orange, and gray. Conductors 1 AWG and larger are generally available in black only.

Grounded Conductor

For alternating-current circuits, the *NEC®* requires that the grounded (identified) conductor have an outer finish of either a continuous white or gray color, or have an outer finish (not green) that has three continuous white stripes along the conductor's entire length. In multiwire circuits, the grounded conductor is also called a *neutral* conductor.

A grounded conductor is often referred to as a neutral conductor. A neutral conductor is always grounded under the requirements of *NEC® 250.20*, but a grounded conductor is not always a neutral conductor, such as in the case of the grounded phase conductor of a three-phase, grounded B-phase system.

NEC® 200.6(A) requires that a grounded conductor, sizes 6 AWG or smaller, be identified by a continuous white or gray color. *NEC® 200.6(B)* permits grounded conductors larger than 6 AWG to be identified at its terminations. For example, this would allow a black conductor to be identified in a disconnect switch or panel with white paint.

When more than one system is installed in the same raceway, such as combinations of 208Y/120-volt wye and 277Y/480-volt wye circuits, you must be able to distinguish which grounded conductor belongs to which set of ungrounded conductors. Generally, white is used for the grounded conductor of 208Y/120-volt circuits and gray is used for the grounded conductor of 277Y/480-volt circuits. The color chosen for identifying the grounded conductors of the different systems shall be posted at each branch-circuit panelboard. See *NEC® 200.6(D)*.

A white or gray conductor shall not be used as a hot phase conductor except when reidentified according to *NEC® 200.7*, which requires it to be a part of a cable or flexible cord.

Three Continuous White Stripes. In commercial and industrial wiring, some conductors have three continuous white stripes. This method of identifying conductors is done by the cable manufacturer and is found on larger size conductors and some service-entrance cables, such as those used for a mobile-home feeder (two black, one black that has 3-white stripes, and one green).

Equipment Grounding Conductor

An equipment grounding conductor may be bare, green, or green with one or more yellow stripes, see *NEC® 250.119*. Never use a green insulated conductor for a hot phase conductor.

Typical Color Coding (for Conduit Wiring)

The following are some examples of color coding quite often used for the conductors of branch-circuits, feeders, and services. These color coding recommendations are not code mandated, other than for the white (or gray) grounded conductor. Certainly, other color combinations are permitted. Just be sure that you do not use white, gray, green, and bare conductors other than for their permitted uses discussed previously.

- Two-wire, single-phase: black, white
- Three-wire, single-phase: black, white, red
- Three-wire, three-phase delta: brown, orange, yellow
- Three-wire, three-phase delta grounded B-phase: black, red, white for the "B" phase
- Four-wire (three-phase, four-wire wye 208Y/120 volt): black, white, red, blue
- Four-wire (three-phase, four-wire wye 480/277 volt): brown, white, yellow, orange
- Four-wire (three-phase, four-wire delta with "high leg"): white, brown, yellow, orange (the high leg). You could also use white, black, blue, orange. See *110.15* for requirement to use the color orange to identify the "high" leg. See *408.3(E)* for phase arrangement of the "high" leg.
- Switch legs: Use a different color than the phase conductors.
- Travelers: Travelers are used for three-way and four-way switch connections. Use conductors that have a color different than the phase conductors or the actual switch leg; for example, a pair of blue conductors, a pair of yellow conductors, or a pair of brown conductors.

CAUTION: For new work, the grounded conductor is permitted to be white or gray. See *NEC® 200.6* and *200.7*. In old installations, a conductor having gray insulation was sometimes used as an ungrounded hot conductor. This is no longer permitted. Be very careful when working on any circuit that has a gray insulated conductor. It might be a hot conductor . . . and could be lethal.

Typical Color Coding (for Cable Wiring)

When wiring with cable (NMC, AC, or MC), the choice of colors is limited to those that are furnished with the cable:

Two-Wire Cable: black, white, bare or green equipment grounding conductor

Three-Wire Cable: black, white, red, bare, or green equipment grounding conductor

Four-Wire Cable: black, white, red, blue, bare, or green equipment grounding conductor

Five-Wire Cable: black, white, red, blue, yellow, bare, or green equipment grounding conductor

Several of the wiring diagrams in this unit illustrate how the white wire is correctly used in switching circuits.

SWITCH AND RECEPTACLE COVERS

The cover that is placed on a recessed box containing a receptacle or a switch is called a faceplate, and a cover placed on a surface-mounted box, such as a 4-in. square, is called a raised cover.

Faceplates come in a variety of colors, shapes, and materials, but for our purposes, they are placed in two categories, metal and insulating. Metal faceplates can become a hazard because they can conduct electricity. Metal faceplates are considered to be effectively grounded through the No. 6-32 screws that fasten the faceplate to the grounded yoke of a receptacle or switch. If the box is nonmetallic, the yoke must be grounded, and both switches and receptacles are available with a grounding screw for grounding the metal yoke.

In the past, it was common practice for a receptacle to be fastened to a raised cover by a single 6–32 screw. *NEC® 406.4(C)* now prohibits this practice and requires that two screws or another approved method be used to fasten a receptacle to a raised cover.

Receptacles that are installed outdoors in damp or wet locations, as stated in *NEC® 406.8(A)* and *(B)*, must be covered with weatherproof covers. If installed in a wet location, the outdoor receptacles must be covered with weatherproof covers that maintain their integrity when the receptacle is in use. The cover would be deep enough to shelter the attachment plug.

REVIEW QUESTIONS

Refer to the *National Electrical Code*® or the working drawings when necessary. Where applicable, responses should be written in complete sentences.

Indicate which of the following switches may be used to control the loads listed in Questions 1 through 7.

A. AC/DC   10 A–125 V / 5 A–250 V
B. AC only 10 A–120 V
C. AC only 15 A–120/277 V
D. AC/DC   20 A–125 V–T / 10 A–250 V–T

1. __D__   A 120-volt incandescent lamp load (tungsten filament) consisting of ten 150-watt lamps

2. __C__   A 120-volt fluorescent lamp load (inductive) of 1500 volt-amperes

3. __C__   A 277-volt fluorescent lamp load of 625 volt-amperes

4. __A__   A 120-volt motor drawing 10 amperes

5. __B__   A 120-volt resistive load of 1250 watts

6. __B__   A 120-volt incandescent lamp load of 2000 watts

7. __A__   A 230-volt motor drawing 2.5 amperes

Using the information in Table 5-2 and Table 5-3, select by number the correct receptacle for the conditions described in Questions 8 through 11.

8. A 20-ampere, 120/208-volt, single-phase load __14-20R__

9. A 50-ampere, 230-volt, three-phase load __15-50R__

10. A 30-ampere, 208-volt, single-phase load __6-30R__

11. A 20-ampere, hospital-grade, 120-volt, single-phase __5-20R Hospital grade__

12. The metal yoke of an isolated grounding-type receptacle (is) (is not) connected to the green equipment grounding terminal of the receptacle. (Circle the correct answer.)

For Questions 13 through 15, check the correct statement.

13. ☐ The recommended method to ground computer/data-processing equipment is to use the grounded metal raceway as the equipment grounding conductor.
    ☑ The recommended method to ground computer/data-processing equipment is to install a separate insulated equipment grounding conductor in the same raceway in which the branch-circuit conductors are installed.

14. When a separate insulated equipment grounding conductor is installed to minimize line disturbances on the circuit supplying an isolated grounding-type receptacle, it

    ☐ shall terminate in the panelboard where the branch-circuit originates.
    ☑ may run through the panelboard where the branch-circuit originates, back to the main disconnect (the source) for the building.

15. To eliminate the overheated neutral problem associated with branch-circuit wiring that supplies computer/data-processing equipment:

    ☑ install a separate neutral for each phase conductor.
    ☐ install a common neutral on multiwire branch-circuits.

An electrical system has been installed for three-way control of a light. The power source is available at one of the three-way switches; the load is located between the three-way switches.

16. Draw a wiring diagram illustrating the connection of the conductors.

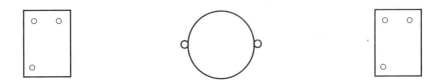

17. Draw the conductors in the raceway diagram that follows. Draw the connections to the switches and the light. Label the conductors for color and function.

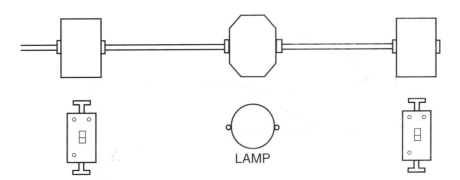

LAMP

18. Following is the same arrangement but using nonmetallic sheathed cable. Draw connections to the switches and the light. Label the conductors for color and function. A two-wire cable would contain black and white conductors; a three-wire cable would contain red, black, and white conductors.

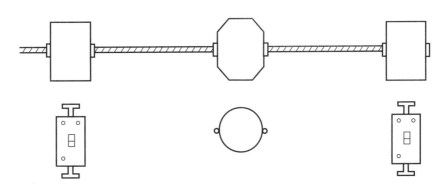

19. This diagram consists of two lamps and two three-way switches. Draw a wiring diagram of the connections. Bring the circuit in from the left.

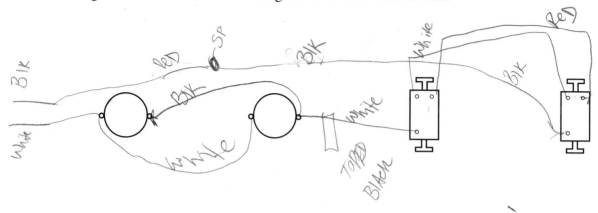

20. Following is a diagram showing the installation of a metallic raceway system connecting two lamps and two three-way switches. Draw the necessary conductors and show their connections.

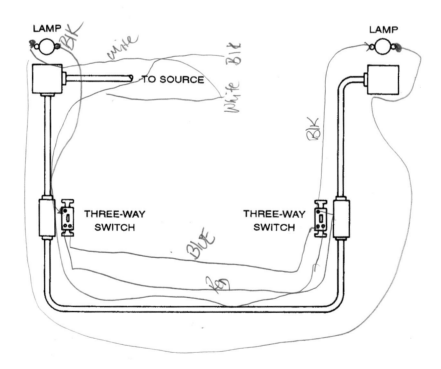

21. Following is a diagram showing the installation of a nonmetallic sheathed cable system connecting two lamps and two three-way switches. Draw the necessary conductors and show their connections.

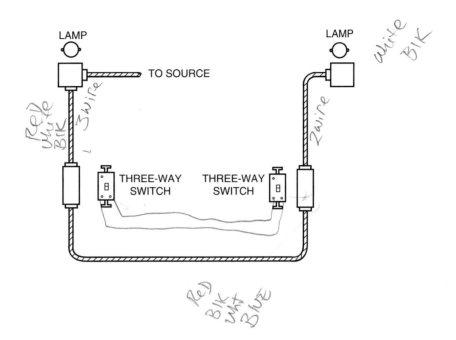

22. This digram consists of one lamp, two three-way switches, and one four-way switch. Draw a wiring diagram of the connections. Bring the circuit in from the top (lamp location first).

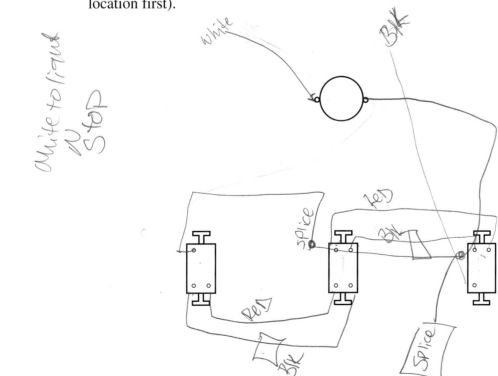

23. Following is a diagram showing the installation of a metallic raceway system connecting a lamp outlet to a four-way switch system. Draw the necessary conductors and show their connections.

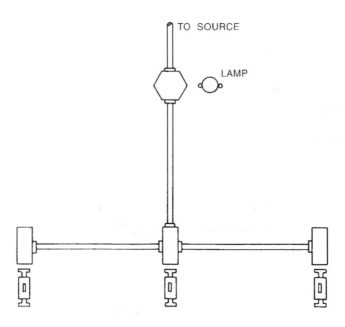

24. Following is a diagram showing the installation of a nonmetallic sheathed cable system connecting a lamp and a four-way switch system. Draw the necessary conductors and show their connections.

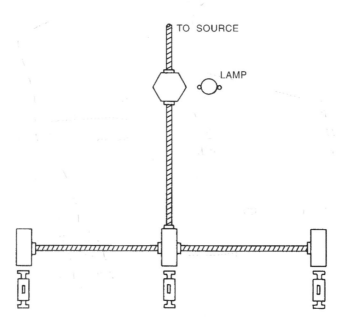

UNIT 6

Branch-Circuit Installation

OBJECTIVES

After studying this unit, the student should be able to

- select the proper raceway for the conditions.
- identify the installation requirements for a raceway.
- select the proper raceway size, dependent upon the conductors to be installed.
- select the proper size of box, dependent upon the fill.
- select the proper size of box, dependent upon entering raceways.

In a commercial building, the major part of the electrical work is the installation of the branch-circuit wiring. The electrician must have the ability to select and install the correct materials to ensure a successful job.

The term *raceway*, which is used in this unit as well as others, is defined by the *NEC®* as a channel that is designed and used expressly for the purpose of holding wires, cables, or busbars.

The following paragraphs describe several types of materials classified as a raceway. The *NEC®* recognizes intermediate metal conduit, rigid metal conduit, flexible metal conduit, liquidtight flexible metal conduit, rigid nonmetallic conduit, high density polyethylene conduit, nonmetallic underground conduit with conductors, liquidtight flexible nonmetallic conduit, electrical metallic tubing, flexible metallic tubing, and electrical nonmetallic tubing.

It will be useful to review some of the terms used in the *NEC®* concerning raceways:

- A ferrous conduit is made of iron or steel; a nonferrous conduit is made of a metal other than iron or steel.
- The common nonferrous raceway is made of aluminum.
- Metal or metallic would include both ferrous and nonferrous.

- Nonmetallic raceways would include PVC and fiberglass.
- Couplings are used to couple sections of a raceway.
- Connectors are used to fasten raceways to boxes or fittings.
- Integral couplings are formed into the raceway and cannot be removed.
- Associated couplings/connectors are separate items.
- A running thread is a double-length thread cut on one conduit. Connection to another conduit is achieved by screwing a coupling on the running thread, butting the conduits together, and then backing the coupling onto the second conduit. Running threads are not permitted for connecting conduits together with a coupling, *NEC® 344.42*.

RIGID METAL CONDUIT (RMC)

Rigid metal conduit, Figure 6-1, is of heavy-wall construction to provide a maximum degree of physical protection to the conductors that run through it. Rigid conduit is available in either steel or aluminum. The conduit can be threaded on the job, or nonthreaded fittings may be used where permitted. See *NEC® Article 344* and Figure 6-2.

Figure 6-1   Rigid metal conduit.

Figure 6-2   Rigid metal conduit fittings.

NEC® 300.6 covers the need for protection of metal raceways against corrosion. When rigid metal conduit is threaded in the field (on the job), the thread must be coated with an approved electrically conductive, corrosion-resistant compound.

RMC bends can be purchased, or they can be made using special bending tools. Bends in trade sizes ½, ¾, and 1 RMC can be made using hand benders or hickeys, Figure 6-3. Hydraulic benders must be used to make bends in larger sizes of conduit.

Figure 6-3   Tubing benders (without handles).

INTERMEDIATE METAL CONDUIT (IMC)

Intermediate metal conduit has a wall thickness that is between that of rigid metal conduit and electrical metallic tubing (EMT). This type of conduit can be installed using either threaded or nonthreaded fittings. *NEC® Article 342* defines IMC and describes the uses permitted and not permitted, the installation requirements, and the construction specifications.

ELECTRICAL METALLIC TUBING (EMT)

EMT is a thin-wall metal raceway that is not to be threaded, Figure 6-4. The specifications for commercial building permit the use of EMT for all branch-circuit wiring. *NEC® Article 358* should be consulted for exact installation requirements.

Fittings

Electrical metallic tubing is nonthreaded, thin-wall conduit. Because EMT is not to be threaded, conduit sections are joined together and connected to boxes, other fittings, or cabinets by couplings and connectors. Several styles of EMT fittings are available, including set-screw, compression, and indenter styles.

Set-screw.   When used with this type of fitting, the EMT is pushed into the coupling or connector and is secured in place by tightening the set-screws, Figure 6-5. This type of fitting is classified as concrete-tight.

Figure 6-4   Electrical metallic tubing.

Figure 6-5   EMT connector and coupling, set-screw type.

Compression. EMT is secured in these fittings by tightening the compression nuts with a wrench or pliers, Figure 6-6. These fittings are classified as raintight and concrete-tight types.

Indenter. A special tool is used to secure EMT in this style of fitting. The tool places an indentation in both the fitting and the conduit. It is a standard wiring practice to make two sets of indentations at each connection. This type of fitting is classified as concrete-tight. Figure 6-7 shows a straight indenter connector.

Installation of EMT

The efficient installation of EMT requires the use of a bender, Figure 6-8. This tool is commonly available in hand-operated models for EMT in trade sizes ½, ¾, and 1 and in power-operated models for EMT larger than trade size 1.

Three kinds of bends can be made with the use of the bending tool. The stub bend, the back-to-back bend, and the angle bend are shown in Figure 6-9. The manufacturer's instructions that accompany each bender indicate the method for making each type of bend.

Installation of Metallic Raceway

RMC, IMC, and EMT are to be installed according to the requirements of *NEC® Articles 342, 344,* and *358.* The following points summarize the contents of these articles. All conduit runs should be level, straight, plumb, and neat, showing good workmanship. Do not do sloppy work.

The rigid types of metal conduit:

- may be installed in concealed and exposed work.
- may be installed in or under concrete when of the type approved for this purpose.
- must not be installed in or under cinder concrete or cinder fill that is subject to permanent moisture, unless the conduit is encased in at least 2 in. (50 mm) of noncinder concrete, is at least 2 in. (450 mm) under the fill, or is of corrosion-resistant material suitable for the purpose.
- must not be installed where subject to severe mechanical damage. (An exception to this is rigid metal conduit, which may be installed in a location where it is subject to damaging conditions.)

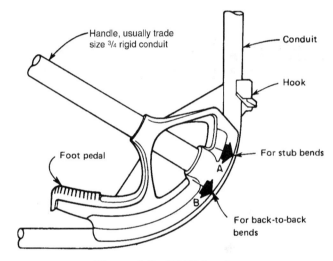

Figure 6-8 EMT bender.

Figure 6-6 EMT connector and coupling, compression type.

Figure 6-7 EMT connector, indenter type.

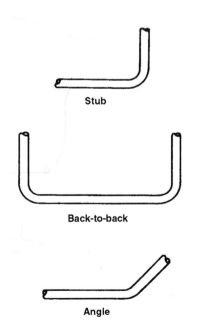

Figure 6-9 Conduit bends.

- may contain up to four quarter bends (for a total of 360 degrees) in any run.

- shall be fastened within 3 ft (900 mm) of each outlet box, junction box, cabinet, or conduit body but may be extended to 5 ft (1.5 m) to accommodate structural members.

- shall be securely supported at least every 10 ft (3 m) except if threaded couplings are used as given in *NEC® Table 344.30(B)(2)*.

- may be installed in wet or dry locations if the conduit is of the type approved for this use.

- shall have the ends reamed to remove rough edges.

- is considered adequately supported when run through drilled, bored, or punched holes in framing members, such as studs or joists.

- Grounding: All listed fittings for metallic raceways are tested for a specified amount of current for a specified length of time, making the fittings acceptable for grounding.

FLEXIBLE CONNECTIONS

The installation of certain equipment requires flexible connections, both to simplify the installation and to stop the transfer of vibrations.

There are three basic wiring methods used for flexible connections:

1. Flexible metal conduit (FMC), Figure 6-10, Figure 6-12, and Figure 6-15. See *NEC® Article 348*.

- Liquidtight flexible metal conduit (LFMC), Figure 6-11 and Figure 6-13. See *NEC® Article 350*.

- Liquidtight flexible nonmetallic conduit (LFNC), Figure 6-14. See *NEC® Article 356*.

- Grounding: All listed fittings for metallic raceways are tested for a specified amount of current for a specified length of time, making the fittings acceptable for grounding.

Figure 6-11   Liquidtight flexible metal conduit and fitting.

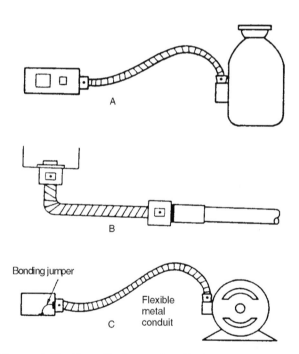

Figure 6-12   Installations using flexible metal conduit.

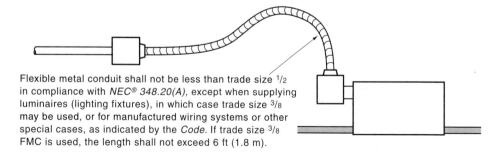

Flexible metal conduit shall not be less than trade size 1/2 in compliance with *NEC® 348.20(A)*, except when supplying luminaires (lighting fixtures), in which case trade size 3/8 may be used, or for manufactured wiring systems or other special cases, as indicated by the *Code*. If trade size 3/8 FMC is used, the length shall not exceed 6 ft (1.8 m).

Figure 6-10   Flexible metal conduit.

NEC® Article 348 regulates the use and installation of FMC. FMC is similar to armored cable, except that the conductors are installed by the electrician. For armored cable, the cable armor is wrapped around the conductors at the factory to form a complete cable assembly.

Some of the more common installations using flexible metal conduit are shown in Figure 6-12. Note that the flexibility required to make the installation is provided by the flexible metal conduit. The figure calls attention to the *National Electrical Code®* and Underwriters Laboratories restrictions on the use of FMC with regard to relying on the metal armor as a grounding means.

Flexible metal and nonmetallic conduit is commonly used to connect recessed luminaires (lighting fixtures). When used in this manner, electricians refer to it as a "fixture whip."

FMC of trade size ¾ or less may be used as a grounding means if:

- It is listed.
- It is not more than 6 ft (1.8 m) in length.
- It is connected with fittings listed for grounding purposes.
- The circuit is rated at 20 amperes or less. See Figure 6-12A.

FMC of a trade size larger than ¾ may be used as a grounding means if:

- It is listed.
- It is not more than 6 ft (1.8 m) in length.
- It is connected with fittings marked "GRND." See Figure 6-12B.

FMC of any size may be installed with a bonding jumper:

- outside the conduit if it is not more than 6 ft (1.8 m) in length.
- inside the conduit. See Figure 6-12C.

LFMC of trade size ½ or smaller may be used as a grounding means if:

- It is listed.
- It is not more than 6 ft (1.8 m) in length.
- It is connected with fittings listed for grounding purposes.
- The circuit is rated at 20 amperes or less. See Figure 6-13.

The use and installation of LFMC is described in *NEC® Article 350*. LFMC has a tighter fit of its spiral turns as compared to FMC. LFMC has a thermoplastic outer jacket that is liquidtight and is commonly used as a flexible connection to central air-conditioning units located outdoors, Figure 6-13.

LFMC of trade size ¾, 1, or 1¼ may be used as a grounding means if:

- It is listed.
- It is not more than 6 ft (1.8 m) in length.
- It is connected with fittings listed for grounding purposes.
- The circuit is rated at 60 amperes or less. See Figure 6-13.

LFMC of any size or length may be installed with a required bonding jumper. See Figure 6-14.

Liquidtight flexible nonmetallic conduit (LFNC) may be used:

- in exposed or concealed locations.
- where flexibility is required.
- for direct burial, when so listed.

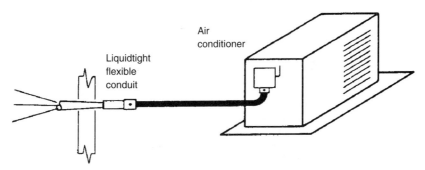

Figure 6-13 An application of liquidtight flexible metal conduit.

Figure 6-14   An application for liquidtight flexible nonmetallic conduit.

- where not subject to damage.
- where the combination of ambient and conductor temperature does not exceed that for which the flexible conduit is approved.
- in lengths of no more than 6 ft (1.8 m) unless certain provisions are adhered to.
- with fittings identified for use with the flexible conduit.
- in trade sizes ½ to 2. For enclosing motor leads, trade size ⅜ is suitable.
- for "fixture whips" not less than 18 in. (450 mm) or over 6 ft (1.8 m) in length permitted or required by *NEC® 410.67(C)*, or for flexible connection to equipment, such as a motor.

When metal fittings are not marked GRND, it can be assumed that they are not approved for equip-

ment grounding purposes. In this case, a separate equipment grounding conductor must be installed. The conductor is sized according to *NEC® Table 250.122*. Figure 6-15 illustrates the application of this table.

The three types of liquidtight flexible nonmetallic conduit are marked

LFNC-A for layered conduit

LFNC-B for integral conduit

LFNC-C for corrugated conduit

Both metal and nonmetallic fittings listed for use with the various types of LFNC will be marked with the "A," "B," or "C" designations.

Liquidtight flexible nonmetallic conduit is restricted to a length of 6 ft (1.8 m), but there are exceptions. See *NEC® 356.10, 356.12*, and *356.20*.

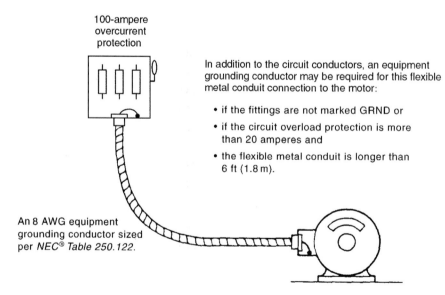

100-ampere overcurrent protection

In addition to the circuit conductors, an equipment grounding conductor may be required for this flexible metal conduit connection to the motor:

- if the fittings are not marked GRND or
- if the circuit overload protection is more than 20 amperes and
- the flexible metal conduit is longer than 6 ft (1.8 m).

An 8 AWG equipment grounding conductor sized per *NEC® Table 250.122*.

Figure 6-15   Equipment grounding conductor in flexible metal conduit.

ARMORED (TYPE AC) AND METAL-CLAD (TYPE MC) CABLES

Two increasingly popular wiring systems are presented in *NEC® Articles 320* and *330*. As they are similar and different in many ways, the information on these systems is presented in the following parallel comparative format.

TYPE AC AND TYPE MC CABLES	
Type AC (*NEC® Article 320*)	**Type MC (*NEC® Article 330*)**
Definition: Type AC cable is a fabricated metallic assembly of insulated conductors in a flexible metallic enclosure.	Definition: Type MC cable is a factory assembly of one or more insulated circuit conductors with or without optical fiber members enclosed in an armor of interlocking metal tape or a smooth or corrugated metallic sheath.
Number of conductors: Two to four current-carrying conductors plus a bonding wire. It may also have a separate equipment grounding conductor.	Number of conductors: Any number of current-carrying conductors. At least one manufacturer has a "Home Run Cable" with conductors for more than one branch-circuit plus the equipment grounding conductors.
Conductor size and type: Copper conductors 14 AWG through 1 AWG; aluminum conductors 12 AWG through 1 AWG, which will be marked "AL (CU-CLAD)" or "Cu-Clad Al" on the carton or reel.	Conductor size and type: Copper conductors 18 AWG through 2000 kcmil; aluminum or copper-clad aluminum 12 AWG through 2000 kcmil, which will be marked "AL (CU-CLAD) or "Cu-Clad Al" on the carton or reel.
Color coding: Two conductors—one black and one white. Three conductors—one black, one white, and one red. Four conductors—one black, one white, one red, and one blue. These are in addition to any equipment grounding and bonding conductors.	Color coding: Two conductors—one black and one white. Three conductors—one black, one white, and one red. Four conductors—one black one white, one red, and one blue. These are in addition to any equipment grounding and bonding conductors.
Bonding and grounding: Has a 16 AWG bare bonding wire in continuous contact with metal armor. These act together to serve as an acceptable equipment ground. The bonding wire does need not be terminated; it may be folded back over the armor or cut off. It shall not be used as a neutral or an equipment grounding conductor.	Bonding and grounding: A green insulated equipment grounding conductor is included and may have two equipment grounding conductors. Unless specifically "listed for grounding" the interlocking jacket of MC cable is not to be used as an equipment grounding conductor. The sheath of smooth or corrugated tube type MC is "listed" as an acceptable equipment grounding conductor.
Insulation: Available with Type AC, thermoset, Type ACTH 167°F (75°C) thermoplastic, Type ACHH 194°F (90°C) thermoset, or Type ACTHH 194°F (90°C) thermoplastic insulations.	Insulation: Available with Type MC 194°F (90°C) thermoplastics or thermoset insulation.
Wrapping: Conductors are individually wrapped with flame retardant, light brown paper.	Wrapping: No individual wrapping. A polyester (Mylar) tape is placed over all the conductors.
Covering over armor: None available.	Covering over armor: A PVC outer covering is available.
Approved locations: Approved only for dry locations.	Approved locations: Approved for wet locations and direct burial if so "listed." May be installed in a raceway.
Insulating bushings: Required to protect conductor insulation from damage. These keep sharp edges of the armor from cutting the conductor insulation. Usually provided with the cable.	Insulating bushings: Recommended but not mandatory. Provides additional protection for conductor insulation. Usually provided with the cable.

Minimum radius of bends:
Five times the diameter of the cable.
	Minimum radius of bends:
	Smooth sheath:
	    Ten times for sizes not more than ¾ in. (19 mm) diameter.
	    Twelve times for sizes more than ¾ in. (19 mm) and not more than 1½ in. (38 mm) in diameter.
	    Fifteen times for sizes more than 1½ in. (38 mm) in diameter.
	Corrugated or interlocked:
	    Seven times diameter of cable.

Type AC (*NEC® Article 320*)	Type MC (*NEC® Article 330*)

Super neutral:
   Not available.
	Super neutral:
	   Available. May be needed where harmonics are high, such as branch-circuits for computers. Example: a 10 AWG neutral with three 12 AWG phase conductors.

Fiber optic cables:
   Not available.
	Fiber optic cables:
	   Available with power conductors.

Supports:
   Not more than 4 ft 6 in. (1.35 m) apart.
   Not more than 1 ft (300 mm) from box or fitting.
   Not required when fished through wall or ceiling or run through holes in studs, joists, or rafters.
   Not required for lengths of no more than 6 ft (1.8 m) when used as a fixture whip in an accessible ceiling.
	Supports:
	   Not more than 6 ft (1.8 m) apart.
	   Not more than 1 ft (300 mm) from box or fitting for cable having not more than four 10 AWG or smaller conductors.
	   Not required when fished through wall or ceiling or run through holes in studs, joists, or rafters.
	   Not required for lengths of no more than 6 ft (1.8 m) when used as a fixture whip in an accessible ceiling.

Voltage rating:
   Not higher than 600 volts.
	Voltage rating:
	   Not higher than 600 volts, except some MC has a rating of not higher than 2000 volts.

Connectors:
   Set-screw-type connectors not permitted with aluminum armor.
   "Listed" connectors are suitable for grounding purposes.
	Connectors:
	   Set-screw-type connectors not permitted with aluminum armor.
	   "Listed" connectors are suitable for grounding purposes.

RIGID NONMETALLIC CONDUIT (RNC)
(*NEC® Article 352*)

▶The most commonly used RNC is made of polyvinyl chloride (PVC), which is a plastic material. Solvent-type cement is used for RNC connections and terminations.

Overview.   The two most popular types RNC are:

- Schedule 40: Permitted underground (direct burial or encased in concrete) and above ground (indoors and outdoors exposed to sunlight) where not subject to physical damage. Some Schedule 40 is marked for underground use only. Schedule 40 has a thinner wall than Schedule 80.

- Schedule 80: Permitted underground (direct burial or encased in concrete) and above ground (indoors and outdoors in sunlight) where subject to physical damage. Schedule 80 has a thicker wall than Schedule 40.

- The main differences between above ground and underground listed RNC are its fire resistance rating and its resistance to sunlight (UV).

- Some RNC is made of fiberglass (Reinforced Thermosetting).

RNC is permitted:

- concealed in walls, floors, and ceilings.

- in corrosive areas.

- in cinder fill.

- in wet and damp locations.

- for exposed work where not subject to physical damage unless identified for such use. (Schedule 80 is so listed.)

- for underground installations.

- in places of assembly if the requirements of *NEC® 518.4(B)*, *518.4(C)*, and *520.5(C)* are met.

RNC is not permitted:

- in hazardous (classified) locations except for limited use as set forth in *NEC® 501.4(B)*, *503.3(A)*, *504.20*, *514.8*, and *515.8*.

- for support of luminaires (fixtures).

- where exposed to physical damage unless so identified. (Schedule 80 is so listed.)

- where subject to ambient temperatures exceeding 122°F (50°C).

- for conductors whose insulation temperature limitations would exceed those for which the conduit is listed. (Check the marking or listing label to obtain this temperature rating.)

- for electrical circuits serving patient care areas in hospitals and health care facilities. The issue here is the need for redundant grounding.

- in places of assembly and theaters except as provided in *NEC® Articles 518.4(B)* and *520.5(C)*.

On an outdoor exposed run of RNC, it is important that expansion fittings be installed to comply with the requirements set forth in *NEC® 352.44*, where expansion characteristics of RNC are given. Also consider expansion fittings where earth movement will be a problem, *300.5(J)*. An equipment grounding conductor must be installed in RNC, and if metal boxes are used, the grounding conductor must be connected to the metal box. See *NEC® Article 352* for additional information for the installation of RNC. ◄

RNC Fittings and Boxes

A complete line of fittings, boxes, and accessories are available for RNC. Where RNC is connected to a box or enclosure, a box adapter is recommended over a male connector because the box adapter is stronger and provides a gentle radius for wire pulling.

ELECTRICAL NONMETALLIC TUBING (ENT)
(NEC® Article 362)

ENT is a pliable, corrugated raceway made of polyvinyl chloride, a plastic material that is also used to make RNC. ENT is hand-bendable and is available in coils and reels to facilitate a single-length installation between pull points. ENT is designed for use within a building or encased in concrete and is not intended for outdoor use. The specification for the commercial building allows the use of ENT for feeders, branch-circuits, and telephone wiring where a raceway trade size 2 or smaller is required. ENT is available in three colors, which allows the color coding of the various systems: blue for electrical power wiring . . . yellow for communication wiring . . . red for fire alarm wiring.

ENT may be used:

- exposed, where not subject to physical damage, in buildings not more than three floors above grade.

- concealed, without regard to building height; in walls, floors, and ceilings of a material with a 15-minute finish rating (½ in. [12.7 mm] UL classified gypsum board has a 15-minute finish rating). Check with the manufacturer for further information.

- subject to corrosive influences (check with manufacturer's recommendations).

- in damp and wet locations, with fittings identified for the use.

- above a suspended ceiling where the ceiling tile has a 15-minute finish rating. The ceiling tile does not need a finish rating if the building is not more than three floors above grade.

- embedded in poured concrete, including slab-on-grade or below grade, where fittings identified for the purpose are used.

ENT shall not be used:

- in hazardous (classified) locations except as permitted by *NEC® 504.20* and *505.15*.

- for luminaire (lighting fixture) or equipment support.

- subject to ambient temperature in excess of 50°C (122°F).

- for conductors whose insulation temperature would exceed those for which ENT is listed. (Check the UL label for the actual temperature rating.)

- for direct earth burial.

- where the voltage is more than 600 volts.

- in places of assembly and theaters except as provided in *NEC® Articles 518* and *520*.

See *NEC® Article 362* for other ENT installation requirements.

ENT Fittings

Two styles of listed mechanical fittings are available for ENT: one piece snap-on and a clamshell variety. Solvent cement PVC fittings for RNC are also listed for use with ENT. The package label will indicate which of these fittings are concrete-tight without tape.

ENT Boxes and Accessories

Wall and ceiling boxes, with knockouts for trade sizes ½, ¾, and 1 are available. ENT can also be connected to any RNC box. ENT mud boxes listed for luminaire (fixture) support are available, with knockouts for trade sizes ½, ¾, and 1.

RACEWAY SIZING

The conduit size required for an installation depends upon three factors:

1. the number of conductors to be installed

2. the cross-sectional area of the conductors

3. the permissible raceway fill

The relationship of these factors is defined in *NEC® Chapter 9, Notes 1* through *9* and *Tables 1, 4, 5, 5A, 8,* and *Annex C* of the *NEC®.* After examining the working drawings and determining the number of conductors to be installed to a certain point, either of the following procedures can be used to find the conduit size. Refer to Table 6-1, Table 6-2, Table 6-3, and Table 6-4.

If all of the conductors have the same insulation, the raceway size can be determined directly by referring to *NEC® Annex C.*

For example, assume that three 8 AWG, Type THWN conductors are to be installed in electrical metallic tubing to an air-conditioning unit. Referring to *NEC® Annex C, Table C1,* it will be determined that a trade size ½ raceway is acceptable.

Next, assume that a short section of the EMT is replaced by liquidtight flexible metal conduit. To comply with *NEC® 350.60,* this change requires the installation of an equipment grounding conductor. Because the rating of the circuit would be greater than 20 amperes, and less than 60 amperes, *NEC® Table 250.122* specifies the use of a 10 AWG. The equipment grounding conductor must be included in the conductor fill calculations in compliance with *NEC® Chapter 9, Tables, Note 3.*

One 10 AWG Type THWN @ 0.0211 in.² (14 mm²)

Three 8 AWG, Type THWN @ 3 × 0.0366 in.² = 0.1098 in.² (3 × 24 mm² = 72 mm²)

Total cross-sectional area = 0.0211 in.² + 0.1098 in.² = 0.1309 in.² (14 mm² + 72 mm² = 86 mm²)

Referring to *NEC® Chapter 9, Table 4*:

- for electrical metallic tubing, the allowable 40 percent fill for more than two conductors in trade size ½ is 0.122 in.² (78 mm²).

- for liquidtight flexible metallic conduit, the allowable 40 percent fill for more than two conductors in trade size ½ is 0.125 in.² (81 mm²).

As both fill values must be 0.1309 in.² (86 mm²) or greater, the use of a trade size ½ is not permitted. Trade size ¾ is the minimum size.

NEC® Chapter 9, Table 1 is based on common conditions of proper cabling and alignment of conductors where the length of the pull and the number of bends are within reasonable limits. It should be recognized that, for certain conditions, a larger size conduit or a lesser conduit fill should be considered.

When pulling three conductors or cables into a raceway, if the ratio of the raceway diameter to the conductor diameter is between 2.8 and 3.2, jamming can occur. With four or more conductors or cables, jamming is highly unlikely.

The limitations of this table apply only to complete conduit or tubing systems and is not intended to apply to sections of conduit or tubing used to protect exposed wiring from physical damage.

Where conduit or tubing nipples not exceeding 24 in. (600 mm) are installed between boxes, cabinets, and similar enclosures, they shall be permitted to be filled to 60 percent of their total cross-sectional area, and *Article 310* and adjustment factors need not be applied.

Before any installation of conductors or cables, the complete set of notes to *NEC® Chapter 9, Table 1* should be consulted.

SPECIAL CONSIDERATIONS

Following the selection of the circuit conductors and the branch-circuit protection, there are a number of special considerations that generally are factors for each installation. The electrician is usually given the responsibility of planning the routing of the conduit to ensure that the outlets are connected properly. As a result, the electrician must determine the length and the number of conductors

Table 4 Dimensions and Percent Area of Conduit and Tubing (Areas of Conduit or Tubing for the Combinations of Wires Permitted in Table 1, Chapter 9)

Article 358 — Electrical Metallic Tubing (EMT)

Metric Designator	Trade Size	Nominal Internal Diameter mm	Nominal Internal Diameter in.	Total Area 100% mm²	Total Area 100% in.²	60% mm²	60% in.²	1 Wire 53% mm²	1 Wire 53% in.²	2 Wires 31% mm²	2 Wires 31% in.²	Over 2 Wires 40% mm²	Over 2 Wires 40% in.²
16	½	15.8	0.622	196	0.304	118	0.182	104	0.161	61	0.094	78	0.122
21	¾	20.9	0.824	343	0.533	206	0.320	182	0.283	106	0.165	137	0.213
27	1	26.6	1.049	556	0.864	333	0.519	295	0.458	172	0.268	222	0.346
35	1¼	35.1	1.380	968	1.496	581	0.897	513	0.793	300	0.464	387	0.598
41	1½	40.9	1.610	1314	2.036	788	1.221	696	1.079	407	0.631	526	0.814
53	2	52.5	2.067	2165	3.356	1299	2.013	1147	1.778	671	1.040	866	1.342
63	2½	69.4	2.731	3783	5.858	2270	3.515	2005	3.105	1173	1.816	1513	2.343
78	3	85.2	3.356	5701	8.846	3421	5.307	3022	4.688	1767	2.742	2280	3.538
91	3½	97.4	3.834	7451	11.545	4471	6.927	3949	6.119	2310	3.579	2980	4.618
103	4	110.1	4.334	9521	14.753	5712	8.852	5046	7.819	2951	4.573	3808	5.901

Article 362 — Electrical Nonmetallic Tubing (ENT)

Metric Designator	Trade Size	Nominal Internal Diameter mm	Nominal Internal Diameter in.	Total Area 100% mm²	Total Area 100% in.²	60% mm²	60% in.²	1 Wire 53% mm²	1 Wire 53% in.²	2 Wires 31% mm²	2 Wires 31% in.²	Over 2 Wires 40% mm²	Over 2 Wires 40% in.²
16	½	14.2	0.560	158	0.246	95	0.148	84	0.131	49	0.076	63	0.099
21	¾	19.3	0.760	293	0.454	176	0.272	155	0.240	91	0.141	117	0.181
27	1	25.4	1.000	507	0.785	304	0.471	269	0.416	157	0.243	203	0.314
35	1¼	34.0	1.340	908	1.410	545	0.846	481	0.747	281	0.437	363	0.564
41	1½	39.9	1.570	1250	1.936	750	1.162	663	1.026	388	0.600	500	0.774
53		51.3	2.020	2067	3.205	1240	1.923	1095	1.699	641	0.993	827	1.282
63	2½	—	—	—	—	—	—	—	—	—	—	—	—
78	3	—	—	—	—	—	—	—	—	—	—	—	—
91	3½	—	—	—	—	—	—	—	—	—	—	—	—

Article 348 — Flexible Metal Conduit (FMC)

Metric Designator	Trade Size	Nominal Internal Diameter mm	Nominal Internal Diameter in.	Total Area 100% mm²	Total Area 100% in.²	60% mm²	60% in.²	1 Wire 53% mm²	1 Wire 53% in.²	2 Wires 31% mm²	2 Wires 31% in.²	Over 2 Wires 40% mm²	Over 2 Wires 40% in.²
12	⅜	9.7	0.384	74	0.116	44	0.069	39	0.061	23	0.036	30	0.046
16	½	16.1	0.635	204	0.317	122	0.190	108	0.168	63	0.098	81	0.127
21	¾	20.9	0.824	343	0.533	206	0.320	182	0.283	106	0.165	137	0.213
27	1	25.9	1.020	527	0.817	316	0.490	279	0.433	163	0.253	211	0.327
35	1¼	32.4	1.275	824	1.277	495	0.766	437	0.677	256	0.396	330	0.511
41	1½	39.1	1.538	1201	1.858	720	1.115	636	0.985	372	0.576	480	0.743
53	2	51.8	2.040	2107	3.269	1264	1.961	1117	1.732	653	1.013	843	1.307
63	2½	63.5	2.500	3167	4.909	1900	2.945	1678	2.602	982	1.522	1267	1.963
78	3	76.2	3.000	4560	7.069	2736	4.241	2417	3.746	1414	2.191	1824	2.827
91	3½	88.9	3.500	6207	9.621	3724	5.773	3290	5.099	1924	2.983	2483	3.848
103	4	101.6	4.000	8107	12.566	4864	7.540	4297	6.660	2513	3.896	3243	5.027

Table 6-1 *(continues)*

Reprinted with permission from NFPA 70-2005

Table 4 *Continued*

Article 342 — Intermediate Metal Conduit (IMC)

Metric Designator	Trade Size	Nominal Internal Diameter		Total Area 100%		60%		1 Wire 53%		2 Wires 31%		Over 2 Wires 40%	
		mm	in.	mm²	in.²	mm²	in.²	mm²	in.²	mm²	in.²	mm²	in.²
12	3/8	—	—	—	—	—	—	—	—	—	—	—	—
16	1/2	16.8	0.660	222	0.342	133	0.205	117	0.181	69	0.106	89	0.137
21	3/4	21.9	0.864	377	0.586	226	0.352	200	0.311	117	0.182	151	0.235
27	1	28.1	1.105	620	0.959	372	0.575	329	0.508	192	0.297	248	0.384
35	1¼	36.8	1.448	1064	1.647	638	0.988	564	0.873	330	0.510	425	0.659
41	1½	42.7	1.683	1432	2.225	859	1.335	759	1.179	444	0.690	573	0.890
53	2	54.6	2.150	2341	3.630	1405	2.178	1241	1.924	726	1.125	937	1.452
63	2½	64.9	2.557	3308	5.135	1985	3.081	1753	2.722	1026	1.592	1323	2.054
78	3	80.7	3.176	5115	7.922	3069	4.753	2711	4.199	1586	2.456	2046	3.169
91	3½	93.2	3.671	6822	10.584	4093	6.351	3616	5.610	2115	3.281	2729	4.234
103	4	105.4	4.166	8725	13.631	5235	8.179	4624	7.224	2705	4.226	3490	5.452

Article 356 — Liquidtight Flexible Nonmetallic Conduit (LFNC-B*)

Metric Designator	Trade Size	Nominal Internal Diameter		Total Area 100%		60%		1 Wire 53%		2 Wires 31%		Over 2 Wires 40%	
		mm	in.	mm²	in.²	mm²	in.²	mm²	in.²	mm²	in.²	mm²	in.²
12	3/8	12.5	0.494	123	0.192	74	0.115	65	0.102	38	0.059	49	0.077
16	1/2	16.1	0.632	204	0.314	122	0.188	108	0.166	63	0.097	81	0.125
21	3/4	21.1	0.830	350	0.541	210	0.325	185	0.287	108	0.168	140	0.216
27	1	26.8	1.054	564	0.873	338	0.524	299	0.462	175	0.270	226	0.349
35	1¼	35.4	1.395	984	1.528	591	0.917	522	0.810	305	0.474	394	0.611
41	1½	40.3	1.588	1276	1.981	765	1.188	676	1.050	395	0.614	510	0.792
53	2	51.6	2.033	2091	3.246	1255	1.948	1108	1.720	648	1.006	836	1.298

*Corresponds to 356.2(2)

Article 356 — Liquidtight Flexible Nonmetallic Conduit (LFNC-A*)

Metric Designator	Trade Size	Nominal Internal Diameter		Total Area 100%		60%		1 Wire 53%		2 Wires 31%		Over 2 Wires 40%	
		mm	in.	mm²	in.²	mm²	in.²	mm²	in.²	mm²	in.²	mm²	in.²
12	3/8	12.6	0.495	125	0.192	75	0.115	66	0.102	39	0.060	50	0.077
16	1/2	16.0	0.630	201	0.312	121	0.187	107	0.165	62	0.097	80	0.125
21	3/4	21.0	0.825	346	0.535	208	0.321	184	0.283	107	0.166	139	0.214
27	1	26.5	1.043	552	0.854	331	0.513	292	0.453	171	0.265	221	0.342
35	1¼	35.1	1.383	968	1.502	581	0.901	513	0.796	300	0.466	387	0.601
41	1½	40.7	1.603	1301	2.018	781	1.211	690	1.070	403	0.626	520	0.807
53	2	52.4	2.063	2157	3.343	1294	2.006	1143	1.772	669	1.036	863	1.337

*Corresponds to 356.2(1)

Table 6-1 *(continued)*

Reprinted with permission from NFPA 70-2005

Table 4 *Continued*

Article 350 — Liquidtight Flexible Metal Conduit (LFMC)

Metric Designator	Trade Size	Nominal Internal Diameter mm	in.	Total Area 100% mm²	in.²	60% mm²	in.²	1 Wire 53% mm²	in.²	2 Wires 31% mm²	in.²	Over 2 Wires 40% mm²	in.²
12	⅜	12.5	0.494	123	0.192	74	0.115	65	0.102	38	0.059	49	0.077
16	½	16.1	0.632	204	0.314	122	0.188	108	0.166	63	0.097	81	0.125
21	¾	21.1	0.830	350	0.541	210	0.325	185	0.287	108	0.168	140	0.216
27	1	26.8	1.054	564	0.873	338	0.524	299	0.462	175	0.270	226	0.349
35	1¼	35.4	1.395	984	1.528	591	0.917	522	0.810	305	0.474	394	0.611
41	1½	40.3	1.588	1276	1.981	765	1.188	676	1.050	395	0.614	510	0.792
53	2	51.6	2.033	2091	3.246	1255	1.948	1108	1.720	648	1.006	836	1.298
63	2½	63.3	2.493	3147	4.881	1888	2.929	1668	2.587	976	1.513	1259	1.953
78	3	78.4	3.085	4827	7.475	2896	4.485	2559	3.962	1497	2.317	1931	2.990
91	3½	89.4	3.520	6277	9.731	3766	5.839	3327	5.158	1946	3.017	2511	3.893
103	4	102.1	4.020	8187	12.692	4912	7.615	4339	6.727	2538	3.935	3275	5.077
129	5	—	—	—	—	—	—	—	—	—	—	—	—
155	6	—	—	—	—	—	—	—	—	—	—	—	—

Article 344 — Rigid Metal Conduit (RMC)

Metric Designator	Trade Size	Nominal Internal Diameter mm	in.	Total Area 100% mm²	in.²	60% mm²	in.²	1 Wire 53% mm²	in.²	2 Wires 31% mm²	in.²	Over 2 Wires 40% mm²	in.²
12	⅜	—	—	—	—	—	—	—	—	—	—	—	—
16	½	16.1	0.632	204	0.314	122	0.188	108	0.166	63	0.097	81	0.125
21	¾	21.2	0.836	353	0.549	212	0.329	187	0.291	109	0.170	141	0.220
27	1	27.0	1.063	573	0.887	344	0.532	303	0.470	177	0.275	229	0.355
35	1¼	35.4	1.394	984	1.526	591	0.916	522	0.809	305	0.473	394	0.610
41	1½	41.2	1.624	1333	2.071	800	1.243	707	1.098	413	0.642	533	0.829
53	2	52.9	2.083	2198	3.408	1319	2.045	1165	1.806	681	1.056	879	1.363
63	2½	63.2	2.489	3137	4.866	1882	2.919	1663	2.579	972	1.508	1255	1.946
78	3	78.5	3.090	4840	7.499	2904	4.499	2565	3.974	1500	2.325	1936	3.000
91	3½	90.7	3.570	6461	10.010	3877	6.006	3424	5.305	2003	3.103	2584	4.004
103	4	102.9	4.050	8316	12.882	4990	7.729	4408	6.828	2578	3.994	3326	5.153
129	5	128.9	5.073	13050	20.212	7830	12.127	6916	10.713	4045	6.266	5220	8.085
155	6	154.8	6.093	18821	29.158	11292	17.495	9975	15.454	5834	9.039	7528	11.663

Article 352 — Rigid PVC Conduit (RNC), Schedule 80

Metric Designator	Trade Size	Nominal Internal Diameter mm	in.	Total Area 100% mm²	in.²	60% mm²	in.²	1 Wire 53% mm²	in.²	2 Wires 31% mm²	in.²	Over 2 Wires 40% mm²	in.²
12	⅜	—	—	—	—	—	—	—	—	—	—	—	—
16	½	13.4	0.526	141	0.217	85	0.130	75	0.115	44	0.067	56	0.087
21	¾	18.3	0.722	263	0.409	158	0.246	139	0.217	82	0.127	105	0.164
27	1	23.8	0.936	445	0.688	267	0.413	236	0.365	138	0.213	178	0.275
35	1¼	31.9	1.255	799	1.237	480	0.742	424	0.656	248	0.383	320	0.495
41	1½	37.5	1.476	1104	1.711	663	1.027	585	0.907	342	0.530	442	0.684
53	2	48.6	1.913	1855	2.874	1113	1.725	983	1.523	575	0.891	742	1.150
63	2½	58.2	2.290	2660	4.119	1596	2.471	1410	2.183	825	1.277	1064	1.647
78	3	72.7	2.864	4151	6.442	2491	3.865	2200	3.414	1287	1.997	1660	2.577
91	3½	84.5	3.326	5608	8.688	3365	5.213	2972	4.605	1738	2.693	2243	3.475
103	4	96.2	3.786	7268	11.258	4361	6.755	3852	5.967	2253	3.490	2907	4.503
129	5	121.1	4.768	11518	17.855	6911	10.713	6105	9.463	3571	5.535	4607	7.142
155	6	145.0	5.709	16513	25.598	9908	15.359	8752	13.567	5119	7.935	6605	10.239

Table 6-1 *(continued)*

Reprinted with permission from NFPA 70-2005

Table 4 *Continued*

Articles 352 and 353 — Rigid PVC Conduit (RNC), Schedule 40, and HDPE Conduit

Metric Designator	Trade Size	Nominal Internal Diameter		Total Area 100%		60%		1 Wire 53%		2 Wires 31%		Over 2 Wires 40%	
		mm	in.	mm²	in.²	mm²	in.²	mm²	in.²	mm²	in.²	mm²	in.²
12	⅜	—	—	—	—	—	—	—	—	—	—	—	—
16	½	15.3	0.602	184	0.285	110	0.171	97	0.151	57	0.088	74	0.114
21	¾	20.4	0.804	327	0.508	196	0.305	173	0.269	101	0.157	131	0.203
27	1	26.1	1.029	535	0.832	321	0.499	284	0.441	166	0.258	214	0.333
35	1¼	34.5	1.360	935	1.453	561	0.872	495	0.770	290	0.450	374	0.581
41	1½	40.4	1.590	1282	1.986	769	1.191	679	1.052	397	0.616	513	0.794
53	2	52.0	2.047	2124	3.291	1274	1.975	1126	1.744	658	1.020	849	1.316
63	2½	62.1	2.445	3029	4.695	1817	2.817	1605	2.488	939	1.455	1212	1.878
78	3	77.3	3.042	4693	7.268	2816	4.361	2487	3.852	1455	2.253	1877	2.907
91	3½	89.4	3.521	6277	9.737	3766	5.842	3327	5.161	1946	3.018	2511	3.895
103	4	101.5	3.998	8091	12.554	4855	7.532	4288	6.654	2508	3.892	3237	5.022
129	5	127.4	5.016	12748	19.761	7649	11.856	6756	10.473	3952	6.126	5099	7.904
155	6	153.2	6.031	18433	28.567	11060	17.140	9770	15.141	5714	8.856	7373	11.427

Article 352 — Type A, Rigid PVC Conduit (RNC)

Metric Designator	Trade Size	Nominal Internal Diameter		Total Area 100%		60%		1 Wire 53%		2 Wires 31%		Over 2 Wires 40%	
		mm	in.	mm²	in.²	mm²	in.²	mm²	in.²	mm²	in.²	mm²	in.²
16	½	17.8	0.700	249	0.385	149	0.231	132	0.204	77	0.119	100	0.154
21	¾	23.1	0.910	419	0.650	251	0.390	222	0.345	130	0.202	168	0.260
27	1	29.8	1.175	697	1.084	418	0.651	370	0.575	216	0.336	279	0.434
35	1¼	38.1	1.500	1140	1.767	684	1.060	604	0.937	353	0.548	456	0.707
41	1½	43.7	1.720	1500	2.324	900	1.394	795	1.231	465	0.720	600	0.929
53	2	54.7	2.155	2350	3.647	1410	2.188	1245	1.933	728	1.131	940	1.459
63	2½	66.9	2.635	3515	5.453	2109	3.272	1863	2.890	1090	1.690	1406	2.181
78	3	82.0	3.230	5281	8.194	3169	4.916	2799	4.343	1637	2.540	2112	3.278
91	3½	93.7	3.690	6896	10.694	4137	6.416	3655	5.668	2138	3.315	2758	4.278
103	4	106.2	4.180	8858	13.723	5315	8.234	4695	7.273	2746	4.254	3543	5.489
129	5	—	—	—	—	—	—	—	—	—	—	—	—
155	6	—	—	—	—	—	—	—	—	—	—	—	—

Article 352 — Type EB, PVC Conduit (RNC)

Metric Designator	Trade Size	Nominal Internal Diameter		Total Area 100%		60%		1 Wire 53%		2 Wires 31%		Over 2 Wires 40%	
		mm	in.	mm²	in.²	mm²	in.²	mm²	in.²	mm²	in.²	mm²	in.²
16	½	—	—	—	—	—	—	—	—	—	—	—	—
21	¾	—	—	—	—	—	—	—	—	—	—	—	—
27	1	—	—	—	—	—	—	—	—	—	—	—	—
35	1¼	—	—	—	—	—	—	—	—	—	—	—	—
41	1½	—	—	—	—	—	—	—	—	—	—	—	—
53	2	56.4	2.221	2498	3.874	1499	2.325	1324	2.053	774	1.201	999	1.550
63	2½	—	—	—	—	—	—	—	—	—	—	—	—
78	3	84.6	3.330	5621	8.709	3373	5.226	2979	4.616	1743	2.700	2248	3.484
91	3½	96.6	3.804	7329	11.365	4397	6.819	3884	6.023	2272	3.523	2932	4.546
103	4	108.9	4.289	9314	14.448	5589	8.669	4937	7.657	2887	4.479	3726	5.779
129	5	135.0	5.316	14314	22.195	8588	13.317	7586	11.763	4437	6.881	5726	8.878
155	6	160.9	6.336	20333	31.530	12200	18.918	10776	16.711	6303	9.774	8133	12.612

Table 6-1 *(continued)*
Reprinted with permission from NFPA 70-2005

Cross-Sectional Area of Commonly Used Conductors			
(Please see *NEC* Chapter 9, Table 5 for complete listing.)			
Type	AWG	Cross-sectional Area	
		Square Inches	Square millimeters
TW	12	0.0181	12
	10	0.0243	16
	8	0.0437	28
	6	0.0726	47
THW	12	0.0260	17
	10	0.0333	21
	8	0.0556	36
	6	0.0726	47
THHN & THWN	12	0.0133	9
	10	0.0211	14
	8	0.0366	24
	6	0.0507	33

Table 6-2

Summary of Conduit and Tubing Types						
	Electrical Nonmetallic Tubing ENT	Intermediate Metal Conduit IMC	Rigid Metal Conduit RMC	Rigid Nonmetallic Conduit RNC	Electrical Metallic Tubing EMT	Flexible Metallic Tubing FMT
NEC	Article 362	Article 342	Article 344	Article 352	Article 358	Article 360
Trade size	½ to 2	½ to 4	½ to 6	½ to 6	½ to 4	⅜ to ¾
Min. radii of bends	Refer to *NEC* Chapter 9, Table 2.					
Min. spacing of supports	3 ft (900 mm) from outlet, 3 ft (900 mm) apart.	3 ft (900 mm) from outlet, 10 ft (3.0 m) apart.	3 ft (900 mm) from outlet, 10 ft (3.0 m) apart. See *NEC* Table 344.30(B)(2).	3 ft (900 mm) from box or outlet. See *NEC* Table 352.30(B).	3 ft (900 mm) from box or outlet. 10 ft (3.0 m) apart.	No stipulation.
Uses allowed	In buildings not over 3 floors. Concealed if 15-minute fire rating. Damp or dry.	All conditions. Under cinder fill if down 18 in. (450 mm) or in 2-in. (50-mm) concrete.	All conditions. Under cinder fill if down 18 in. (450 mm) or in 2-in. (50-mm) concrete.	All conditions.	Under cinder fill if down 18 in. (450 mm) or in 2-in. (50-mm) concrete.	Dry, accessible locations.
Uses not allowed	Hazardous locations. Support of equipment. Direct burial. In theaters unless enclosed in concrete.	See *NEC* 300.6 for protection against corrosion. See Note 1.	See *NEC* 300.6 for protection against corrosion. See Note 1.	In hazardous locations. In theaters.	Where subject to damage. Hazardous locations. See *NEC* 300.6 for protection against corrosion.	Hazardous locations. Underground in concrete. Lengths more than 6 ft (1.8 m).
Miscellaneous	Pliable. Corrugated. Nongrounding.	Grounding. Threaded with taper die.	Grounding. Threaded with taper die.	Expansion joints required. Nongrounding.	Not threaded.	Permitted for grounding per *NEC* 250.118(7).

Note 1: ►Where corrosion protection is necessary, all field-cut threads shall be coated with an approved electrically conductive corrosion-resistant compound. ◄

Table 6-3

Table 1 Percent of Cross Section of Conduit and Tubing for Conductors

Number of Conductors	All Conductor Types
1	53
2	31
Over 2	40

FPN No. 1: Table 1 is based on common conditions of proper cabling and alignment of conductors where the length of the pull and the number of bends are within reasonable limits. It should be recognized that, for certain conditions, a larger size conduit or a lesser conduit fill should be considered.

FPN No. 2: When pulling three conductors or cables into a raceway, if the ratio of the raceway (inside diameter) to the conductor or cable (outside diameter) is between 2.8 and 3.2, jamming can occur. While jamming can occur when pulling four or more conductors or cables into a raceway, the probability is very low.

Table 6-4

Reprinted with permission from NFPA 70-2005

in each raceway. In addition, the electrician must derate the conductors' ampacity as required, make allowances for voltage drops, recognize the various receptacle types, and be able to install these receptacles correctly on the system.

BOX STYLES AND SIZING

The style of box required on a building project is usually established in the specifications. However, the sizing of the boxes is usually one of the decisions made by the electrician. (See Table 6-5 and Table 6-6.)

Device Boxes

Device boxes, Figure 6-16, are available with depths ranging from 1½ to 3½ in. (38 to 90 mm). These boxes can be purchased for trade size ½ or ¾ conduit or with cable clamps. Each side of the box has holes through which a 20-penny nail can be inserted for nailing to wood studs. The boxes can

Table 314.16(A) Metal Boxes

mm	in.		Minimum Volume cm³	in.³	18	16	14	12	10	8	6
100 × 32	(4 × 1¼)	round/octagonal	205	12.5	8	7	6	5	5	5	2
100 × 38	(4 × 1½)	round/octagonal	254	15.5	10	8	7	6	6	5	3
100 × 54	(4 × 2⅛)	round/octagonal	353	21.5	14	12	10	9	8	7	4
100 × 32	(4 × 1¼)	square	295	18.0	12	10	9	8	7	6	3
100 × 38	(4 × 1½)	square	344	21.0	14	12	10	9	8	7	4
100 × 54	(4 × 2⅛)	square	497	30.3	20	17	15	13	12	10	6
120 × 32	(4¹¹⁄₁₆ × 1¼)	square	418	25.5	17	14	12	11	10	8	5
120 × 38	(4¹¹⁄₁₆ × 1½)	square	484	29.5	19	16	14	13	11	9	5
120 × 54	(4¹¹⁄₁₆ × 2⅛)	square	689	42.0	28	24	21	18	16	14	8
75 × 50 × 38	(3 × 2 × 1½)	device	123	7.5	5	4	3	3	3	2	1
75 × 50 × 50	(3 × 2 × 2)	device	164	10.0	6	5	5	4	4	3	2
75 × 50 × 57	(3 × 2 × 2¼)	device	172	10.5	7	6	5	4	4	3	2
75 × 50 × 65	(3 × 2 × 2½)	device	205	12.5	8	7	6	5	5	4	2
75 × 50 × 70	(3 × 2 × 2¾)	device	230	14.0	9	8	7	6	5	4	2
75 × 50 × 90	(3 × 2 × 3½)	device	295	18.0	12	10	9	8	7	6	3
100 × 54 × 38	(4 × 2⅛ × 1½)	device	169	10.3	6	5	5	4	4	3	2
100 × 54 × 48	(4 × 2⅛ × 1⅞)	device	213	13.0	8	7	6	5	5	4	2
100 × 54 × 54	(4 × 2⅛ × 2⅛)	device	238	14.5	9	8	7	6	5	4	2
95 × 50 × 65	(3¾ × 2 × 2½)	masonry box/gang	230	14.0	9	8	7	6	5	4	2
95 × 50 × 90	(3¾ × 2 × 3½)	masonry box/gang	344	21.0	14	12	10	9	8	7	4
min. 44.5 depth	FS — single cover/gang (1¾)		221	13.5	9	7	6	6	5	4	2
min. 60.3 depth	FD — single cover/gang (2⅜)		295	18.0	12	10	9	8	7	6	3
min. 44.5 depth	FS — multiple cover/gang (1¾)		295	18.0	12	10	9	8	7	6	3
min. 60.3 depth	FD — multiple cover/gang (2⅜)		395	24.0	16	13	12	10	9	8	4

*Where no volume allowances are required by 314.16(B)(2) through (B)(5).

Table 6-5

Reprinted with permission from NFPA 70-2005

Table 314.16(B) Volume Allowance Required per Conductor

Size of Conductor (AWG)	Free Space Within Box for Each Conductor	
	cm³	in.³
18	24.6	1.50
16	28.7	1.75
14	32.8	2.00
12	36.9	2.25
10	41.0	2.50
8	49.2	3.00
6	81.9	5.00

Table 6-6

Reprinted with permission from NFPA 70-2005

Available in 1-, 2-, 3-, 4-, and 5-gang sizes.

Figure 6-17   Four-gang masonry box.

Figure 6-16   Device (switch) box.

Figure 6-18   Handy box and receptacle cover.

be ganged by removing the common sides of two or more boxes and connecting the boxes together. Plaster ears may be provided for use on plasterboard or for work on old installations.

Securing Boxes with Screws

There are many ways to secure a box to framing members such as boxes with brackets, nails, screws, bolts, etc. ▶Make sure that if you use screws that will pass through the box, that exposed threads on the screws are protected in an approved manner so as not to damage conductor insulation, *NEC® 314.23(B)(1).*◀

Approved means *"Acceptable to the authority having jurisdiction."*

Masonry Boxes

Masonry boxes are designed for use in masonry block or brick. These boxes do not require an extension cover but will accommodate devices directly, Figure 6-17. Masonry boxes are available with depths of 2½ in. (65 mm) and 3½ in. (90 mm). Boxes of this type are available with knockout sizes up to trade size 1.

Handy Boxes

Handy boxes, Figure 6-18, are generally used in exposed installations and are available with trade size ½, ¾, and 1 knockouts. These boxes range in depth from 1¼ to 2½ in. (32 to 58 mm) and can accommodate most devices without the use of an extension cover.

Trade Size 4 × 4 Square Boxes

A trade size 4 × 4 square box, Figure 6-19, is used for surface or concealed installations. Extension covers of various depths are available to accommodate devices where the box is surface mounted. This type of box is available with knockouts up to trade size 1.

Octagonal Boxes

Boxes of this type are used primarily to install ceiling outlets. Octagonal boxes are available either for mounting in concrete or for surface or concealed mounting, Figure 6-20. Extension covers are available, but are not always required. Octagonal boxes are commonly used in depths of 1¼, 1½, and 2⅛ in. (32, 38, and 54 mm). These boxes are available with knockouts up to trade size 1.

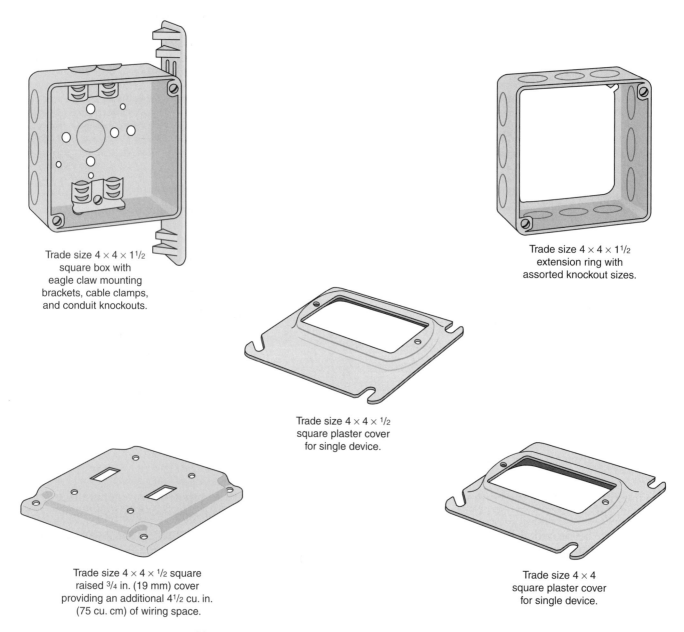

Trade size 4 × 4 × 1½ square box with eagle claw mounting brackets, cable clamps, and conduit knockouts.

Trade size 4 × 4 × 1½ extension ring with assorted knockout sizes.

Trade size 4 × 4 × ½ square plaster cover for single device.

Trade size 4 × 4 × ½ square raised ¾ in. (19 mm) cover providing an additional 4½ cu. in. (75 cu. cm) of wiring space.

Trade size 4 × 4 square plaster cover for single device.

Figure 6-19   Trade size 4 square boxes and covers.

Figure 6-20   Octagonal box on telescopic hanger and box extension.

Trade Size 4¹¹⁄₁₆ Square Boxes

These spacious boxes are used where the larger size is required, usually the result of larger size conductors, many conductors, or wiring devices in the box. Available in 1½ in. and 2⅛ in. deep sizes, with trade size ½ and ¾ knockouts. These boxes require an extension ring or a raised plaster cover to permit the attachment of wiring devices, Figure 6-21.

Box Sizing

NEC® 314.16 states that outlet boxes, switch boxes, and device boxes must be large enough to provide ample room for the wires in that box without having to jam or crowd the wires in the box.

Figure 6-21   Trade size 4¹¹⁄₁₆ square box.

When conductors are the same size, the proper box size can be determined by referring to *NEC® Table 314.16(A)*. When conductors are of different sizes, refer to *NEC® Table 314.16(B)*.

NEC® Table 314.16(A) does not consider fittings or devices such as studs for luminaires (fixtures), cable clamps, hickeys, switches, pilot lights, or

receptacles that may be in the box. Table 6-7 provides a list of conditions that may arise when evaluating box fill and the appropriate response to each of those conditions. Figure 6-22 covers the special case presented when transformer leads enter a box. Figure 6-23 addresses the common situation of luminaire (fixture) wires in a box. If there are not more than four luminaire (fixture) wires, and they are smaller than 14 AWG, they may be omitted from the box fill calculations. See *NEC® 314.16(B)(1), Exception.*

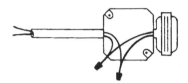

Figure 6-22   Box containing four conductors. Transformer leads 18 AWG or larger are to be included, *NEC® 314.16(B)(1).*

Box Fill According to *NEC® 314.16*		
When a box contains no fittings, devices, studs for luminaires (fixtures), cable clamps, hickeys, switches, receptacles, or equipment grounding conductors.	⇨	Refer directly to *NEC® Table 314.16(A)* or *Table 314.16(B)*.
When a box contains one or more internal cable clamps.	⇨	Add a single-volume based on the largest conductor in the box.
When a box contains one or more fixture studs for luminaires (fixtures) or hickeys.	⇨	Add a single-volume for each type based on the largest conductor in the box.
When a box contains one or more wiring devices on a yoke.	⇨	Add a double-volume for each yoke based on the largest conductor connected to a device on that yoke.
When a box contains one or more equipment grounding conductors.	⇨	Add a single-volume based on the largest conductor in the box.
When a box contains one or more additional isolated (insulated) equipment grounding conductors as permitted by *NEC® 250.146(D)* for noise reduction.	⇨	Add a single-volume based on the largest conductor in the box
▶For conductors less than 12 in. (300 mm) long between the raceway entries for the conductor that is looped through the box without being spliced.◀	⇨	▶Add a single-volume for each conductor that is looped through the box.◀
▶For conductors 12 in. (300 mm) or longer between the raceway entries for the conductor that is looped through the box without being spliced.◀	⇨	▶Add a double-volume for each conductor that is looped through the box.◀
When conductors originate outside of the box and terminate inside the box.	⇨	Add a single-volume for each conductor that originates outside the box and terminates inside the box.
When no part of the conductor leaves the box—for example, a jumper wire used to connect three wiring devices on one yoke, or pigtails as illustrated in Figure 6-23.	⇨	Do not count this (these). No additional volume is required.
When an equipment grounding conductor has not more than four conductors smaller than 14 AWG that originate from a luminaire (fixture) canopy or similar canopy (like a fan) and terminate in the box.	⇨	Do not count this (these). No additional volume is required.
When small fittings such as lock-nuts and bushings are present.	⇨	Do not count this (these). No additional volume is required.

Table 6-7

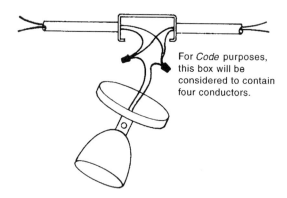

For *Code* purposes, this box will be considered to contain four conductors.

Figure 6-23 Box containing four conductors. Not more than four luminaire (fixture) wires that are smaller than 14 AWG may be omitted from the box fill calculations, *NEC® 314.16(B)(1), Exception*.

SELECTING THE CORRECT SIZE BOX

Box Fill When All Conductors Are the Same Size

A box contains one stud for luminaire (fixture) and two internal cable clamps. Four 12 AWG conductors enter the box.

Four 12 AWG conductors	4
One stud for luminaire (fixture)	1
Two cable clamps (count only one)	1
Total	6

Referring to *NEC® 314.16(A)*, we find that a $4 \times 1\frac{1}{2}$ in. $(100 \times 38$ mm) octagonal box is suitable for this example.

Box Fill When the Conductors Are Different Sizes

When a box contains different size wires, refer to *NEC® 314.16* and do the following:

- size the box based upon the volume required for the conductors according to *NEC® 314.16(B)*.

- then make the adjustment necessary due to devices, cable clamps, luminaires (fixtures), studs, etc. This volume adjustment must be based on the volume allowance for the largest conductor in the box as determined by *NEC® Table 314.16(B)*.

- when conductors of different sizes are connected to a device(s) on one yoke or strap, the cross-sectional area of the largest conductor connected to the device(s) shall be used in computing the box fill.

- when more than one equipment grounding conductor is in a box, the cross-sectional area of the largest grounding conductor shall be used in computing the box fill.

Box Fill When a Box Contains Devices

A box is to contain two devices—a duplex receptacle and a toggle switch. Two 12 AWG conductors are connected to the receptacle, and two 14 AWG conductors are connected to the toggle switch. Two grounding conductors are in the box, a 12 AWG and a 14 AWG. Two cable clamps are in the box.

The minimum box size would be calculated as follows:

Item	Customary Volume	SI Volume
Two 12 AWG	4.50 in.³	73.8 cm³
Two 14 AWG	4.00 in.³	65.6 cm³
One cable clamp	2.25 in.³	36.9 cm³
One equipment grounding conductor	2.25 in.³	36.9 cm³
One duplex receptacle	4.50 in.³	73.8 cm³
One toggle switch	4.00 in.³	65.6 cm³
Minimum volume	21.50 in.³	352.6 cm³

Therefore, select a box having a *minimum* volume of 21.5 in.³ (352.6 cm³). The volume may be marked on the box; otherwise, refer to the second column of *NEC® Table 314.16(A)*, titled "Min. Cu. In. Cap."

The *Code* requires that all boxes *other* than those listed in *NEC® Table 314.16(A)* be durable and legibly marked by the manufacturer with their cubic-inch capacity. When sectional boxes are ganged together, the volume to be filled is the total cubic-inch volume of the assembled boxes. Fittings may be used with the sectional boxes, such as plaster rings, raised covers, and extension rings.

When these fittings are marked with their volume in cubic inches or have dimensions comparable to those boxes shown in *NEC® Table 314.16(A)*, their volume may be considered in determining the total cubic-inch volume to be filled. See *NEC® 314.16(A)(1)*. The following example illustrates how the volume of a plaster ring or of a raised cover, Figure 6-19, is added to the volume of the box to increase the total volume.

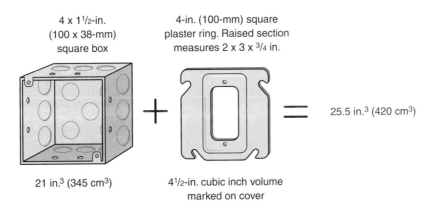

4 x 1½-in.
(100 x 38-mm)
square box

4-in. (100-mm) square
plaster ring. Raised section
measures 2 x 3 x ¾ in.

25.5 in.³ (420 cm³)

21 in.³ (345 cm³)

4½-in. cubic inch volume
marked on cover

Figure 6-24   Box and raised plaster ring.

Box Fill When There Is a Plaster Ring

How many 12 AWG conductors are permitted in this box and raised plaster ring as shown in Figure 6-24?

See *NEC® 314.16* and *NEC® Tables 314.16(A)* and *314.16(B)* for the volume required per conductor. This box and cover will take

$$\frac{25.5 \text{ in.}^3 (420 \text{ cm}^3)}{2.25 \text{ in.}^3 (36.9 \text{ cm}^3)} = \begin{array}{l}\text{Eleven 12 AWG conductors}\\\text{maximum, less the required}\\\text{deductions per } NEC®\\314.16(B)(2)\end{array}$$

Many devices such as GFCI receptacles, dimmers, and timers are much larger than the conventional receptacle or switch. The *Code* has recognized this problem by requiring that one or more devices on one strap require a conductor reduction of two when determining maximum box capacity. It is always good practice to install a box that has ample room for the conductors, devices, and fittings rather than forcibly crowding the conductors into the box.

The sizes of equipment grounding conductors are shown in *NEC® Table 250.122*. The grounding conductors are the same size as the circuit conductors in cables having 14, 12, or 10 AWG circuit conductors. Thus, box sizes can be calculated using *NEC® Table 314.16(A)*.

Another example of the procedure for selecting the box size is the installation of a four-way switch in the bakery. Four 12 AWG conductors and one device are to be installed in the box. Therefore, the box selected must be able to accommodate six conductors; a 3 × 2 × 2¼ in. (75 × 50 × 57 mm) switch box is adequate.

When the box contains fittings such as luminaire (fixture) studs, clamps, or hickeys, the number of conductors permitted in the box is one less for *each* type of device. For example, if a box contains two clamps, deduct one conductor from the allowable number shown in *NEC® Table 314.16(A)* (Figure 6-25).

An example of this situation occurs in the bakery at the box where the conduit leading to the four-way switch is connected.

- There are two sets of travelers (one from each of the three-way switches) passing through the box on the way to the four-way switch (four conductors).

- A neutral, from the panelboard, is spliced to a neutral that serves the luminaire (fixture) attached to the box and to neutrals entering the two conduits leaving the box that go to other luminaires (fixtures) (four conductors).

- A switch return is spliced in the box to a conductor that serves the luminaire (fixture) that is attached to the box and to a switch return that enters the conduits leaving the box to serve other luminaires (fixtures) (four conductors).

- A luminaire (fixture) stud-stem assembly is used to connect the luminaire (fixture) directly to the box (one conductor).

This is a total of 13 conductors.

From *NEC® Table 314.16(A)*, a 4 × 2⅛ in. (100 × 54 mm) square or a 4¹¹⁄₁₆ × 1½ in. (100 × 38 mm) square box would qualify. Because these boxes require plaster rings, a smaller size box could be used if the plaster ring volume added to the box volume were sufficient for the 13 conductors.

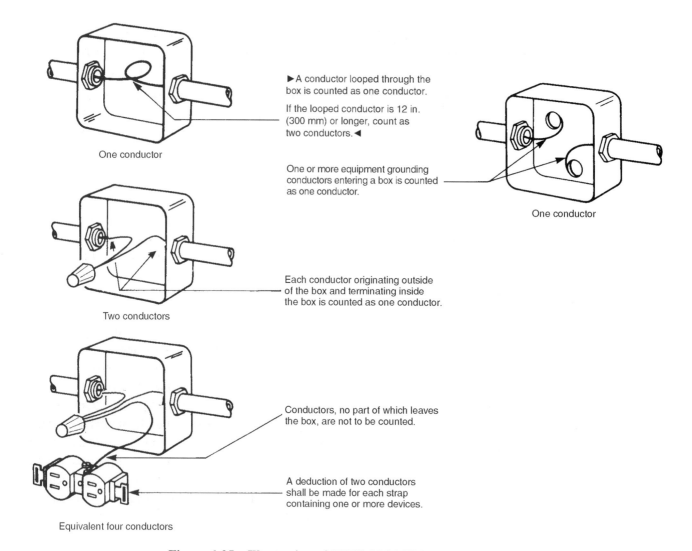

►A conductor looped through the box is counted as one conductor.

If the looped conductor is 12 in. (300 mm) or longer, count as two conductors.◄

One or more equipment grounding conductors entering a box is counted as one conductor.

Each conductor originating outside of the box and terminating inside the box is counted as one conductor.

Conductors, no part of which leaves the box, are not to be counted.

A deduction of two conductors shall be made for each strap containing one or more devices.

Figure 6-25   Illustration of *NEC*® 314.16(B)(1) requirements.

Pull and Junction Boxes and Conduit Bodies

When necessary, conduit bodies (Figure 6-26) or pull boxes (Figure 6-27) are used. They must be sized according to *NEC*® 314.28 when they contain conductors 4 AWG or larger.

- For straight pulls, the box must be at least eight times the diameter of the largest raceway. See Figures 6-27 and 6-28.

- For angle pulls or U pulls, the distance between the raceways and the opposite wall of the box must be at least six times the diameter of the largest raceway. To this value, it is necessary to add the diameters of all other raceways entering in any one row of the same wall of the box. See Figures 6-29 and 6-30. When more than one row exists, the row with the maximum diameters must be used in sizing the box.

Figure 6-26   Conduit bodies.

Figure 6-27   Threaded pull box.

Minimum distance
8d
d
Use largest conduit

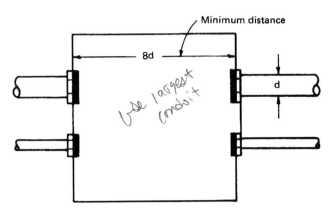

Figure 6-28 Straight pulls.

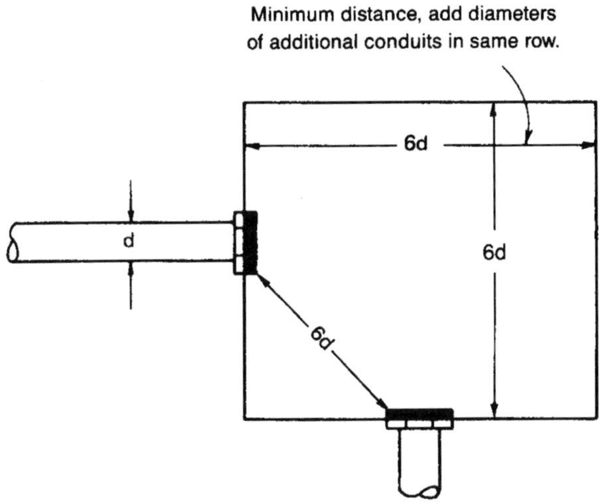

Minimum distance, add diameters of additional conduits in same row.
6d
d
6d
6d

Figure 6-29 Angle pull.

- Boxes may be smaller than the preceding requirements when approved and marked with the size and number of conductors permitted.
- Conductors must be racked or cabled if any dimension of the box exceeds 6 ft (1.83 m).
- Boxes must have an approved cover.
- Boxes must be accessible.

Depth of Pull Boxes. The depth of a pull box when a simple angle pull is made is not clearly addressed in the *NEC*.®

In reality, the depth would have to be at least the diameter of the bushing for the raceway entering the side of the pull box *plus* a little more to be able to install the bushing and tighten it.

For example, a locknut and bushing for a 3-in. rigid metal conduit has a diameter of approximately 3.8 to 4 in. If this were a grounding/bonding bushing, the diameter (including the huge grounding/bonding lug) would be much more than that of a simple locknut and bushing. To install the grounding/bonding lug, 5 to 6 in. may be needed to turn the bushing onto the conduit. With a little space to spare, the pull box would need to be at least 8 in. deep.

For multiple raceways, you have to take into consideration the diameters of the bushings as discussed previously.

Minimum distance
20 in. (500 mm)
Trade Size 3
Trade Size 2
Minimum distance
20 in. (500 mm)
Minimum distance
18 in. (450 mm)
Minimum distance
12 in. (300 mm)
Trade Size 2
Trade Size 3

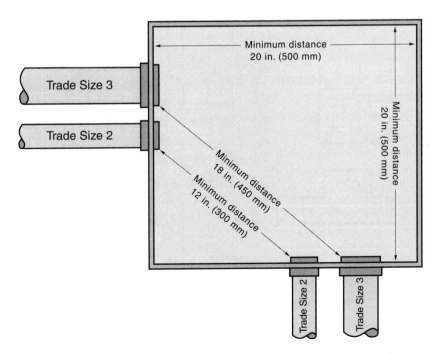

Figure 6-30 Angle pulls with side-by-side conduits.

RACEWAY SUPPORT

Many available devices are used to support conduit. The more popular devices are shown in Figure 6-31 and Figure 6-32.

The *NEC*® article that deals with a particular type of raceway also gives the requirements for supporting that raceway. Figure 6-33 shows the support requirements for rigid metal conduit, intermediate metal conduit, electrical metallic tubing, flexible metal conduit, and liquidtight flexible metal conduit.

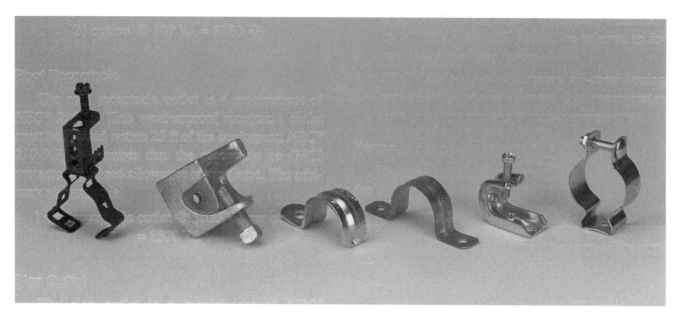

Figure 6-31   Raceway support devices.

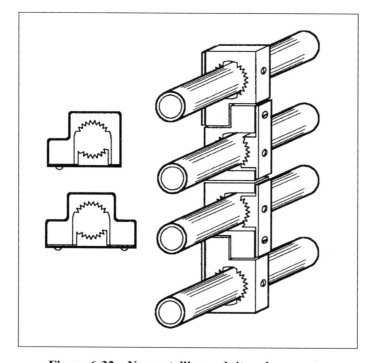

Figure 6-32   Nonmetallic conduit and supports.

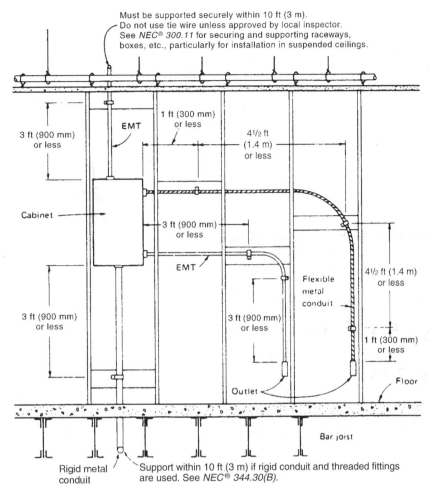

Figure 6-33   Securing and supporting requirements for raceways, tubing, and cable is found in the specific wiring method article, in the section number *XXX.30*.

REVIEW QUESTIONS

Refer to the *National Electrical Code®* or the working drawings when necessary. Where applicable, responses should be written in complete sentences.

When responding to Questions 1 through 4, select from the following list of raceway types: (check the letter[s] representing the correct response[s])

a. electrical metallic tubing

b. electrical nonmetallic tubing

c. flexible metallic tubing

d. rigid metal conduit

e. intermediate metal conduit

f. rigid nonmetallic conduit

1. Which raceway(s) may be used where cinder fill is present?
   ☐ a.   ☐ b.   ☐ c.   ☒ d.   ☐ e.   ☐ f.

2. Which raceway(s) may be used in hazardous locations?
☐ a. ☐ b. ☐ c. ☒ d. ☐ e. ☐ f.

3. Which raceway(s) may be used in wet locations?
☐ a. ☐ b. ☐ c. ☐ d. ☐ e. ☐ f.

4. Which raceway(s) may be used for service entrances?
☐ a. ☐ b. ☐ c. ☐ d. ☐ e. ☐ f.

Give the raceway size required for each of the following conductor combinations:

 # 10
.0333 × 5 = .1665

5. Five, 10 AWG, Type THW in SCH80 RNC   _____ 1"

.486 × 3 = *1.458*

6. Three, 1/0 AWG, Type THHN in EMT   _____ 2½"

7. Three, 1/0 AWG and one, 1 AWG, Type THW in EMT   2½"
   .366 × 3 = 1.098 +   .299 = 1.397

Determine the correct box size for each of the following conditions:

8. Two nonmetallic sheathed cables with two 12 AWG conductors, a grounding conductor, and a switch in a metal box.

12 awg   4.5 in³ × 2 = 9.0 in³
1 grn   2.25 in³   + 2.25
Switch   4.0 in³   + 4.0
_____
15.25 in³

4 × 1½

where do you get in³?
Switch in³?

9. A conduit run is serving a series of luminaires (fixtures), connected to a total of three circuits. The luminaires (fixtures) are supplied by 120 volts from a 3-phase, 4-wire system. Each box will contain two circuits running through the box and a third circuit connected to a luminaire (fixture), which is hung from a luminaire (fixture) stud. Use #12 THHN conductors.

2.25 × 4 = .9 in³

10. What is the minimum cubic in. volume that is permissible for the following outlet box? Show your calculations.

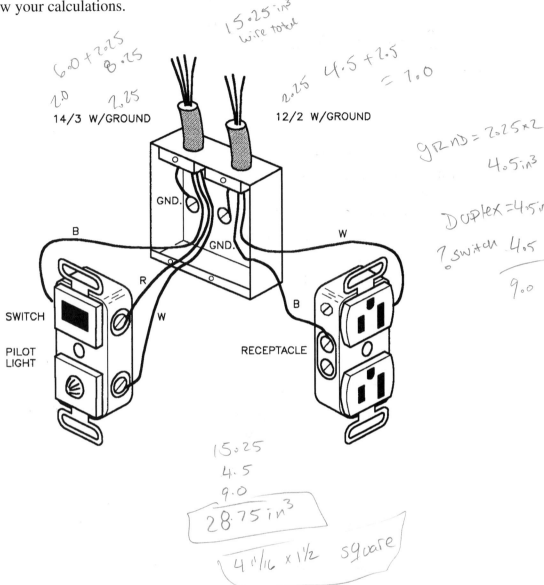

6.0 + 2.25
8.25
2.0   2.25

15.25 in³
wire total

2.25   4.5 + 2.5
= 7.0

14/3 W/GROUND   12/2 W/GROUND

GRND = 2.25 × 2
4.5 in³

DUPLEX = 4.5 in³

? switch 4.5
9.0

B

GND.

GND.

W

R

SWITCH

W

B

PILOT
LIGHT

RECEPTACLE

15.25
4.5
9.0
28.75 in³

4 11/16 × 1½ square

Two trade size 3 raceways enter a box directly across from each other. No other raceways enter the box. What are the minimum dimensions of the box?

11. Length ___24___   3 × 8 = 24

12. Width ___6___ ?

13. Depth ___6___ ?

Two trade size 3 raceways enter a box at right angles to each other. No other raceways enter the box. What are the minimum dimensions of the box?

14. Length ___18___

15. Width ___18___

16. Depth ___6___

17. If you had the freedom of choice, what type of raceway would you select for installation under the sidewalks and driveways of the commercial building? Support your selection. *I would use schedual 80 PVC conduit code 350.10G*

? *Using Table 300.5 To Meet minimum cover requirement*
6

Determine the required EMT raceway size for the following combination of conductors:

18. Four 8 AWG Type THW and four 12 AWG Type THW

8AWG .0437 × 4 = .1748

12AWG .0181 × 4 = + .0724

.2472 Total

1 inch EMT

19. Three 350 kcmil and one 250 kcmil Type TW conductors and a 4 AWG bare conductor

TW

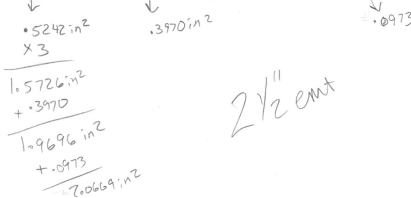

.5242 in²   *.3970 in²*   *.0973*

× 3

1.5726 in²

+ .3970

1.9696 in²

+ .0973

2.0669 in²

2½" emt

Determine the required box size for the following situations:

20. In a nonmetallic sheathed cable installation, a 10/3 with ground is installed in a metal octagonal box to supply two 12/2 with ground branch-circuit cables. What is the minimum size box? The box contains cable clamps.

4 wires   *Clamps 1 wire*

6 wires

1

11

4 11/16 × 1½

Two 3-in. raceways enter a box, one through a side and the other in the back. Four 500 kcmil, type THHN conductors will be installed in the raceway. No other raceways enter the box. What are the minimum dimensions of the box?

21. Length ____18____

22. Width ____18____

23. Depth ____6____

Calculate the answers using information shown in the drawing.

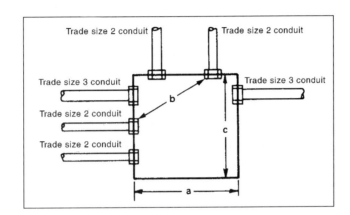

24. Dimension *a* must be at least ____28____ inches.

25. Dimension *b* must be at least ____12____ inches.

26. Dimension *c* must be at least ____17____ inches.

27. Do you foresee any difficulties in installing the conductors in these raceways? Explain.

IF you Moved 3on leFt To bottom would not interFear with diagnoe Pulls.

UNIT 7

Motor and Appliance Circuits

OBJECTIVES

After studying this unit, the student should be able to

- use and interpret the word *appliance*.
- use and interpret the term *utilization equipment*.
- choose an appropriate method for installing electrical circuits to appliances.
- identify appropriate grounding for appliances.
- determine branch-circuit ratings, conductor sizes, and overcurrent protection for appliances and motors.
- understand the meaning of the terms *Type 1* and *Type 2* protection.

APPLIANCES

Most of the *Code* rules that relate to the installation and connection of electrical appliances are found in *NEC® Article 422*. For motor-operated appliances, reference will be made to *NEC® Article 430* (*Motors, Motor Circuits, and Controllers*). Where the appliance is equipped with hermetic refrigerant motor compressor(s), such as refrigeration and air-conditioning equipment, *NEC® Article 440* (*Air-Conditioning Equipment*) applies.

The *National Electrical Code®* defines an appliance as utilization equipment, generally other than industrial, of standardized sizes or types, that is installed or connected as a unit to perform such functions as air-conditioning, washing, food mixing, and cooking, among others.

The *Code* defines utilization equipment as equipment that uses electric energy for mechanical, chemical, heating, lighting, or similar purposes.

In commercial buildings, one of the tasks performed by the electrician is to make the electrical connections to appliances. The branch-circuit sizing may be specified by the consulting engineer in the specifications or on the working drawings. In other

cases, it is up to the electrician to do all of the calculations, and/or to make decisions based upon the nameplate data of the appliance.

The bakery in the commercial building provides examples of several different types of electrical connection methods that can be used to connect electrical appliances.

Appliance Branch-Circuit Overcurrent Protection

- For those appliances that are marked with a maximum size and type of overcurrent protective device, this rating shall not be exceeded, see *NEC® 422.11(A)*.
- If the appliance is not marked with a maximum size overcurrent protective device, *NEC® 422.11(A)* stipulates that *NEC® 240.4* applies. This section states that the fundamental rule is to protect conductors at their ampacities, as specified in *NEC® Table 310.16*.

There are several variations allowed to this fundamental rule, three of which relate closely to motor-operated appliances, air-conditioning, and refrigeration equipment.

NEC® 240.4(B) allows the use of the next standard higher ampere rating overcurrent device when the ampacity of the conductor does not match a standard ampere rating overcurrent device. There are two exceptions to this—if the conductors are part of a multioutlet branch-circuit supplying cord- and plug-connected portable loads or if the next standard rating is greater than 800-ampere, then the next lower standard rating overcurrent device must be chosen. Standard ratings of overcurrent devices are listed in *NEC® 240.6.*

NEC® 422.11 lists overcurrent protection requirements for several types of appliances. Commercial appliances include:

- infrared lamp heating
- surface heating elements
- single nonmotor-operated appliances
- resistance-type heating elements
- sheathed-type heating elements
- water heaters and steam boilers
- motor-operated appliances

For a single nonmotor-operated appliance, *NEC® 422.11(E)* states that the overcurrent protection shall not exceed

a. the overcurrent device rating if so marked on the appliance.
b. 150 percent of the appliance's rated current, if the appliance is rated at more than 13.3 amperes. If the 150 percent results in a nonstandard ampere overcurrent device value, then it is permitted to use the next higher standard rating overcurrent device.

For example, an appliance is rated at 4500 volt-ampere, 240 volts. What is the maximum size fuse permitted?

$$\frac{4500 \text{ VA}}{240 \text{ V}} = 18.75 \text{ A}$$

$$18.75 \text{ A} \times 1.5 = 28.125 \text{ A}$$

Therefore, the *Code* permits the use of a 30-ampere fuse for the overcurrent protection.

Appliance Grounding

- When appliances are to be grounded, the requirements are given in *NEC® Article 250.* Grounding is discussed in several places in this

text. It is suggested that the reader consult the *Code* Index in the Appendix where all *NEC®* references are listed.

Appliance-Disconnecting Means

- Permanently connected appliances, *NEC® 422.31:*
  a. If the appliance rating does not exceed ⅛ horsepower or does not exceed 300 volt-amperes, the branch-circuit overcurrent device may serve as the appliance's disconnecting means, *NEC® 422.31(A).*
  b. If the appliance rating exceeds 300 volt-amperes or exceeds ⅛ horsepower, the branch-circuit disconnect switch or breaker can serve as the appliance's disconnecting means, but only if they are within sight of the appliance or are capable of being locked in the off position. ▶The lock-off provision must be permanently installed on or at the switch or circuit breaker, *NEC® 422.31(B).*◀

- For motor-driven, permanently connected appliances rated at more than ⅛ horsepower, the disconnecting means must be within sight of the motor controller, *NEC® 422.32.*

- Many appliances have an integral unit switch. Where this unit switch has a marked off position and disconnects all of the ungrounded conductors, this unit switch can be considered to be the appliance's disconnecting means, but only, in nondwelling occupancies, if the branch-circuit switch or circuit breaker supplying the appliance is readily accessible, *NEC® 422.34(D).* The term *readily accessible* is covered in the *Definitions* in the *NEC®.* Simply stated, this means that there must be easy access to the disconnect switch without the need to *climb over or remove obstacles or to resort to portable ladders, and so forth.*

- The term *in sight* is covered in the *Definitions* in the *NEC®.* Further discussion is found in *NEC® 430.102.*

- *NEC® 422.33* permits the use of a cord- and plug-connected arrangement to serve as the appliance's disconnecting means. Such cord-and-plug assemblies may be furnished and attached to the appliance by the manufacturer to meet the requirements of specific Underwriters Laboratories standards.

THE BASICS OF MOTOR CIRCUITS

Motor circuits can be simple or complex depending on factors such as the type, size, and characteristics of the motor or motors, how each motor is to be operated and the electrical requirements. Given this information, the electrician usually begins by locating the power source, then planning the circuit between there and the motor. The first component in this circuit is usually the disconnecting means.

Disconnecting Means

The *Code* requires that all motors be provided with a means to disconnect the motor from its electrical supply. These requirements are found in *NEC® Article 430, Part IX*.

For most applications, the disconnect switch for a motor must be horsepower rated, *NEC® 430.109*. The most common exceptions are:

- for motors ⅛ horsepower or less where the branch-circuit overcurrent device can serve as the disconnecting means.

- for stationary motors of 2 horsepower or less, rated at 300 volts or less, a general-use switch is permitted.

- for stationary ac motors greater than 100 horsepower, a general-use disconnect switch, or an isolating switch is permitted when the switch is marked "Do Not Open Under Load."

- an attachment plug is permitted to serve as the disconnecting means if the attachment plug cap is horsepower-rated not less than that of the load. Refer to *NEC® 430.109(F)*. The horsepower rating of an attachment plug cap is considered to be the corresponding horsepower rating for 80 percent of the ampere rating of the attachment plug cap.

Example.   A two-wire, 125-volt, 20-ampere attachment plug cap:

$$20 \times 0.80 = 16 \text{ amperes}$$

Checking *NEC® Table 430.148* for single-phase, 115-volt, motor, we determine that a 1-horsepower motor has a full-load ampere rating of 16 amperes. Thus, this attachment plug cap is permitted to serve as the disconnecting means for this motor.

NEC® 430.110(A) requires that the disconnecting means for motors rated at 600 volts or less must have an ampere rating of not less than 115 percent of the motor's full-load ampere rating.

The nameplate on all disconnect switches, as well as the manufacturer's technical data, furnishes the horsepower rating, the voltage rating, and the ampere rating of the disconnect switch.

The *NEC®* requires that a disconnecting means be provided for motors and motor controllers, *NEC® 430.101*. This requirement is broken into two specific rules, and both rules must be met. Figures 7-1A, 7-1B, 7-1C, and 7-1D illustrate the location of the disconnecting means in relation to the motor's controller, the motor, and the driven machinery.

Rule #1.   See Figure 7-1A. *NEC® 430.102(A):* An individual disconnect must be in sight of the controller and must disconnect the controller. This is the basic rule. The *NEC®* definition of "In Sight" means that the controller must be visible and not more than 50 ft (15 m) from the disconnect.

Rule #2.   See Figure 7-1B. *NEC®430.102(B):* Disconnect must be in sight of motor and driven machinery. If the disconnect as required in *430.102(A)* is in sight of the controller, the motor, and driven machinery, then that disconnect meets the requirements of both *430.102(A)* and *430.102(B)*.

If, because of the nature of the installation, the disconnect is in sight of the controller, but not in sight of the motor and the driven machinery, then another disconnect must be installed, Figure 7-1C.

As with many *NEC®* requirements, there are exceptions. Figure 7-1D shows the exceptions to the disconnect requirements for motors, controllers, and driven machinery.

The reason the *Code* uses the term *motor location* instead of *motor* is that, in many instances, the motor is inside an enclosure and is out of sight until an access panel is removed. If the term *motor* were to be used, then it would be mandatory to install the disconnect inside of the enclosure. This is not always practical. The commercial building's rooftop air-conditioning units are examples.

NEC® 440.14 does permit the disconnecting means to be installed on or within the air-conditioning unit, as might be the case for large air-conditioning units.

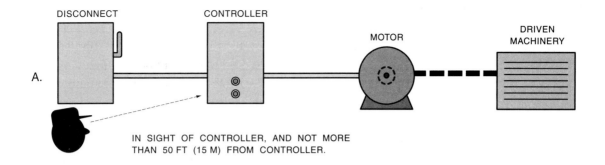

A.

IN SIGHT OF CONTROLLER, AND NOT MORE
THAN 50 FT (15 M) FROM CONTROLLER.

B.

IN SIGHT OF CONTROLLER, MOTOR, AND DRIVEN MACHINERY, AND NOT MORE
THAN 50 FT (15 M) FROM CONTROLLER, MOTOR, AND DRIVEN MACHINERY.

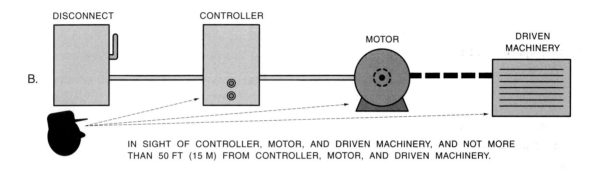

C.

THIS DISCONNECT IS IN SIGHT OF
MOTOR AND DRIVEN MACHINERY.

THIS DISCONNECT IS IN SIGHT OF CONTROLLER, BUT NOT IN SIGHT OF THE
MOTOR AND DRIVEN MACHINERY, OR IS MORE THAN 50 FT (15 M) FROM THE
MOTOR AND DRIVEN MACHINERY; THEN ANOTHER DISCONNECT IS REQUIRED.

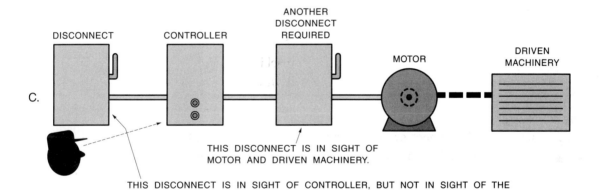

D.

EXCEPTION: Disconnect need not be in sight of motor and driven machinery if the disconnect can be individually locked
in the open (off) position. The locking provision must be of a permanent type installed on the switch or circuit breaker.
Most disconnect switches, combination starters, and motor control centers have this "lock-off" feature. "Listed" circuit breaker
"lock-off" devices that fit over the top of the circuit breaker handle also meet this requirement.

To apply the EXCEPTION, *either* of two conditions must be met:

1. If the disconnect location is impracticable . . . or might cause additional or increased hazards to persons or property.

2. In industrial installations where there is a *written* safety in the workplace program . . . and where conditions of
   maintenance and supervision ensure that only qualified persons will service the equipment.

Figure 7-1   Location of motor disconnect means.

Exhaust Fan

To turn the exhaust fan on and off, an AC general-use, single-pole snap (toggle) switch is installed on the wall below the fan. The electrician will install an outlet box approximately 1 ft from the ceiling (see the electrical plan) from which he will run a short length of flexible metal conduit to the junction box on the fan. The fan exhausts the air from the bakery. The fan is installed through-the-wall, approximately 18 in. below the ceiling, Figure 7-2.

NEC® 430.111 permits this switch to serve as both the controller and the required disconnecting means.

Motor Branch-Circuit Conductors

Conductors for a motor branch-circuit shall not be less than 125 percent of the full-load current draw of the motor, *NEC® 430.22(A)*.

Example. Determine the minimum size Type THHN conductors for a 7½-horsepower, 208-volt, three-phase motor, Service Factor 1.15. The panelboard, circuit breakers, and the motor conductors have been checked to confirm that all are rated for 167°F (75°C).

1. Refer to *NEC® Table 430.250* to obtain the motor's full-load current rating. The motor's full-load current draw is 24.2 amperes.

2. Multiply the motor's full load amperes by 1.25.

    24.2 amperes × 1.25 = 30.25 amperes

3, Check *NEC® 110.14(C)* to find limitations on terminations. For the size of the conductors

used for this motor, the 167°F (75°C) ampacity values found in *NEC® Table 310.16* can be used. If it is unknown whether the panelboard and breakers are suitable for 167°F (75°C), the 140°F (60°C) column for circuits 100 amperes or less, or 14 AWG through 1 AWG conductors, is required to be used.

4. *NEC® Table 310.13* shows that Type THHN conductors have a temperature rating of 194°F (90°C). *NEC® Table 310.16* shows that a 167°F (75°C):

    - 12 AWG Type THHN conductor has an allowable ampacity of 25 amperes.

    - 10 AWG Type THHN conductor has an allowable ampacity of 35 amperes.

The conclusion for this example is that at minimum 10 AWG THHN conductors are needed to meet the minimum conductor ampacity of 30.25 amperes for the motor.

Motor Overload Protection

NEC® Article 430, Part III addresses motor overload protection. This part discusses continuous and intermittent duty motors, motors that are larger than 1 horsepower, motors that are automatically started, and motors that are marked with a service factor that have integral (built-in) overload protection.

The exhaust fan in the bakery draws 2.9 amperes, is less than 1 horsepower, and is nonautomatically started. According to *NEC® 430.32(D)*, second sentence, *Any motor rated at 1 horsepower or less that is permanently installed shall be protected in accordance with* NEC® 430.32(B), which in part states that the overload protection shall be as shown in Table 7-1.

The same values indicated in Table 7-1 are also valid for continuous duty motors of more than 1 horsepower, see *NEC® 430.32(A)*.

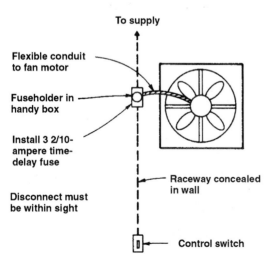

Figure 7-2   Exhaust fan installation.

Overload Protection	
Motor Nameplate Rating	**Overload Protection as a Percentage of Motor Nameplate Full-load Current Rating**
Service factor not less than 1.15	125%
Temperature rise not over 104°F (40°C)	125%
All other motors	115%

Table 7-1

Unit 17, "Overcurrent Protection: Fuses and Circuit Breakers," covers in detail the subject of overcurrent protection using fuses and circuit breakers. Dual-element, time-delay fuses are an excellent choice for overload protection of motors. They can be sized close to the ampere rating of the motor. When the motor has integral overload protection or where the motor controller has thermal overloads, then dual-element, time-delay fuses are selected to provide backup overload protection to the thermal overloads.

The starting current of an electrical motor is very high for a short period of time as is illustrated in Figure 7-3. The current decreases to its rated value as the motor reaches full speed. Dual-element, time-delay fuses will carry five times their ampere rating for at least 10 seconds. This allows the motor to start and then provides good overload protection while it is running. For example, a 4-ampere, full-load rated motor could have a starting current as high as 24 amperes. The fuse must not open needlessly when it sees this inrush.

In this situation, a 15-ampere ordinary fuse or breaker might be required to allow the motor to start. This would provide the branch-circuit overcurrent protection, but it would not provide overload protection for the appliance.

If the motor does not have built-in overload protection, sizing the branch-circuit overcurrent protection as stated would allow the motor to destroy itself if the motor were to be subjected to continuous overloaded or stalled conditions.

Abuse, lack of oil, bad bearings, and jammed V-belts are a few of the conditions that can cause a motor to draw more than normal current. Too much current results in the generation of too much heat within the windings of the motor. For every 50°F (10°C) above the maximum temperature rating of the motor, the expected life of the motor is reduced by 50 percent. This is sometimes referred to as the "half-life rule."

The solution is to install a dual-element, time-delay fuse that permits the motor to start, yet opens on an overload before the motor is damaged. To provide this overload protection for single-phase, 120- or 240-volt equipment, fuses and fuseholders as illustrated in Figure 7-4 are often used.

Type S fuses and the corresponding Type S adapter are inserted into the fuseholder, Figure 7-5. Once the adapter has been installed, it is extremely difficult to remove it. This makes it virtually impossible to replace a properly sized Type S fuse with a fuse having an ampere rating too large for the specific load.

Table 7-2 compares the load responses of ordinary fuses and dual-element, time-delay fuses.

Table 7-3A shows how to size dual-element, time-delay fuses for motors that are marked with a service factor of not less than 1.15 and for motors marked with a temperature rise of not over 104°F (40°C).

These charts show one manufacturer's recommendations for the selection of dual-element, time-delay fuses for motor circuits. The Design B energy efficient motors have a higher starting current than the older style motors. The nameplate should be checked to determine the design designation; then refer to *NEC® Table 430.52* to determine proper branch-circuit, short-circuit, and ground-fault protection. Finally, check the protective device time-current curve to determine whether it is capable of carrying the starting current until the motor reaches full speed.

Table 7-3B shows how to size dual-element, time-delay fuses for motors that have a service

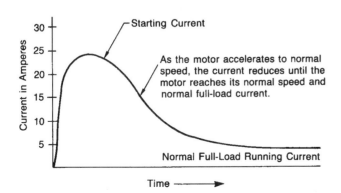

Figure 7-3 Typical motor starting current.

Figure 7-4 Type S plug fuse and adapter.

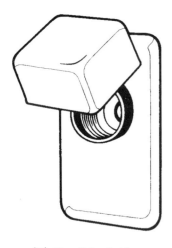

(A) Type S fuseholder

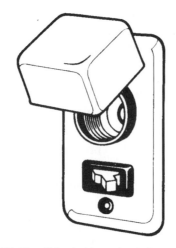

(B) Type S fuseholder and switch

(C) Type S fuseholder with switch and pilot light

Figure 7-5   Fuseholders.

Load Response Time for Four Types of Fuses (Approximate Time in Seconds for Fuse to Blow)				
Load (amperes)	4-ampere Dual-element Fuse (time-delay)	4-ampere Ordinary Fuse (nontime-delay)	8-ampere Ordinary fuse (nontime-delay)	15-ampere Ordinary Fuse (nontime-delay)
5	More than 300 sec.	More than 300 sec.	Won't blow	Won't blow
6	250 sec.	5 sec.	Won't blow	Won't blow
8	60 sec.	1 sec.	Won't blow	Won't blow
10	38 sec.	Less than 1 sec.	More than 300 sec.	Won't blow
15	17 sec.	Less than 1 sec.	5 sec.	Won't blow
20	9 sec.	Less than 1 sec.	Less than 1 sec.	300 sec.
25	5 sec.	Less than 1 sec.	Less than 1 sec.	10 sec.
30	2 sec.	Less than 1 sec.	Less than 1 sec.	4 sec.

Table 7-2

factor of less than 1.15 and for motors with a temperature rise of more than 104°F (40°C).

For abnormal installations, dual-element, time-delay fuses larger than those shown in Table 7-3A and Table 7-3B may be required. These larger-rated fuses (or circuit breakers) will provide short-circuit protection only for the motor branch-circuit. *NEC®* *430.52* and *Table 430.52* show the maximum size fuses and breakers permitted. These percentages have already been discussed in this unit.

Abnormal conditions might include:

* situations where the motor is started, stopped, jogged, inched, plugged, or reversed frequently.
* high inertia loads such as large fans and centrifugal machines, extractors, separators, pulverizers, etc. Machines having large flywheels fall into this category.
* motors having a high *Code* letter and full-

voltage start as well as some older motors without *Code* letters.

Remember, the higher the *Code* letter, the higher the starting current. For example, two motors have exactly the same full-load ampere rating. One motor is marked *Code* letter J. The other is marked *Code* letter D. The *Code* letter J motor will draw more momentary starting inrush current than the *Code* letter D motor. The *Code* letter of a motor is found on the motor nameplate. See *NEC®* *430.7* and *Table 430.7(B)*.

Refer to the *National Electrical Code®* and to Bussmann's *Electrical Protection Handbook* for more data relating to motor overload and motor branch-circuit protection, disconnecting means, controller size, conductor size, conduit size, and voltage drop.

Selection of Dual-Element Fuses for Motor Overload Protection

A. MOTORS MARKED WITH NOT LESS THAN 1.15 SERVICE FACTOR OR TEMP. RISE NOT OVER 104°F (40°C)			
Motor (40°C or 1.15 S.F.) Ampere Rating	Dual-element Fuse Amp Rating Max. 125%	Motor (40°C or 1.15 S.F.) Ampere Rating	Dual-element Fuse Amp Rating Max. 125%
1.00 to 1.11	1 1/4	20.0 to 23.9	25
1.12 to 1.27	1 4/10	24.0 to 27.9*	30
1.28 to 1.43	1 6/10	28.0 to 31.9	35
1.44 to 1.59	1 8/10	32.0 to 35.9	40
1.60 to 1.79	2	36.0 to 39.9	45
1.80 to 1.99	2 1/4	40.0 to 47.9	50
2.00 to 2.23	2 1/2	48.0 to 55.9	60
2.24 to 2.55	2 8/10	56.0 to 63.9	70
2.56 to 2.79	3 2/10	64.0 to 71.9	80
2.80 to 3.19	3 1/2	72.0 to 79.9	90
3.20 to 3.59	4	80.0 to 87.9*	100
3.60 to 3.99	4 1/2	88.0 to 99.9	110
4.00 to 4.47	5	100 to 119	125
4.48 to 4.99	5 6/10	120 to 139	150
5.00 to 5.59	6 1/4	140 to 159	175
5.60 to 6.39	7	160 to 179*	200
6.40 to 7.19	8	180 to 199	225
7.20 to 7.99	9	200 to 239	250
8.00 to 9.59	10	240 to 279	300
9.60 to 11.9	12	280 to 319	350
12.0 to 13.9	15	320 to 359*	400
14.0 to 15.9	17 1/2	360 to 399	450
16.0 to 19.9	20	400 to 480	500

Table 7-3A

B. ALL OTHER MOTORS (i.e., LESS THAN 1.15 SERVICE FACTOR OR GREATER THAN 104°F [40°C] RISE)			
All Other Motors–Ampere Rating	Dual-element Fuse Amp Rating Max. 115%	All Other Motors–Ampere Rating	Dual-element Fuse Amp Rating Max. 115%
1.00 to 1.08	1 1/8	17.4 to 20.0	20
1.09 to 1.21	1 1/4	21.8 to 25.0	25
1.22 to 1.39	1 4/10	26.1 to 30.0*	30
1.40 to 1.56	1 6/10	30.5 to 34.7	35
1.57 to 1.73	1 8/10	34.8 to 39.1	40
1.74 to 1.95	2	39.2 to 43.4	45
1.96 to 2.17	2 1/4	43.5 to 50.0	50
2.18 to 2.43	2 1/2	52.2 to 60.0	60
2.44 to 2.78	2 8/10	60.9 to 69.5	70
2.79 to 3.04	3 2/10	69.6 to 78.2	80
3.05 to 3.47	3 1/2	78.3 to 86.9	90
3.48 to 3.91	4	87.0 to 95.6*	100
3.92 to 4.34	4 1/2	95.7 to 108	110
4.35 to 4.86	5	109 to 125	125
4.87 to 5.43	5 6/10	131 to 150	150
5.44 to 6.08	6 1/4	153 to 173	175
6.09 to 6.95	7	174 to 195*	200
6.96 to 7.82	8	196 to 217	225
7.83 to 8.69	9	218 to 250	250
8.70 to 10.0	10	261 to 300	300
10.5 to 12.0	12	305 to 347	350
13.1 to 15.0	15	348 to 391*	400
15.3 to 17.3	17 1/2	392 to 434	450
		435 to 480	500

*Note: Disconnect switch must have an ampere rating at least 115 percent of motor ampere rating, NEC® 430.110(A). Next larger size switch with fuse reducers may be required.

Table 7-3B

Dual-element, time-delay fuses are designed to withstand motor-starting inrush currents and will not open needlessly. However, they will open if a sustained overload occurs. Selecting dual-element, time-delay fuses of the Type S plug type, Figure 7-4, or of the cartridge type, as illustrated in Unit 18 for motors, should be sized in the range of 115 percent, but not to exceed 125 percent of the motor full-load ampere rating.

For the exhaust fan motor that has a 2.9-ampere full-load current draw, select a dual-element, time-delay fuse in the range of

$$2.9 \times 1.15 = 3.34 \text{ amperes}$$
$$2.9 \times 1.25 = 3.62 \text{ amperes}$$

Referring to Table 7-3A and Table 7-3B, find dual-element, time-delay fuses rated at 3½ amperes. If the motor has built-in or other overload protection, the 3½-ampere size will provide backup overload protection for the motor.

Motor Branch-Circuit, Short-Circuit, and Ground-Fault Protection

It is beyond the scope of this text to cover every aspect of motor circuit design, as an entire text could be devoted to this subject. The basics of the typical motor installations will be presented, but for the not-so-common installations, refer to *NEC® Article 430*.

- Motor branch-circuit conductors, controllers, and the motor must be provided with short-circuit and ground-fault overcurrent protection, *NEC® 430.51*.

- The short-circuit and ground-fault overcurrent protection must have sufficient time-delay to permit the motor to be started, *NEC® 430.52*.

- *NEC® 430.52* and *Table 430.52* provide the maximum percentages permitted for fuses and circuit breakers, for different types of motors.

- When applying the percentages listed in *NEC® Table 430.52*, and the resulting ampere rating or setting does not correspond to the standard ampere ratings found in *NEC® 240.6*, we are permitted to round up to the next higher standard rating. See *NEC® 430.52(C)(1), Exception 1*.

- If the size selected by rounding up to the next higher standard size will not allow the motor to start, such as might be the case on Design B (energy-efficient motors) motor circuits, then we are permitted to size the motor branch-circuit fuse or breaker even larger, *NEC® 430.52(C)(1), Exception 2*. See Table 7-4.

- Always check the manufacturer's overload relay table in a motor controller to see whether the manufacturer has indicated a maximum size or type of overcurrent device, for example,

Sizing of Motor Branch-Circuit, Short-Circuit, and Ground-Fault Protection	Maximum Rating	Absolute Maximum
Nontime-delay fuses not over 600 amperes	300%	400%
Nontime-delay fuses over 600 amperes	300%	300%
Time-delay dual-element fuses not over 600 amperes	175%	225%
Fuses over 600 amperes	300%	300%
Inverse-time circuit breakers Motor FLA 100 amperes or less	250%	400%
Inverse-time circuit breakers Motor FLA over 100 amperes	250%	300%
Instant-trip circuit breakers for other than Design B energy-efficient motors	800%*	1300%**
Instant-trip circuit breakers for Design B energy-efficient motors	1100%*	1700%**

*Only permitted when part of a listed combination motor controller.
**Only permitted where necessary after doing an engineering evaluation to prove that the 800 percent and 1100 percent sizing will not perform satisfactorily.

Table 7-4

"Maximum Size Fuse 25-ampere." Do not exceed this ampere rating, even though the percentage values listed in *NEC® Table 430.52* result in a higher ampere rating. Do not use a circuit breaker. To do so would be a violation of *NEC® 110.3(B)*.

Example.   Determine the rating of dual-element, time-delay fuses for a Design B, 7½-horsepower, 208-volt, three-phase motor, Service Factor 1.15. First refer to *NEC® Table 430.250* to obtain the motor's full-load current rating, and then refer to *NEC® Table 430.52*. The first column shows the type of motor, and the remaining columns show the maximum rating or setting for the branch-circuit, short-circuit, and ground-fault protection using different types of fuses and circuit breakers. Also refer to *NEC® 430.52*.

$$24.2 \text{ amperes} \times 1.75 = 42.35 \text{ amperes}$$

NEC® 240.6 lists 45-ampere as the next higher standard size.

If it is determined that 45-ampere, dual-element, time-delay fuses will not hold the starting current of the motor, we are permitted to increase the ampere rating of the fuses, but not to exceed 225 percent of the motor's full-load current draw.

$$24.2 \text{ amperes} \times 2.25 = 54.45 \text{ amperes}$$

The next lower standard OCPD size is rated 50-ampere.

For an easy-to-use computer version for doing motor circuit calculations, visit the Bussmann Web site at http://www.bussmann.com. Point to "Application Info" in the left column, and in the drop-down window click on "Software." In the following drop-down window click on "Motor Circuit Selection Guide."

Motor Starting Currents/*Code* Letters

Motor branch-circuit short-circuit and ground-fault protection is based upon the type of motor and the *Code* letter found on the motor's nameplate. The higher the *Code* letter, the higher the starting current.

NEC® Table 430.7(B) lists *Code* letters for electric motors. Because this table is based upon kVA per horsepower, it is easy to calculate the range of the motor's starting current in amperes.

Example.   Calculate the starting current for a 240-volt, three-phase, Code Letter J, 7½-horsepower motor. The minimum starting current would be:

$$I = \frac{kVA \times 1000}{E \times 1.73} = \frac{7.1 \times 7.5 \times 1000}{240 \times 1.73} = 128 \text{ amperes}$$

The maximum starting current would be

$$I = \frac{kVA \times 1000}{E \times 1.73} = \frac{7.99 \times 7.5 \times 1000}{240 \times 1.73} = 144 \text{ amperes}$$

Motor Design Designations

NEC® 430.52 and *Tables 430.52* and *430.251(B)* reference *design* types of motors. The *design* types refer to the National Electrical Manufacturers Association (NEMA) designation for alternating current, polyphase induction electric motors. Different design types of motors have different applications as shown in Table 7-5.

Type 1 and Type 2 Coordination

Because these terms might be found on the nameplate of a motor controller, familiarity with their meanings is necessary.

NEMA Design Application			
NEMA Design	Starting Current	Torque	Application
A	High	Normal	A variation of Design B, having a higher starting current.
B	Normal	Normal	For normal starting torque for fans, blowers, rotary pumps, unloaded compressors, some conveyors, metal cutting machine tools, misc. machinery. Very common for general-purpose across-the-line starting. Slight change of speed as load varies.
C	Normal	High	For high inertia starts such as large centrifugal blowers, fly wheels, and crusher drums. Loaded starts, such as piston pumps, compressors, and conveyors, where rapid acceleration is needed. Slight change of speed as load varies.
D	Low	Very high	For very high inertia and loaded starts such as punch presses, shears and forming machine tools, cranes, hoists, elevators, oil well pumping jacks. Considerable change of speed as load varies.

Table 7-5

These terms are used by manufacturers of motor controllers based on the International Electrotechnical Commission (IEC) Standard 947–4-1. This is a European standard. Electrical equipment manufactured overseas to European standards is being shipped into the United States, and electrical equipment manufactured in this country to U.S. standards (NEMA and UL Standard 508) is being shipped overseas. Conformance to one standard does not necessarily mean that the equipment conforms to the other standard.

In the United States, we prefer to use the term *protection* instead of *coordination* because coordination has a different meaning as discussed in Unit 18, "Short Circuit Calculations and Coordination of Overcurrent Protection Devices."

Type 1 short-circuit protection means that under short-circuit conditions, the contactor and starter shall not cause danger for persons working on or near the equipment. However, a certain amount of damage is acceptable to the controller, such as welding of the contacts or burning out the overload relays. Replacing the contacts of a controller does not constitute a hazard to personnel. Type 1 protection and UL 508 are similar in this respect. The steps necessary to get Type 1 protected controllers back on line after a short circuit or ground-fault occurs are:

1. Disconnect the power.
2. Locate and repair the fault.
3. Replace the contacts in the controller if necessary.
4. Replace the branch-circuit fuses or reset the circuit breaker.
5. Restore power.

Type 2 short-circuit protection means that under short-circuit conditions, the contactor and starter shall not cause danger for persons working on or near the equipment and, in addition, the contactor or starter shall be suitable for further use. No damage to the controller is acceptable. The steps necessary to get Type 2 protected controllers back on line after a short circuit or ground-fault occurs are:

1. Disconnect the power.
2. Locate and repair the fault.
3. Replace the branch-circuit fuses.
4. Restore power.

Type 2 protection requires extremely current-limiting overcurrent devices, such as Class J and Class CC fuses. Standard circuit breakers or older style fuses are typically not fast enough under short-circuit conditions to protect the motor starter from damage.

Although Type 2 protected controllers are marked by the manufacturer with the maximum size and type of fuse, the maximum size might be larger than that permitted for motor branch-circuit protection when applying the rules of *NEC® 430.52* and *Table 430.52* for a particular motor installation. For example, the controller might be marked "Maximum Ampere Rating Class J Fuse: 8 amperes." Yet, the proper size Class J time-delay fuse for a particular motor circuit, applying the maximum values according to *NEC® 430.52* and *Table 430.52* might be 4½ amperes. In this example, the proper maximum size fuse is 4½ amperes.

EQUIPMENT INSTALLATION

There are a number of topics that must be addressed.

Motor Controllers

▶The *2005 NEC®* contains a totally new *NEC® Article 409*. This article covers industrial control panels operating at 600 volts or less. Industrial control panels cover a pretty wide range of equipment. Used in both industrial and commercial applications, these panels might include the motor controller, overload relays, fused disconnect switches, circuit breakers, control devices, push-button stations, selector switches, timers, control relays, terminal blocks, pilot lights, and similar components. The equipment and the field wiring of the equipment must conform to *NEC® Article 409* relative to such things as conductor sizing, grounding, disconnecting means, rating or setting of the overcurrent protective device, construction, phase arrangement, wire space, wire bending, and marking.

Motor controllers must be marked with the manufacturer's name, voltage, current or horsepower, and the short-circuit current rating, *NEC® 430.8.*◀

Equipment and Lighting on a Branch-Circuit

The exhaust fan in the bakery is an example of an equipment load being connected on a branch-circuit with lighting. Circuit 1 serves 1305 VA of lighting and the exhaust fan. This fan is a permanently connected motor-operated appliance. Permanently connected is often called hard wired, differentiating it from cord- and plug-connected.

NEC® 210.23(A) permits, on 15- and 20-ampere branch-circuits, the supply of lighting and equipment provided the hard-wired equipment does not exceed 50 percent of the circuit rating.

This particular exhaust fan has a $1/10$ horsepower, 120-volt motor with a full-load rating of 2.9 amperes. The load would be

$$1305 \text{ VA} + (120 \text{ V} \times 2.9 \text{ A} \times 1.25) = 1740 \text{ VA}$$

$$\frac{1740 \text{ VA}}{120 \text{ V}} = 14.5 \text{ A}$$

In selecting the proper size conductor, remember what has been previously discussed. *NEC® Table 310.16* requires the use of the 140°F (60°C) column. Accordingly, a 12 AWG Type THHN conductor has an allowable ampacity of 25 amperes, even though the allowable ampacity at the conductors 194°F (90°C) rating is 30 amperes.

According to *NEC® 240.4(D)*, the maximum overcurrent protection for a 12 AWG copper conductor is 20 amperes.

Considering Circuit 1 to be a "continuous load," the 80 percent maximum loading factor on the branch-circuit overcurrent device and branch-circuit conductors is:

$$20 \times 0.80 = 16 \text{ amperes}$$

Therefore, 12 AWG Type THHN conductors protected by a 20-ampere circuit breaker meets the requirements of the *NEC®.*

The fan motor is under the 50 percent limitation of *NEC® 210.23(A)*.

Combination Load on Individual Branch-Circuit

When a cord- and plug-connected appliance is supplied by an individual branch-circuit, the load is limited to 80 percent of the circuit rating by *NEC® 210.23(A)(1)*. An example of this is Circuit 7 serving the dough machine:

Motor load 792 volt-amperes

Heater load 2000 volt-amperes

Total load 2792 volt-amperes

Conversion to amperes

$$\frac{2792 \text{ VA}}{360} = 7.8 \text{ A}$$

A 20-ampere-rated circuit, the minimum allowed, is assigned to supply this load.

Supplying a Specified Load

The bake oven installed in the bakery is an electrically heated commercial-type bake oven, Figure 7-10. The oven has a marked nameplate indicating a load of 16,000 volt-amperes, at three-phase, 208-volt. This load includes all electrical heating elements, drive motors, timers, transformers, controls, operating coils, lights, and all other electrically powered apparatus. The nameplate also indicates that all necessary protective devices are installed.

The instructions furnished with the bake oven, and on the nameplate, specify that the supply conductors have a minimum ampacity of 56 amperes and that the terminations are rated for 167°F (75°C).

If the panelboard were to have terminations rated at 140°F (60°C), the conductor selection would be made in the 140°F (60°C) column of *NEC® Table 310.16*. A minimum size conductor 4 AWG having the lowest allowable ampacity higher than 56 is required. If the panelboard terminations are rated at 167°F (75°C), then the selection is made from the 167°F (75°C) column and a 6 AWG having an allowable ampacity of 65 amperes would qualify. See *NEC® 110.14(C)*. These sizes are required even though 194°F (90°C), Type THHN conductors are being used. A disconnect means, without overcurrent protection, is installed at the site to allow maintenance and adjustments to be made to the internal electrical system. This disconnect is not required by the *NEC®* if the branch-circuit panelboard is in sight of the appliance and can be locked. ▶The lock-off provision must be permanently installed on or at the switch or circuit breaker, *NEC® 422.31(B)*.◀

Conductors Supplying Several Motors

NEC® 430.24 requires that the conductors supplying several motors on one circuit shall have an ampacity of not less than 125 percent of the largest motor plus the full-load current ratings of all other motors on that circuit.

For example, consider the bakery's multi-mixers and dough divider, which are supplied by a single branch-circuit and have full load ratings of 7.48, 2.2, and 3.96 amperes as determined by referring to the equipment's nameplate and installation instructions.

Therefore:

$$1.25 \times 7.48 = \quad 9.35 \text{ amperes}$$
$$\text{plus} \qquad 2.20 \text{ amperes}$$
$$\text{plus} \qquad \underline{3.96 \text{ amperes}}$$
$$\text{Total} \qquad 15.51 \text{ amperes}$$

The branch-circuit conductors must have an ampacity of 15.51 amperes minimum. Specifications for this commercial building call for 12 AWG minimum. Checking NEC® Table 310.16, a trade size 12 AWG Type THHN copper conductor has an ampacity of 25 amperes, more than adequate to serve the three appliances. The ampacity of 25 amperes is from the 167°F (75°C) column. When adjusting ampacities for more than three conductors in one raceway, or when correcting ampacities for high temperatures, we would begin the derating by using the 194°F (90°C) column ampacity to determine the 194°F (90°C) Type THHN conductor's ampacity.

Several Motors and Other Loads on One Branch-Circuit

This type of installation is not very common. This situation may occur on equipment that has multiple motors and other integral loads. In just about all encounters with this kind of equipment, it will be marked with the required conductor and overcurrent protection sizing.

This is covered in NEC® 430.53. This kind of group installation does not have individual motor branch-circuit, short-circuit, and ground-fault protection. The individual motors do have overload protection. The individual motor controllers, circuit breakers of the inverse-time type, and overload devices must be listed for group installation. The label on the controller indicates the maximum rating of fuse or circuit breaker suitable for use with the particular overload devices in the controller(s).

The branch-circuit fuses or circuit breakers are sized using these steps:

1. Determine the fuse or circuit-breaker ampere rating or setting per NEC® 430.52 for the highest rated motor in the group.

2. To this, add the full-load current ratings for all other motors in the group.

3. To this, add the current ratings for the "other loads."

Example. To determine the maximum overcurrent protection using dual-element, time-delay fuses for a branch-circuit serving three motors plus an additional load, the FLA rating of the motors are 27, 14, and 11 amperes. There is 20 amperes of "other load" connected to the branch-circuit. First determine the maximum time-delay fuse for the largest motor:

$$27 \times 1.75 = 47.15 \text{ amperes}$$

The next higher standard ampere rating is 50 amperes; next add the remaining ampere ratings to this value:

$$50 + 14 + 11 + 20 = 95 \text{ amperes}$$

For these four loads the maximum ampere rating of dual-element, time-delay fuses is 95 amperes, or rounded up to the next higher standard rating of 100 amperes. The disconnect switch would be a 100-ampere rating.

It can be readily seen that the rating of the branch-circuit overcurrent device is a variable, depending upon the type of overcurrent device used. For example, if time-delay fuses sized at 225 percent are chosen because of the electrician's familiarity with high inrush current during start-up:

$$27 \times 2.25 = 60.75 \text{ amperes (round down to 60)}$$

$$\text{Then } 60 + 14 + 11 + 20 = 105 \text{ amperes}$$

Again, using common sense with full knowledge of the variables in sizing the branch-circuit overcurrent protection, you would install 100-ampere time-delay fuses in a 100-ampere horsepower-rated disconnect switch.

Several Motors on One Feeder

NEC® 430.24 and 430.62 set forth the requirements for this situation. The individual motors will have branch-circuit, short-circuit, and ground-fault protection sized according to NEC® 430.52. The individual motors do have overload protection.

The feeder fuses or circuit breakers are sized using these steps:

1. Determine the fuse or circuit-breaker rating or setting from *NEC® Table 430.52* for the highest rated motor in the group.

2. To this, add the full-load current ratings for all other motors in the group.

Example.   To determine the maximum overcurrent protection using dual-element, time-delay fuses for a branch-circuit serving three motors, the FLA ratings of the motors are 27, 14, and 11 amperes.

First determine the maximum time-delay fuse for the largest motor:

$$27 \times 1.75 = 47.15 \text{ amperes}$$

The next higher standard ampere rating overcurrent device is 50 amperes.

Next add the remaining ampere ratings to this value:

$$50 + 14 + 11 = 75 \text{ amperes}$$

For these three motors, the maximum ampere rating of dual-element, time-delay fuses is 75 amperes.

Because 75-ampere, dual-element, time-delay fuses are not a standard size as listed in *NEC® 240.6*, it would be permissible to install 80-ampere fuses. You could also install 70-ampere, dual-element, time-delay fuses if you are confident that they would have sufficient time-delay to allow the motors to start. In either case, the size of the disconnect switch would be 100 amperes.

Sometimes there is confusion when one or more of the motors are protected with instant-trip breakers, sized at 800 percent to 1700 percent of the motor's full-load ampere rating. This would result in a very large (and possibly unsafe) feeder overcurrent device. *NEC® 430.62(A), Exception No. 1* states that in such cases, make the feeder calculation as though the motors protected by the instant-trip breakers were protected by dual-element, time-delay fuses.

For an easy-to-use, computer version for doing a group-motor circuit calculation, visit the Bussmann Web site at http://www.bussmann.com. Point to "Application Info" in the left column and in the drop-down window click on "Software." In the following drop-down window click on "Group Motor Protection Guide."

DISCONNECTING MEANS

Each of the three appliances is furnished with a four-wire cord-and-plug (three phases plus equipment ground). This meets the requirements for the disconnecting of cord- and plug-connected appliances, *NEC® 422.33*.

The disconnecting-means requirement for permanently connected appliances is covered in *NEC® 422.31*. *NEC® 422.32* requires that the disconnect means for motor-driven appliances be located within sight of the appliance or be capable of being locked in the off position. ▶The lock-off provision must be permanently installed on or at the switch or circuit breaker, *NEC® 422.31(B)*.◀

The disconnecting means shall have an ampere rating not less than 115 percent of the motor full-load current rating, *NEC® 430.110*.

The disconnecting means for more than one motor shall not be less than 115 percent of the sum of the full-load current ratings of all of the motors supplied by the disconnecting means, *NEC® 430.110(C)(2)*.

GROUNDING

WHY?	*NEC® 250.4(A)(1) and (2)* explains *why* equipment must be grounded.
WHERE?	*NEC® 250.110* explains locations *where* equipment fastened in place or connected by permanent wiring methods (fixed) must be grounded.
WHAT?	*NEC® 250.112* explains *what* equipment that is fastened in place or connected by permanent wiring methods (fixed) must be grounded. *NEC® 250.114* explains *what* equipment that is cord- and plug-connected must be grounded.
HOW?	*NEC® 250.134* explains *how* to ground equipment that is fastened in place and is connected by permanent wiring methods. *NEC® 250.138* explains *how* to ground equipment that is cord- and plug-connected.
SIZE?	*NEC® Table 250.122* lists the size of an equipment grounding conductor based on the rating of the OCPD protecting the circuit.

OVERCURRENT PROTECTION

Overcurrent protection for appliances is covered in *NEC® 422.11*. If the appliance is motor-driven, then *NEC® Article 430, Part III* applies. In most cases, the overload protection is built into the appliance. The appliance meets Underwriters Laboratories standards.

A complete discussion for individual motors and appliances can be found in the "Exhaust Fan" section earlier in this unit.

THE BAKERY EQUIPMENT

A bakery can have many types and sizes of food-preparing equipment, such as blenders, choppers, cutters, disposers, mixers, dough dividers, grinders, molding and patting machines, peelers, slicers, wrappers, dishwashers, rack conveyors, water heaters, blower dryers, refrigerators, freezers, hot-food cabinets, and others, Figure 7-6 and Figure 7-7. All of this equipment is designed and manufactured by companies that specialize in food-preparation equipment. These manufacturers furnish specifications that clearly state such things as voltage, current, wattage, phase (single- or three-phase), minimum required branch-circuit rating, minimum supply circuit conductor ampacity, and maximum

Figure 7-7   Dough divider.

overcurrent protective device rating. If the appliance is furnished with a cord-and-plug arrangement, it will specify the NEMA size and type, plus any other technical data that is required in order to connect the appliance in a safe manner as required by the *Code*.

NEC® Article 422, Part V sets forth the requirements for the marking of appliances.

The Mixers and Dough Dividers (Three Appliances Connected to One Circuit)

When more than one appliance is to be supplied by one circuit, a load calculation must be made. Note on the bakery electrical plans that three receptacle outlets are supplied by one three-phase, 208-volt branch-circuit, Table 7-6.

Figure 7-6   Cake mixer.

Appliance Load Data				
Type of Appliance	Voltage	Amperes	Phase	HP
Multimixer	208	3.96	3	¾
Multimixer	208	7.48	3	1½
Dough divider	208	2.2	3	½
All appliances are connected with NEMA 15–20P on four-wire cord.				

Table 7-6

The data for these food-preparation appliances has been taken from the nameplate and installation instruction manuals of the appliances.

A NEMA 15–20R receptacle outlet, Figure 7-8, is provided at each appliance location. These are part of the electrical contract.

Figure 7-8   NEMA 15–20R receptacle and plate.

Each of these appliances is purchased as a complete unit. After the electrician provides the proper receptacle outlets, the appliances are ready for use as soon as they are moved into place and plugged in.

The Doughnut Machine

An individual branch-circuit provides power to the doughnut machine. This machine consists of (1) a 2000-watt heating element that heats the liquid used in frying and (2) a driving motor that has a full-load rating of 2.2 amperes.

As with most food-preparation equipment, the appliance is purchased as a complete, prewired unit. This particular appliance is equipped with a four-wire cord to be plugged into a receptacle outlet of the proper configuration.

Figure 7-9 is the control-circuit diagram for the doughnut machine. The following components are listed in Figure 7-9:

S   A manual switch used to start and stop the machine.

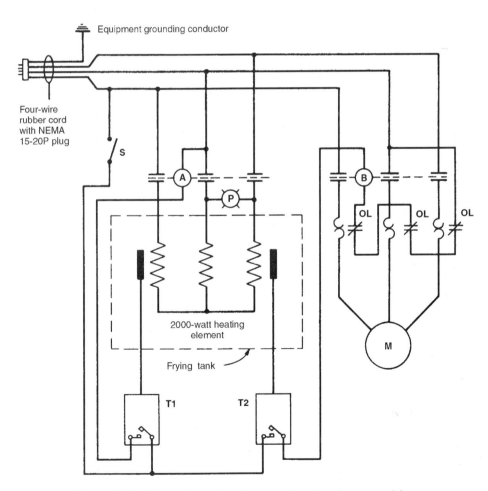

Figure 7-9   Control-wiring diagram for doughnut machine.

T1   A thermostat with its sensing element in the frying tank; this thermostat keeps the oil at the correct temperature.

T2   Another thermostat with its sensing element in the drying tank; this thermostat controls the driving motor.

A   A three-pole contactor controlling the heating element.

B   A three-pole motor controller operating the drive motor.

M   A three-phase motor.

OL   Overload units that provide overload protection for the motor; note one thermal overload unit in each phase.

P   Pilot light to indicate when power to the heating elements is "ON."

Because this appliance is supplied by an individual circuit, its current is limited to 80 percent of the branch-circuit rating according to *NEC® 422.10(A)*. The branch-circuit supplying the doughnut machine must have sufficient ampacity to meet the minimum load requirements as indicated on the appliance nameplate, or in accordance with *NEC® 430.24*.

Doughnut Machine Load Data	
Heater load	2000 VA
Motor load	792 VA
25% motor load	198 VA
Total	2990 VA

The maximum continuous load permitted on a 20-ampere, three-phase branch-circuit is

16 amperes × 208 volts × 1.73 = 5760 VA

The load of 2990 volt-amperes is well within the 5760 volt-amperes permitted loading of the 20-ampere branch-circuit.

The Bake Oven

As previously discussed, the bake oven installed in the bakery is an electrically heated commercial-type bake oven. See Figure 7-10.

According to the Panel Schedule, the bake oven is fed with a 60-ampere, three-phase, three-wire feeder consisting of three 6 AWG Type THHN copper conductors. The metal raceway is considered acceptable for the equipment grounding conductor.

A three-pole, 60-ampere, 250-volt disconnect switch is mounted on the wall near the oven. From

Figure 7-10   Bake oven.

this disconnect switch, a conduit (flexible, rigid, intermediate metal conduit, or electrical metallic tubing as recommended by the oven manufacturer) is run to the control panel on the oven, Figure 7-11.

The bake oven is installed as a complete unit. All overcurrent protection is an integral part of the circuitry of the oven. The internal control circuit of the oven is 120 volts, which is supplied by an integral control transformer.

All of the previous discussion in this unit relating to circuit ampacity, conductor sizing, grounding, etc., also applies to the bake oven, dishwasher, and food-waste disposer. Additional information relating to appliances is found in Unit 3 and Unit 21.

For the bake oven:

1. Ampere rating = 44.5 amperes

2. Minimum conductor ampacity =
   $44.5 \times 1.25 = 55.6$ amperes

3. Branch-circuit overcurrent protection =
   $44.5 \times 1.25 = 55.6$ amperes

It is permitted to install 55- or 60-ampere fuses in a 60-ampere disconnect switch.

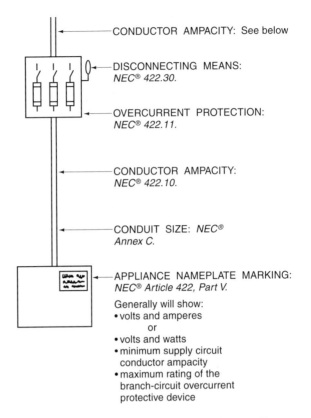

Figure 7-11   Motor branch-circuit for appliance.

REVIEW QUESTIONS

Refer to the *National Electrical Code* or the working drawings when necessary. Where applicable, responses should be written in complete sentences.

Cite the appropriate *NEC* Article, Section, or Table for each of the following:

1. Where does the *NEC* refer specifically to air-conditioning equipment?

   _____

2. Where does the *NEC* refer specifically to appliances? _____

3. Where does the *NEC* refer specifically to disconnecting means for a motor?

   _____

4. Where does the *NEC* refer specifically to grounding exposed noncurrent-carrying metal parts of motor circuits?

   _____

The following questions refer to cord- and plug-connected appliances.

5. In your own words, define an *appliance* and give examples.

   _____

   _____

6. A 20-ampere branch-circuit shall not serve a single appliance with a rating greater than _____

_____ .

7. An appliance is fastened in place and other cord- and plug-connected appliances are connected to a 20-ampere branch-circuit. What is the maximum allowable load for the appliance that is fastened in place? _____

_____

_____

The following questions refer to motors and motor circuits.

8. Where shall the disconnect means for a 208-volt, three-phase motor controller be located? _____

_____

9. Where shall the disconnect means for a motor be located? _____

_____

10. When selecting conductors, from where is the motor full-load ampere rating to be taken? _____

_____

11. When selecting motor overload protective devices, from where is the full-load ampere rating taken?_____

_____

12. A motor with a service factor of 1.25 shall be protected by an overload device set to trip at not more than _____ of the motor full-load current rating.

13. The full-load current rating of a 5-horsepower, 208-volt, three-phase motor is 16.7 amperes. The minimum ampacity of the branch-circuit conductors supplying this motor would be not less than _____ .

A 5-horsepower, three-phase motor has a nameplate current rating of 16 amperes. Dual-element, time-delay fuses are to be used for the motor branch-circuit protection. Show your calculations.

14. The preferred rating of the fuses is _____ amperes.

15. If the fuse with the preferred rating will not permit the starting of the motor, the fuse rating may be increased to _____ amperes.

16. If the motor is to be started and stopped repeatedly, possibly resulting in nuisance opening of the circuit, the fuse rating may be increased to a maximum of _____ amperes.

The following questions apply to alternating-current motors:

17. Define *in sight* as it applies to a motor and its disconnect switch.

_____

_____

_____

18. A motor has a full-load current rating of 74.8 amperes. What is the minimum allowable ampacity of the branch-circuit conductors?

_____

_____

19. A motor has a full-load current rating of 74.8 amperes and the terminations are rated for 167°F (75°C). What is the minimum conductor size and type?

_____

_____

20. Three three-phase motors are supplied by a single set of conductors. The motors are rated 208 volts and 5, 10, and 15 horsepower. The minimum allowable ampacity of the conductors is _____ amperes.

21. A ½-horsepower, 115-volt motor (Service Factor 1.15) is to be installed, and a type S fuseholder and switch is provided to control and protect the motor. The proper size fuse will have a rating of _____ amperes.

22. For a 25-horsepower, 480-volt, *Code* letter G motor, the minimum starting current will be _____ amperes, and the maximum starting current will be _____ amperes.

23. *Type 1* and *Type 2 Protection* are terms that might be found on the nameplate of a motor controller. Explain in your own words what these terms mean.

_____

_____

_____

_____

UNIT 8

Feeders

OBJECTIVES

After studying this unit, the student should be able to

- calculate the feeder loading.
- determine the minimum feeder overcurrent protective device rating.
- determine the minimum feeder conductor size.
- determine the appropriate correction and adjustment factors.
- calculate voltage drops.
- reduce the neutral size as appropriate.
- determine the minimum raceway size.

Feeders are the part of the electrical system that connects the branch-circuit panelboards to the electrical service equipment. In the commercial building a feeder is installed to each of the five occupancies and one to the panelboard for the owner's circuit. (The supply to the boiler is a branch-circuit.)

The feeder layout is shown in a riser diagram on working drawing E4. Specific requirements for feeders are given in *NEC® Article 215* concerning installation requirements and in *NEC® Article 220 Part II*, concerning calculated loads and demand factors.

Some of the information presented in this unit has been introduced previously. This redundancy is intentional because of its application to both feeders and branch-circuits.

FEEDER REQUIREMENTS

Feeder Ampacity

The ampacity of a feeder must comply with the following requirements:

- A feeder must have an ampacity no less than the sum of the volt-amperes of the coincident

loads of the branch-circuits supplied by the feeder as reduced by demand factors. See *NEC® Article 220 Part II*.

- Demand factors are allowed in selected cases where it is unlikely that all the loads would be energized at the same time. See *NEC® 220.60*.
- Only the larger of any noncoincident loads (loads unlikely to be operated simultaneously such as the air-conditioning and the heating) need be included. See *NEC® 220.60*.
- Referring to *NEC® 220.61*, a feeder neutral may be reduced in size in certain situations. It may not be reduced in feeders consisting of two-phase wires and a neutral of a three-phase, four-wire, wye-connected system. The feeder to the doctor's office is an example of this condition.

The neutral of a three-phase, four-wire, wye-connected system may be reduced in size provided it remains of a size sufficient to:

- carry the maximum unbalanced load.
- carry the nonlinear load.

Overcurrent Protection

The details of selecting overcurrent protective devices are covered in Units 17, 18, 19, and elsewhere in the text as indicated by the *Code* index included in the Appendix.

As previously stated, the rating of a branch-circuit is based upon the rating of the circuit overcurrent protective device, *NEC® 210.3*. The rating of the overcurrent device for a feeder shall not be less than the noncontinuous loading plus 125 percent of the continuous loading, *NEC® 215.2(A)(1)*.

The basic requirement for overcurrent protection is that a conductor shall be protected in accordance with its ampacity, but several conditions exist where the rating of the overcurrent protective device (OCPD) can exceed the ampacity of the conductor being protected.

For example, if the ampacity of a feeder does not match a standard OCPD rating, then the next higher standard rating may be used, see *NEC® 240.4(B)*, provided the rating is 800-ampere or less. Feeder conductors with an ampacity of 130 amperes may be protected with a 150-ampere protective device.

Temperature Limitations

When selecting the components for a circuit, remember that no circuit is better than its terminations. *NEC® 110.14(C)* ensures that the consideration of terminations is included in the selection of the circuit components.

Terminations, like other components in a circuit, have temperature ratings. The terminations on breakers, switches, or panelboards with a rating of 100-ampere or less may be rated for 140°F (60°C) or 167°F (75°C). *NEC® 110.14(C)* stipulates that unless equipment is marked with a 167°F (75°C) temperature rating (for example, 140°F/167°F [60°C/75°C] or 167°F [75°C]), the circuit ampacity shall not exceed the value given in the 140°F (60°C) column of *NEC® Table 310.16*. The reason for this is that the heat generated by the current (amperes) in a conductor is dependent on the conductor size. The consequence of *NEC® 110.14(C)* is that it establishes a conductor size that is compatible with the temperature rating of the terminations. For example, the minimum conductor size for a noncontinuous load of 60 amperes would be a 4 AWG conductor unless the terminations are marked 167°F (75°C) in

which case a 6 AWG would satisfy the requirements. The resistance of the 4 AWG is less than the 6 AWG; thus the heat generated would be less and, subsequently, the termination rating may be less.

Wire Selection

The process of selecting a conductor begins with identifying the wire material. An examination of *NEC® Table 310.16* reveals that only two classifications of wire are listed. They are:

1. copper

2. aluminum or copper-clad aluminum

The table indicates that for the same allowable size and type, a copper wire will have the higher allowable ampacity. *NEC® 110.5* stipulates that the wire will be assumed to be copper unless stated otherwise.

Aluminum wire is seldom used in branch-circuits, but, because of weight and cost factors, it is frequently selected for feeders and services. When installing aluminum conductors, the electrician must follow the directions and be especially careful to tighten all terminals to the recommended torque. It is also important to remember that listed connectors are required whenever it is necessary to join copper and aluminum conductors.

Some common problems associated with aluminum conductors when not properly connected may be summarized as follows:

- A corrosive action is set up when copper and aluminum wires come in contact with one another if moisture is present.

- The surface of an aluminum conductor oxidizes as soon as it is exposed to air. If this oxidized surface is not broken through, a poor connection results. When installing aluminum conductors, particularly in larger sizes, the electrician brushes an inhibitor onto the aluminum conductor and then scrapes the conductor with a stiff brush where the connection is to be made. The process of scraping the conductor breaks through the oxidation, and the inhibitor keeps the air from coming in contact with the conductor. Thus, further oxidation is prevented. Aluminum compression-type connectors usually have an inhibitor paste installed inside the connector.

- Aluminum wire expands and contracts to a greater degree than does copper for an equal load. This characteristic is another possible cause of a poor connection. Crimp connections for aluminum conductors are usually longer than those for comparable copper conductors, resulting in greater contact surface of the conductor in the connector.

FEEDER COMPONENT SELECTION

Previously in this unit, the information necessary for selecting feeder and branch-circuit components has been discussed in detail. Here, that information is first summarized and then presented in sequential steps. While studying these steps, it may be useful to review the detailed discussions and peruse the examples presented following the selection procedure.

Before the selection process can begin, it is necessary that the following circuit parameters be known:

- The continuous and noncontinuous loadings, in amperes, calculated in accordance with *NEC® Article 220*.

- The type of conductors that are to be installed. See *NEC® Table 310.13*.

- The ambient temperature of the environment where the conductors are to be installed and the number of current-carrying conductors that will be in the cable or raceway. See *NEC® 310.10*.

Step 1: OCPD Selection

- If the sum of the noncontinuous load plus 125 percent of the continuous load is 800 amperes or less, an OCPD shall be selected that has a rating equal to or next greater than that sum, see *NEC® Sections 240.4(B)* and *240.6*.

- If the sum of the noncontinuous load plus 125 percent of the continuous load is greater than 800 amperes, an OCPD shall be selected that has a rating equal to or less than that sum, see *NEC® Sections 240.4(C)* and *240.6*.

Step 2: Minimum Conductor Size Determination

- If the selected OCPD has a rating of 100-ampere or less, or if any of the terminations are marked 1 AWG or smaller, or are rated at 140°F (60°C), the minimum size conductor is deter-

mined by entering the 140°F (60°C) column of *NEC® Table 310.16* and selecting the conductor size that has an allowable ampacity not less than the sum of the noncontinuous load plus 125 percent of the continuous load.

- If the selected OCPD has a rating greater than 100-ampere, or if all the terminations are rated for 167°F (75°C), the minimum size conductor is determined by entering the 167°F (75°C) column of *NEC® Table 310.16* and selecting the conductor size that has an allowable ampacity equal to or greater than the sum of the noncontinuous load plus 125 percent of the continuous load.

Step 3: Conductor Type Determination

- If the selected OCPD has a rating of 100-ampere or less, the conductors to be installed may be selected from the 140°F (60°C), 167°F (75°C), or 194°F (90°C) columns of *NEC® Table 310.16*.

- If the selected OCPD has a rating greater than 100-ampere, the conductors to be installed shall be selected from either the 167°F (75°C) or 194°F (90°C) column of *NEC® Table 310.16*.

Step 4: Conductor Size Determination

To make the feeder conductor size determination:

From *NEC® Table 310.16* for the type of conductor selected in Step 3, choose a conductor that has an ampacity:

- equal to or greater than the minimum calculated load (noncontinuous load plus 125 percent of the continuous load)

  1. that permits the use of the OCPD selected in Step 1.

  2. is not less than the calculated load (noncontinuous load plus 125 percent of the continuous load).

Step 5: Neutral Size Determination

NEC® 220.61 sets forth conditions for reducing the size of the neutral conductor. This can result in a reduction in wire and raceway cost. The procedure set forth is straightforward except when the supply is a three-phase, four-wire, wye-connected system such as is commonly specified for commercial buildings. No precise method of computing the minimum size is identified by the *NEC®* for this situation. To determine a reasonable minimum size, two types of loads must be considered—linear and nonlinear. Nonlinear loads are defined in *NEC® Article 100*. In the FPN following the definition, the following items are listed: *Electronic equipment, electronic/electric discharge lighting and adjustable speed drives.*

Linear Loads.   Linear loads may be calculated by the following formula:

$$N = \sqrt{A^2 + B^2 + C^2 - AB - BC - AC}$$

Where: N = Neutral current
         A = Phase A current = 30 amperes
         B = Phase B current = 40 amperes
         C = Phase C current = 50 amperes

$$N = \sqrt{30^2 + 40^2 + 50^2 - (30 \times 40) - (40 \times 50) - (30 \times 50)}$$

$$N = \sqrt{900 + 1600 + 2500 - 1200 - 2000 - 1500}$$

$$N = \sqrt{300} = 17.32 \text{ amperes}$$

Assuming phase loads of 30, 40, and 50 amperes, the neutral load would be 17.32 amperes.

Additional calculation would indicate that if the 30-ampere load were disconnected, the neutral load would be 43.6 amperes, and if all but the 50-ampere load were disconnected, the neutral load would be 50 amperes. For the neutral to comply with *NEC® 220.61*, it must be sized for this maximum unbalanced condition.

Nonlinear Loads.   Beginning in 1947 and continuing in each succeeding edition, the *NEC®* has addressed the various problems of neutral currents from nonlinear/harmonic sources.

The following quotes are from the 2005 *National Electrical Code®*:

NEC® Article 100, Nonlinear load *A load where the wave shape of the steady-state current does not follow the wave shape of the applied voltage. (FPN): Electronic equipment, electronic/electric-discharge lighting, adjustable speed drive systems, and similar equipment may be nonlinear loads.* *

NEC® 220.61(C) states that *There shall be no reduction of the neutral or grounded conductor capacity applied to the amount in 220.61(C)(1), or portion of the amount in 220.61(C)(2), from that determined by the basic calculation:* *

(1) Any portion of a 3-wire circuit consisting of 2-phase wires and the neutral of a 4-wire, 3-phase, wye connected system.

(2) That portion consisting of nonlinear loads supplied from a 4-wire, wye-connected, 3-phase system.

NEC® 220.61, FPN 2: A 3-phase, 4-wire, wye connected power system used to supply power to nonlinear loads may necessitate that the power system design allow for the possibility of high harmonic neutral currents. *

Harmonics.   In the past, most connected loads were linear, such as resistive heating and lighting, and motors. In a linear circuit, current changes in proportion to a voltage increase or decrease in that circuit. The voltage and current sine waves are sinusoidal.

The arrival of electronic equipment such as UPS systems, AC-to-DC converters (rectifiers), inverters (AC-to-DC to adjustable frequency AC), computer power supplies, programmable controllers, data processing, electronic ballasts, and similar equipment brought about problems in electrical systems. Unexplained events started to happen such as overheated neutrals, overheating and failure of transformers, overheated motors, hot bus bars in switchboards, unexplained tripping of circuit breakers, incandescent lights blinking, fluorescent lamps flickering, malfunctioning computers, and hot lugs in switches and panelboards, even though the connected loads were found to be well within the conductor and equipment rating. All of the preceding electronic equipment is considered to be nonlinear, when the current in a given circuit does not increase or decrease in proportion to the voltage in that circuit. The resulting distorted voltage and current sine waves are nonsinusoidal. These problems can feed back into an electrical system and affect circuits elsewhere in the building, not just where the nonlinear loads are connected.

The root of the problem can be traced to electronic devices such as thyristors (silicon-controlled rectifiers [SCRs]) that can be switched on and off for durations that are extremely small fractions of a cycle. Another major culprit is switching-mode power supplies that switch at frequencies of 20,000 to more than 100,000 cycles per second. This rapid switching causes distortion of the sine wave. Harmonic frequencies are superimposed on top of the

---

*Reprinted with permission from NFPA 70-2005.

fundamental 60 Hz frequency, creating harmonic distortion.

Harmonics are multiples of a fundamental frequency. In this country, 60-Hz (60 cycles per second) is standard. Other frequencies superimposed on the fundamental frequency can be measured, such as:

Odd Harmonics		Even Harmonics	
3rd	180 cycles	2nd	120 cycles
5th	300 cycles	4th	240 cycles
7th	420 cycles	6th	360 cycles
9th	540 cycles	8th	480 cycles
etc.		etc.	

Currents of the 3rd harmonic and odd multiples of the 3rd harmonic (9th, 15th, 21st, etc.) add together in the common neutral conductor of a three-phase system, instead of canceling each other. These odd multiples are referred to as triplens. For example, on an equally balanced three-phase, four-wire, wye-connected system in which each phase carries 100 amperes, the neutral conductor might be called upon to carry 200 amperes or more. If this neutral conductor had been sized to carry 100 amperes, it would become severely overheated. This is illustrated in Figure 8-1.

Unusually high neutral current will cause overheating of the neutral conductor. There might also be excessive voltage drop between the neutral and the ground. Taking a voltage reading at a receptacle between the neutral conductor and ground with the loads turned on and finding more than two volts present probably indicates that the neutral conductor is overloaded because of the connected nonlinear load.

Attempting to use a low-cost, average-reading, clamp-on meter will not give a true reading of the total current in the neutral conductor because the neutral current is made up of current values from many frequencies. The readings will be from 30 percent to 50 percent lower. To get accurate current readings where harmonics are involved, ammeters referred to as "true rms" must be used.

At least one major manufacturer of Type-MC cable manufactures a cable that has an oversized neutral conductor. For cables in which the phase conductors are 12 AWG, the neutral conductor is an 8 AWG. For cables in which the phase conductors are 10 AWG, the neutral conductor is a 6 AWG. They also provide a Type-MC cable that has a separate neutral conductor for each of the phase conductors. There is still much to be learned about the effects of nonlinear loads. Many experienced electrical consulting engineers are now specifying that three-phase, four-wire branch-circuits that supply nonlinear loads have a separate neutral for each phase conductor, instead of using a typical three-phase four-wire branch-circuit using a common neutral. (See Figure 8-3.) They also specify that the neutral be sized double that of the phase conductor. The Computer and Business Equipment Manufacturers Association (CBEMA) offers the following recommendation:

Run a separate neutral to 120 volt outlet receptacles on each phase. Avoid using shared neutral conductors for single-phase 120-volt outlets on different phases. Where a shared neutral conductor for a 208Y/120 volt system must be used for multiple phases use a neutral conductor having at least 200% of the phase conductors.

What does this mean to the electrician? It simply means that if overheated neutral conductors, overheated transformers, overheated lugs in switches and panels, or other unexplained heating problems are encountered, they might be caused by nonlinear loads. Where a commercial or industrial building contains a high number of nonlinear loads, it is advisable to have the electrical system analyzed and redesigned by an electrical engineer specifically trained and qualified on the subject of nonlinear loads.

The effects of nonlinear loads, such as electric discharge lighting, electronic/data-processing equipment, and/or variable speed motors, is illustrated in Figure 8-1, Figure 8-2, Figure 8-3, and Figure 8-4. In each case, the power is supplied from a three-

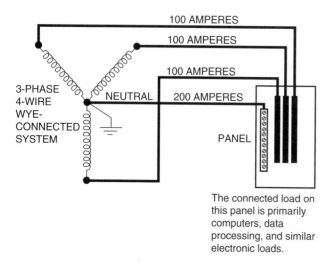

The connected load on this panel is primarily computers, data processing, and similar electronic loads.

Figure 8-1   Feeder neutral current resulting from nonlinear loads.

phase, four-wire 208Y/120-volt system. Figure 8-1 illustrates the possible effect the load could have on a feeder. The neutral could be carrying 200 percent of the phase conductors. Figure 8-2 illustrates a branch-circuit where the load is fluorescent lighting using core and coil ballast. If the ballast were electronic, the neutral current would be even greater. Figure 8-3 illustrates the recommended arrangement for serving three nonlinear loads from a four-wire wye system. If the neutral is shared, as is illustrated in Figure 8-4, then the neutral should have at least 200 percent of the carrying capacity as the individual phase conductors. Using separate neutrals is the preferred arrangement.

It is clear from the information given that nonlinear loads can create neutral currents in excess of the phase currents in a three-phase, four-wire wye system. Considering what is known and unknown, it is prudent to take a conservative approach consider-

ing the minimal savings involved.

Three stipulations are clear:

1. The neutral must be able to carry the maximum unbalanced load.

2. On a three-phase, four-wire, wye system, no reduction is allowed for nonlinear loads.

3. In *NEC® 220.61, FPN No. 2* a warning is issued concerning the currents created by nonlinear loads.

Three conditions of failure may occur with a three-phase, four-wire wye system:

1. If all three phases are operable: In this, the normal situation, by definition the neutral is part of the feeder and must be sized for the same load and environmental requirements. The minimum size shall comply with *NEC® 110.14(C)*, and the adjustment and corrections must be applied as with the phase conductors.

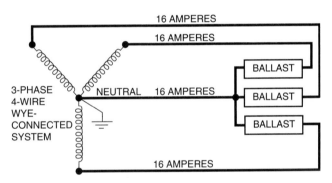

Figure 8-2 Branch-circuit current resulting from nonlinear loads.

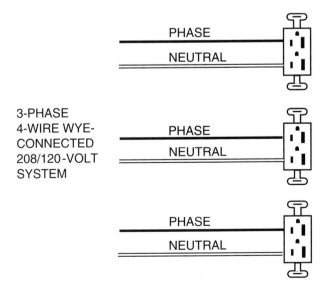

Figure 8-3 Recommended: separate neutrals serving nonlinear loads.

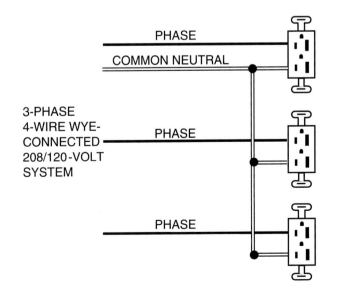

Figure 8-4 Not recommended: shared neutral serving nonlinear loads.

2. If two phases are operable: In essence this creates a situation similar to a feeder consisting of two phases of a three-phase system, a condition where the *NEC*® specifically prohibits a reduction in neutral size.

3. If only a single phase is operable (worst-case scenario): It is a tendency for nonlinear loads to be continuous loads, and thus likely that in this condition, the load in the single phase may be largely a nonlinear load. As previously stated, this is a situation in which the neutral load could exceed the phase load.

Giving full consideration to the worst case scenario, in this commercial building the following strategy will be followed to determine the minimum neutral conductor size:

• The 208-volt, single- and three-phase loads will be subtracted from the OCPD selection load (the noncontinuous load plus 125 percent of the continuous load).

• A load equal to the nonlinear load will be added to the neutral load. This doubles the neutral capacity for the nonlinear load.

Summary of Nonlinear Loads

Because of the complexity of calculating nonlinear loads, most consulting engineers will simply do load calculations per the *NEC*®—double the size of the neutral.

You will also find that for installations where the nonlinear load is huge, most consulting engineers will specify K-rated transformers. A K-rated transformer can handle the heat generated by harmonic currents. A K-rated transformer is manufactured with heavier gauge wire for the windings and has double-size neutral terminals to accommodate the larger size neutral installed by the electrician.

VOLTAGE DROP

NEC® 215.2(A)(3), FPN No. 2 sets forth a recommendation for limiting the voltage drop on feeders. A similar recommendation is made for branch-circuits. The recommendation is for the total voltage drop to be not more than 5 percent, and not more than 3 percent for either the feeder or the branch-circuit. The author recommends that the feeder be limited to 2 percent, thus leaving 3 percent for the branch-circuit.

As an example, the circuit to the drugstore window display receptacle outlets is to be loaded to 1500 volt-amperes or 12.5 amperes as illustrated in Figure 8-5. The distance from the panelboard to the center outlet is 85 ft and the minimum wire size is 12 AWG. Three different methods will be demonstrated to estimate the voltage drop in this circuit.

The first method requires that the resistance per foot of the conductor be known. According to *NEC*® *Chapter 9, Table 8*, the resistance of a 12 AWG

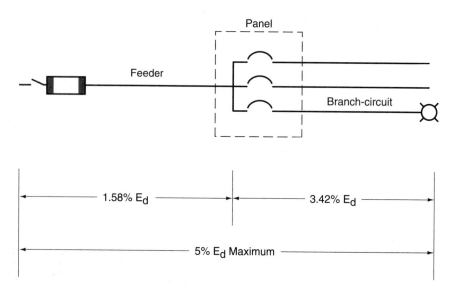

Figure 8-5   Voltage drop calculations.

uncoated copper conductor is 1.93 ohms per 1000 ft (300 m). Therefore, the resistance of the circuit is

$$R = \text{Unit resistance} \times \text{distance}$$

$$R = \left(\frac{1.93 \text{ ohm}}{1000 \text{ ft}}\right) \times (2 \times 85 \text{ ft}) = 0.3281 \text{ ohm}$$

$$R = \left(\frac{1.93 \text{ ohm}}{300 \text{ m}}\right) \times (2 \times 25.5 \text{ m}) = 0.3281 \text{ ohm}$$

The factor of two in the equation is required because both the phase and neutral conductors carry the current. The voltage drop is calculated using Ohm's Law:

$$VD = I \times R$$
$$= 12.5 \text{ amperes} \times 0.3281 \text{ ohms}$$
$$= 4.1 \text{ volts}$$

This is a voltage drop of

$$\frac{4.1}{120} = 0.0342 \text{ or } 3.42\%$$

This would limit the drop allowable for the feeder to 1.58 percent (5 − 3.42 = 1.58).

The second method requires that the circular mil area of the conductor be known. From *NEC® Chapter 9, Table 8,* the kcmil area of a 12 AWG is 6530. The equation is:

$$VD = \frac{K \times I \times L \times 2}{\text{kcmil}}$$

Where K for copper is 39.4 when the length is in meters and 12 when the length is in feet. For aluminum, K is 68 and 20. Therefore

$$VD = \frac{12 \times 12.5 \text{ A} \times 85 \text{ ft} \times 2}{6530 \text{ kcmil}} = 3.9 \text{ volts}$$

$$VD = \frac{39.4 \times 12.5 \text{ A} \times 26 \text{ m} \times 2}{6530 \text{ kcmil}} = 3.9 \text{ volts}$$

Another method is one that considers the power factor, the type of raceway, and whether it is a single- or three-phase circuit. Abbreviated Table 8-1A and Table 8-1B show values appropriate for use with the circuits in the commercial building. The values given are for installation in steel conduit. A complete table is provided in Appendix Table A-2 and Table A-3.

To calculate the voltage drop, multiply the current, the distance, and the proper factor from the table and then move the decimal point six places to the left. The following is the calculation for the branch-circuit to the show window receptacle outlets, assuming a power factor of 90 percent:

$$VD = \frac{12.5 \text{ A} \times 85 \text{ ft} \times 3659}{10^6} = 3.88 \text{ V}$$

$$VD = \frac{12.5 \text{ A} \times 25.9 \text{ m} \times 12,000}{10^6} = 3.88 \text{ V}$$

Installation in Steel Conduit—Meters									
Wire size	12	10	8	6	4	3	2	1	1/0
90% pf, 1-phase	12,000	7260	4790	3070	2000	1643	1340	1105	860
90% pf, 3-phase	10,390	6288	4144	2662	1731	1422	1160	957	748
Wire size	2/0	3/0	4/0	250	300	350	400	500	600
90% pf, 1-phase	744	613	515	465	410	370	344	308	282
90% pf, 3-phase	642	531	445	403	354	321	298	265	245

Table 8-1A

Installation in Steel Conduit—Feet									
Wire size	12	10	8	6	4	3	2	1	1/0
90% pf, 1-phase	3659	2214	1460	937	610	501	409	337	263
90% pf, 3-phase	3169	1918	1264	812	528	434	354	292	228
Wire size	2/0	3/0	4/0	250	300	350	400	500	600
90% pf, 1-phase	227	187	157	142	125	113	105	94	86
90% pf, 3-phase	196	162	136	123	108	98	91	81	75

Table 8-1B

This is a voltage drop of

$$\frac{3.88 \text{ V}}{120 \text{ V}} = 0.032 \text{ or } 3.2\%$$

It is recommended that 10 AWG conductors be used.

Parallel Conductors

Often the load is such that parallel conductors should be considered. This usually occurs with loads over 400 amperes. In the commercial building, the branch-circuit to the boiler consists of two sets of conductors, and the service is three sets. The major reasons for this are the required size of the conductors and the raceways. Conductors larger than 500 kcmil are difficult to install, and often contractors do not have the equipment to use conduit sizes greater than trade size 4.

NEC® 310.4 sets forth the requirements for installing parallel conductors. They shall:

- be 1/0 AWG or larger.
- be the same length.
- be of the same material.
- have the same insulation.
- be the same size in circular mil area.
- be terminated in the same manner.

It is critical that the conductors be the same length. A knowledgeable electrician will mark out the required length on a clean flat surface and, thus, ensure that the conductors have identical lengths. The following example will illustrate how a small difference in length results in a significant difference in current in the conductor sets.

Problem:

A 1600-ampere service consists of four conduits, each containing four 600-kcmil Type THHN/THWN conductors. The allowable ampacity of this conductor is 420 amperes. When the service is operating at 1600-ampere, what is the current in each of the four parallel sets of conductors, if their lengths are 20 ft, 21 ft, 22 ft, and 23 ft?

Solution:

The total length of these conductors is

$$20 + 21 + 22 + 23 = 86 \text{ ft}$$

The current in each set is inversely proportional to the resistance, which is proportional to the conductor's length. The proportional resistance of each set is

$$\frac{20}{86} \times 1600 = 372.09 \text{ amperes}$$

$$\frac{21}{86} \times 1600 = 390.6 \text{ amperes}$$

$$\frac{22}{86} \times 1600 = 409.3 \text{ amperes}$$

$$\frac{23}{86} \times 1600 = 427.9 \text{ amperes}$$

As current is inversely proportional to resistance, the currents would be 428, 409, 390, and 372 amperes. With only one foot difference in each set, the current exceeds the rating in one set of the conductors.

PANELBOARD WORKSHEET SUMMARY

After all loads are identified, it is suggested that the loads be summarized, first to determine that no single phase is carrying a majority of the load and second according to the classifications needed for sizing the feeder. The sum of these two summaries should be equal.

Phase Connection Summary

The volt-ampere loading on each phase is tabulated and compared (Table 8-2). The values should be essentially equal; however, it is unlikely that they will be exactly equal.

Drugstore—Load Summary	
Connection Summary	**VA Total**
Connected Load Phase A	4503
Connected Load Phase B	5343
Connected Load Phase C	4272
Balanced Loads	8604
Connection Total	22,722
Loads Summary	**VA Total**
Continuous	8838
Noncontinuous + Receptacle	5280
Highest Motor	8604
Other Motors	0
Loads Summary Total	22,722

Table 8-2

Loads Summary

The values shown in Table 8-2 are needed for the design loading calculations. The groups are totaled to ensure that no loads were omitted. Refer to Table A-17, Table A-18, and Table A-19 in the Appendix.

FEEDER DETERMINATION, DRUGSTORE

Following is a table, Table 8-3, and an outline illustrating the process of selecting the feeder components for the drugstore.

Load Summary

- Referring to line 1 of Table 8-3, the continuous calculated load is taken from the panelboard summary. It is multiplied by 1.25.

$$8838 \text{ VA} \times 1.25 = 11{,}048 \text{ VA}$$

- Referring to line 2 of Table 8-3, the noncontinuous and receptacle loads are taken from the panelboard schedule. The value remains unchanged.

$$5280 \text{ VA} = 5280 \text{ VA}$$

- Referring to line 3 of Table 8-3, the highest motor load is taken from the panelboard summary. It is multiplied by 1.25.

$$8604 \text{ VA} \times 1.25 = 10{,}755 \text{ VA}$$

- Referring to line 4 of Table 8-3, there were no other motor loads.

- Referring to line 5 of Table 8-3, the four preceding loads are added, and the sum is multiplied by 0.25 to determine the growth. The growth is increased by 25 percent on the assumption that it will all be a continuous load.

$$(8838 \text{ VA} + 5280 \text{ VA} + 8604 \text{ VA}) \times 0.25$$
$$= 5680 \text{ VA}$$

$$5680 \text{ VA} \times 1.25 = 7100 \text{ VA}$$

- Referring to line 6 of Table 8-3, the values in the two columns are summed.

$$8838 \text{ VA} + 5280 \text{ VA} + 8604 \text{ VA} + 5680 \text{ VA}$$
$$= 28{,}402 \text{ VA}$$

$$11{,}048 \text{ VA} + 5280 \text{ VA} + 10{,}755 \text{ VA} +$$
$$7100 \text{ VA} = 34{,}183 \text{ VA}$$

OCPD Selection

- Divide the OCPD selection volt-amperes by 360 to convert to amperes.

$$\frac{34{,}183 \text{ VA}}{360 \text{ V}} = 95 \text{ A}$$

Drugstore—Feeder Selection		
Load Summary, 208Y/120-Volt, 3-Phase, 4-Wire	**Calculated**	**OCPD**
Continuous Load	8838	11,048
Noncontinuous + Receptacle Load	5280	5280
Highest Motor Load	8604	10,755
Other Motor Load	0	0
Growth	5680	7100
Calculated Load & OCPD Load	28,402	34,183
OCPD Selection	**Input**	**Output**
OCPD Load Volt-Amperes & Amperes	34,183	95
OCPD Rating		100
Minimum Conductor Size		1 AWG
Minimum Ampacity		91
Phase Conductor Selection	**Input**	**Output**
Ambient Temperature & Correction Factor	28°C	1
Current Carrying Conductors & Adjustment Factor	4	0.8
Derating Factor		0.8
Minimum & Allowable Ampacity	91	114
Conductor Size		1 AWG
Conductor Type, Allowable Ampacity	THHN	150
Ampacity		120
Voltage Drop, 0.9 pf, 3 ph, per 100 ft	2.3	1.1%
Neutral Conductor Selection	**Input**	**Output**
OCPD Load Volt-Amperes	34,183	
Balanced Load Volt-Amperes	−8604	
Nonlinear Load Volt-Amperes	4962	
Total Load Volt-Amperes & Amperes	30,541	85
Minimum Neutral Size		3 AWG
Minimum Allowable Ampacity		106
Neutral Conductor Type & Size	THHN	3 AWG
Allowable Ampacity	110	88
Raceway Size Determination	**Input**	**Output**
Feeder Conductors Size & Total Area	1 AWG	0.4686
Neutral Conductor Size & Area	3 AWG	0.0973
Grounding Conductor Size & Area		0
Total Conductor Area		0.5659
Raceway Type & Trade Size	RMC	1¼

Table 8-3

- From *NEC® 240.6*, the OCPD ampere rating next higher than 95 is 100-ampere. This becomes the OCPD ampere rating.

- As the OCPD rating is 100-ampere, a 1 AWG is the conductor size with an ampacity equal to or next greater than 100 in the 140°F (60°C) column of *NEC® Table 310.16.*

The minimum feeder conductor ampacity must meet two criteria. It must be high enough to carry the calculated load. In this case that is 28,402 ÷ 360 = 79 amperes. It also must be high enough to allow the use of the selected OCPD. The next lower rating given in *NEC® 240.6* under 100 is 90. An ampacity of 91 would permit the use of an OCPD with a rating of 100 amperes. The higher of these is 91-ampere, thus, that is the minimum ampacity.

Phase Conductor Selection

- The ambient temperature is determined to be 82°F (28°C). As this is within the standard range, the correction factor is 1.

- As the harmonic load (fluorescent lighting, etc.) is significant, the neutral will be considered as a current-carrying conductor. The adjustment factor for four conductors is 0.8.

- The correction and adjustment factors are multiplied to yield a reduction factor of 0.8.

- The minimum allowable ampacity is determined by dividing the minimum ampacity by the derating factor. In this example 91 ÷ 0.8 = 114. This is the lowest ampacity that after derating will be equal to or greater than the minimum value.

- The conductor size selection must also meet two criteria. It must be at least the size of the minimum conductor size, and it must have an allowable ampacity that is equal to or greater than the minimum allowable ampacity. In this case, a 1 AWG qualifies.

- The conductor type is compliant with the construction specifications and the allowable ampacity is taken from *NEC® Table 310.16* for the conductor size.

- The ampacity is the allowable ampacity multiplied by the derating factor. This is checked against the minimum allowable ampacity, but if the steps have been followed correctly, it should never fail.

The voltage drop is calculated for a 90 percent power factor and a distance of 100 ft by the proce-dure presented earlier. By using 100 ft, the drop for the actual distance can be easily calculated. If, after applying the actual length, the drop is in excess of 3 percent, serious consideration should be given to increasing the conductor size.

$$\frac{28{,}402 \text{ volt-amperes}}{360 \text{ volts}} = 79 \text{ amperes}$$

$$\frac{79 \text{ amperes} \times 292 \times 100 \text{ ft}}{10^6} = 2.3 \text{ VD}$$

$$\frac{2.3 \text{ V}}{208 \text{ V}} \times 100 = 1.1\%$$

Neutral Size Determination

As is shown in the feeder selection schedule, the neutral size is determined to be

Total load – Balanced load + Nonlinear load
= Neutral load

34,183 VA – 8604 VA + 4962 VA = 30,541 VA

$$\frac{30{,}541 \text{ VA}}{360 \text{ V}} = 85 \text{ A}$$

From the (140°F) 60°C column, the minimum wire size is a 3 AWG.

The minimum allowable ampacity is

$$\frac{85 \text{ A}}{0.8} = 106 \text{ A}$$

A 3 AWG Type THHN conductor has an allowable ampacity of 110 amperes and qualifies for the neutral conductor.

Raceway Size Determination

It has been concluded that the feeder to the drug-store will consist of three 1 AWG and one 3 AWG. No equipment grounding conductor is necessary if rigid metal conduit is installed. The specification also permits the use of schedule 80 rigid nonmetallic conduit, which would require the installation of an equipment grounding conductor.

From *NEC® Chapter 9, Table 5:*

AWG	Count	Area in.²	Total	Area mm²	Total
1	3	0.1562	0.4686	100.8	302.5
3	1	0.0973	0.0973	62.77	62.77
Area Totals			0.5650		365.27

Consulting *NEC® Chapter 9, Table 4,* for rigid metal conduit the smallest allowable size, for more than two conductors, is trade size 1¼.

REVIEW QUESTIONS

Refer to the *National Electrical Code*® or the working drawings when necessary. Where applicable, responses should be written in complete sentences.

1. Determine the conductor sizes for a feeder to a panelboard. It is a 120/240-volt, single-phase system. The OCPD has a rating of 100-ampere. The calculated load is 15,600 VA. All the loads are 120 volts.

$$\frac{15600}{240} = 65$$

$$\times 125\% = 81$$

For the next questions, you may use either a graphical or mathematical means to determine the answers.

2. Calculate the neutral current in a 120/240-volt, single-phase system when the current in Phase A is 20 amperes and the current in Phase B is 40 amperes.

20amp

3. Calculate the neutral current in a 208Y/120-volt, three-phase, four-wire system when the current in Phase A is 0, in Phase B is 40, and in Phase C is 60 amperes.

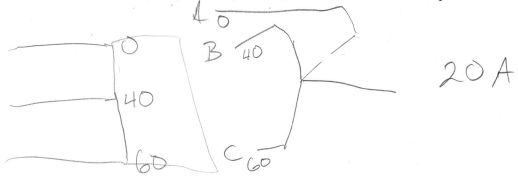

20A

4. Calculate the neutral current in a 208Y/120-volt, three-phase, four-wire system when the current in Phase A is 20, in Phase B is 40, and in Phase C is 60 amperes.

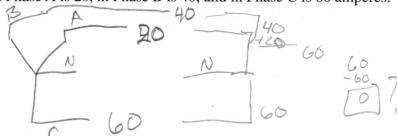

5. A balanced electronic ballast load connected to a three-phase, four-wire multiwire branch-circuit can result in a neutral current of   (zero)   (half)   (two times)   that of the current in the phase conductors. (Circle the correct answer.)

6. Harmonic currents associated with electronic equipment   (add together)   (cancel out) a shared common neutral on multiwire branch-circuits and feeders. (Circle the correct answer.)

7. Check the correct statement:
   For circuits that supply computer/data-processing equipment, it is recommended that
   ☐ a separate neutral be installed for each hot-phase branch-circuit conductor.
   ☐ a common neutral be installed for each multiwire branch-circuit.

8. Determine the feeder size and other information requested in the following table. The load data is:

Continuous	20,000 VA
Noncontinuous/receptacle	12,000 VA
Highest motor	8000 VA
Other motor	16,000 VA
Nonlinear	20,000 VA
Balanced	24,000 VA

A three-phase, four-wire feeder is to be installed. From a distribution panelboard, the feeder is installed underground in rigid PVC, schedule 80, to an adjacent building where intermediate metal conduit (IMC) is used.

• The IMC is installed in a room where the ambient temperature will be 110°F (43°C).

• Type THHN/THWN conductors are used.

• The growth allowance will be 10 percent.

• The power factor will be maintained at 90 percent or higher.

- The conductor length is 180 ft.

- The voltage drop at maximum load shall not exceed 2 percent.

Load Summary, 208Y/120, 3 Phase, 4-Wire	Calculated	OCPD
Continuous Load		
Noncontinuous + Receptacle Load		
Highest Motor Load		
Other Motor Load		
Growth		
Calculated Load & OCPD Load		
OCPD Selection	**Input**	**Output**
OCPD Load Volt-Amperes & Amperes		
OCPD Rating		
Minimum Conductor Size		
Minimum Ampacity		
Phase Conductor Selection	**Input**	**Output**
Ambient Temperature & Correction Factor		
Current-Carrying Conductors & Adjustment Factor		
Derating Factor		
Minimum Allowable Ampacity		
Conductor Size		
Conductor Type, Allowable Ampacity		
Ampacity		
Voltage Drop, 0.9 pf, 3 ph, per 100 ft		
Neutral Conductor Selection	**Input**	**Output**
OCPD Load Volt-Amperes		
Balanced Load Volt-Amperes		
Nonlinear Load Volt-Amperes		
Total Load Volt-Amperes & Amperes		
Minimum Neutral Size		
Minimum Allowable Ampacity		
Neutral Conductor Type & Size		
Allowable Ampacity		
Raceway Size Determination	**Input**	**Output**
Feeder Conductors Size & Total Area		
Neutral Conductor Size & Area		
Grounding Conductor Size & Area		
Total Conductor Area		
Raceway Type & Minimum Trade Size		

UNIT 9

Special Systems

OBJECTIVES

After studying this unit, the student should be able to

- select and install a surface metal raceway.
- select and install multioutlet assemblies.
- calculate the loading allowance for multioutlet assemblies.
- select and install a floor outlet system.
- install a branch-circuit for a computer.

A number of electrical systems are found in almost every commercial building. Although these systems usually are a minor part of the total electrical work to be done, they are essential systems, and it is recommended that the electrician be familiar with the installation requirements of these special systems.

SURFACE METAL RACEWAYS

Surface metal raceways, either metal or nonmetallic, are generally installed as extensions to an existing electrical raceway system, and where it is impossible to conceal conduits, such as in desks, counters, cabinets, and modular partitions. The installation of surface metal raceways is governed by *NEC® Article 386. NEC® Article 388* governs the installation of surface nonmetallic raceways. The number and size of the conductors to be installed in surface raceways is limited by the design of the raceway. Catalog data from the raceway manufacturer will specify the permitted number and size of the conductors for specific raceways. Conductors to be installed in raceways may be spliced at junction boxes or within the raceway if the cover of the raceway is removable. See *NEC® 386.56* and *388.56*. It should be noted that the combined size of the conductors, splices, and taps shall not fill more than 75 percent of the raceway at the point where the splices and taps occur.

Surface raceways are available in various sizes, Figure 9-1, and a wide variety of special fittings, like the supports in Figure 9-2, makes it possible to use surface metal or nonmetallic raceways in almost any dry location. Two examples of the use of surface raceways are shown in Figure 9-3 and Figure 9-4.

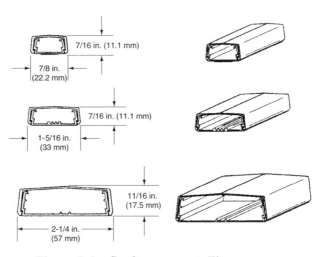

Figure 9-1   Surface nonmetallic raceways.

Figure 9-2   Surface metal raceway supports.

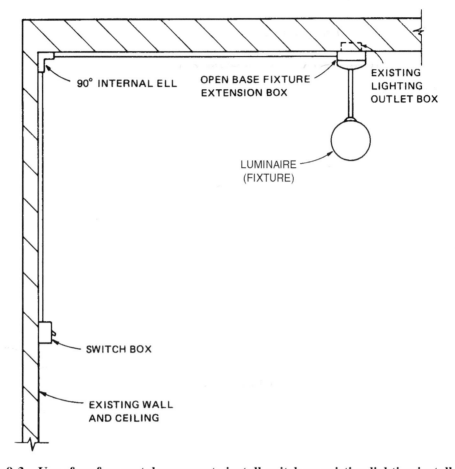

Figure 9-3   Use of surface metal raceway to install switch on existing lighting installation.

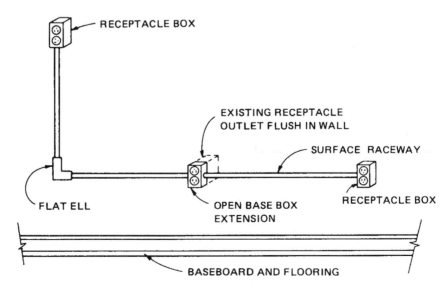

Figure 9-4   Use of surface metal raceway to install additional receptacle outlets.

MULTIOUTLET ASSEMBLIES

Multioutlet assembly is defined in *NEC® Article 100*. The installation requirements are specified in *NEC® Article 380*. Multioutlet assemblies, as illustrated in Figure 9-5, are similar to surface raceways and are designed to hold both conductors and devices. These assemblies offer a high degree of flexibility to an installation and are particularly suited to heavy-use areas where many outlets are required or where there is a likelihood of changes in the installation requirements. The plans for the insurance office specify the use of a multioutlet assembly that will accommodate power, data, telecommunications, security, and audio/visual

systems; see Figure 9-6. This installation will allow the tenant in the insurance office to revise and expand the office facilities as the need arises.

Receptacles for multioutlet assemblies are available with GFCI, isolated ground, and surge suppression features.

The covers are available in steel, aluminum, and PVC with vinyl laminates of different colors and wood veneers such as maple, cherry, mahogany, and oak.

Loading Allowance

The load allowance for a multioutlet assembly is specified by *NEC® 220.14(H)* as:

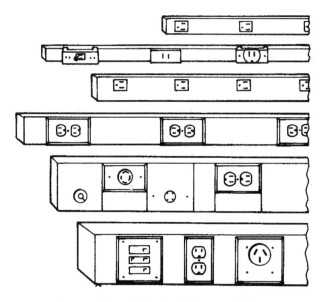

Figure 9-5   Multioutlet assemblies.

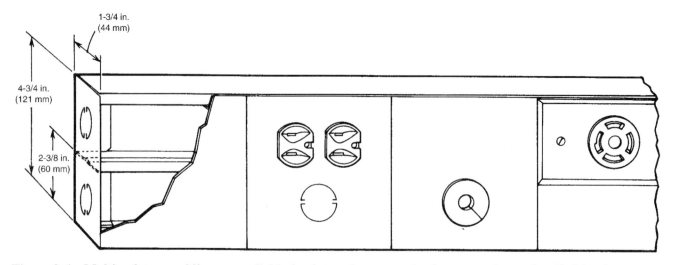

Figure 9-6   Multioutlet assemblies are available for data, telecommunications, security, and audio/visual systems.

- 180 volt-amperes for each 5 ft (1.5 m) of assembly, or fraction thereof, when normal loading conditions exist.

- 180 volt-amperes for each 1 ft (300 mm) of assembly, or fraction thereof, when heavy loading conditions exist.

The usage in the insurance office is expected to be intermittent and made up of only small appliances; thus, it would qualify for the allowable minimum of 180 volt-amperes per 5 ft. However, the contractor is required to install a duplex receptacle every 18 in. (450 mm) for a total of 56 receptacles. The number of receptacles is multiplied by 180 volt-amperes to determine the connected load, which would be 10,080 volt-amperes. This load is connected to five branch-circuits with a total noncontinuous capacity of 12,000 volt-amperes. This allows

for some growth, and additional circuits can be installed easily.

When determining the feeder conductor size, the requirement for the receptacle load can be reduced by the application of the demand factors given in *NEC® Table 220.44*. The total receptacle loading in the insurance office is 12,780 volt-amperes. The first 10,000 volt-amperes is included at 100 percent. The remaining load is included at 50 percent. The calculated load on the feeder due to receptacles is 11,390 amperes.

Receptacle Wiring

The plans indicate that the receptacles to be mounted in the multioutlet assembly must be spaced 18 in. (450 mm) apart. The receptacles may be connected in either of the arrangements shown in Figure 9-7. That is, all of the receptacles on a phase

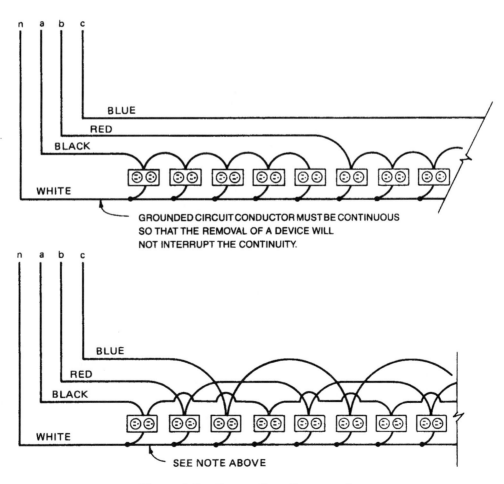

Figure 9-7 Connection of receptacles.

can be connected in a continuous row, or the receptacles can be connected on alternate phases.

COMMUNICATIONS SYSTEMS

The installation of the telephone system in the commercial building will consist of two separate installations. The *electrical contractor* will install an empty conduit system according to the specifications for the commercial building and in the locations indicated on the working drawings. In addition, the installation will meet the rules, regulations, and requirements of the communications company that will serve the building. When the conduit system is complete, the electrical contractor, telephone installing company, or telephone company will install a complete telephone system.

Power Requirements

An allowance of 2500 volt-amperes is made for the installation of the telephone equipment to provide the power required to operate the special switching equipment for a large number of telephones. Because of the importance of the telephone as a means of communication, the receptacle outlets for this equipment are connected to the emergency power system.

The Communications System

Communication/telephonic requirements vary widely according to the type of business and the extent of the communication convenience desired. Details on communication systems are given in *NEC® Article 800.*

Because it is difficult to install the system wiring after the building construction is completed, it is often a requirement that conduits be installed from each of the occupancies to a designated place or area where they are accessible to the installers of the system. The materials used in the installation and the method of installing the raceway lines are the same as those used for light and power wiring with the following modifications:

- Because of small openings and limited space, junction boxes are used rather than the standard conduit fittings, such as ells and tees.

- Because multipair conductors are used extensively in telephone installations, the size of the conduit should be trade size ¾ or larger.

- The number of bends and offsets is kept to a minimum; when possible, bends and offsets are made using greater minimum radii or larger sweeps (the allowable minimum radius is 5 in. [127 mm]).

- A fishwire is installed in each conduit for use in pulling in the cables.

- When basement communication wiring is to be exposed, the conduits are dropped into the basement and terminated with a bushing (junction boxes are not required). No more than 2 in. (50 mm) of conduit should project beyond the joists or ceiling level in the basement.

- The inside conduit drops need not be grounded unless they can become energized; see *NEC® 250.104(B).*

- The service conduit carrying communications cables from the exterior of a building to the interior must be permanently and effectively grounded; see *NEC® 250.104(A).*

Installing the Communications System Raceways

The plans for the commercial building indicate that each of the occupancies requires access to a communication system. Access is achieved by installing a trade size ¾ EMT or ENT to the basement from each occupancy, Figure 9-8. As previously indicated, the insurance office uses a multioutlet assembly as a distribution means. In the other areas, the distribution is the responsibility of the occupant. All the cables and equipment will be installed by others.

FLOOR OUTLETS

In an area the size of the insurance office, it may be necessary to place equipment and desks where wall outlets are not available. Floor outlets may be installed to provide the necessary electrical supply to such equipment. Two methods can be used to provide floor outlets: (1) installing underfloor raceway or (2) installing floor boxes.

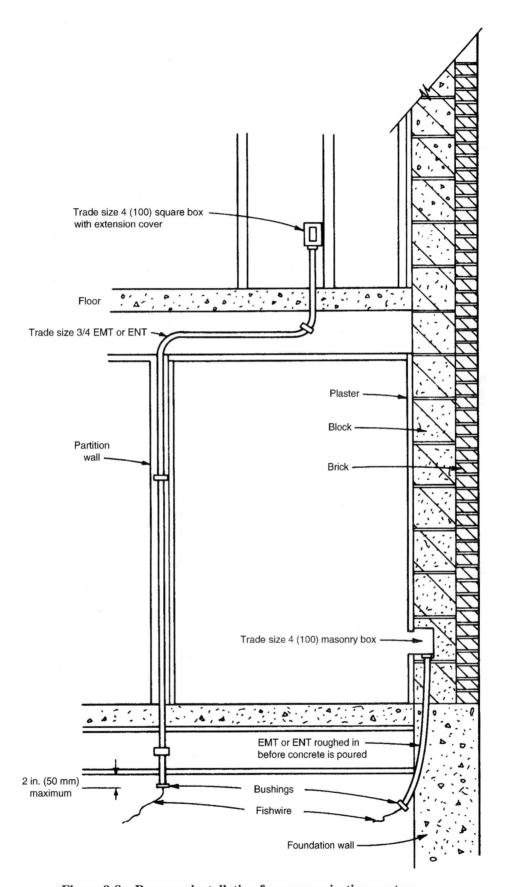

Trade size 4 (100) square box
with extension cover

Floor

Trade size 3/4 EMT or ENT

Plaster

Block

Brick

Partition
wall

Trade size 4 (100) masonry box

EMT or ENT roughed in
before concrete is poured

2 in. (50 mm)
maximum

Bushings

Fishwire

Foundation wall

Figure 9-8   Raceway installation for communications system.

Underfloor Raceway

The installation requirements for underfloor raceway are set forth in *NEC® Article 390*. It is common for this type of raceway to be installed to provide both power and communication outlets in a dual-duct system similar to the one shown in Figure 9-9. The junction box is constructed so that the power and communications systems are always separated from each other. Service fittings are available for the outlets, Figure 9-10.

Floor Boxes

Floor boxes, either metallic or nonmetallic, can be installed using any approved raceway such as rigid conduit, rigid nonmetallic conduit, or electrical nonmetallic tubing. Some boxes must be installed to the correct height by adjusting the leveling screws before the concrete is poured, Figure 9-11. One nonmetallic box can be installed without adjustment and cut to the desired height after the concrete is poured, Figure 9-12 and Figure 9-13.

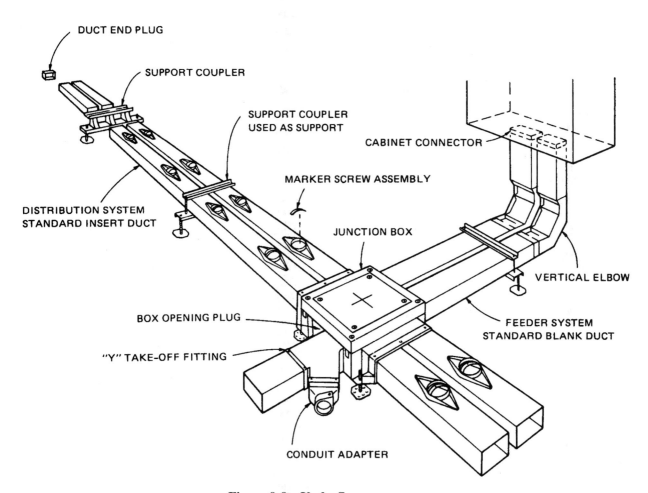

Figure 9-9   Underfloor raceway.

FOR HIGH-POTENTIAL SERVICE

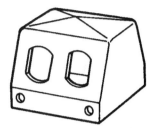 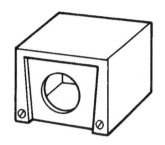 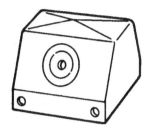

Dimensions: 4-1/8 in. (105 mm) long; 4-1/8 in. (105 mm) wide; 2-15/16 in. (74.6 mm) high.

Figure 9-10   Service fittings.

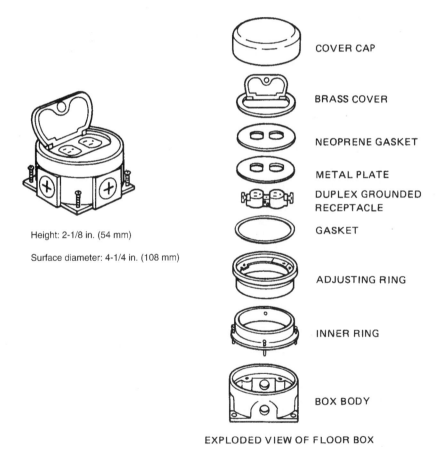

COVER CAP

BRASS COVER

NEOPRENE GASKET

METAL PLATE

DUPLEX GROUNDED
RECEPTACLE

GASKET

ADJUSTING RING

INNER RING

BOX BODY

Height: 2-1/8 in. (54 mm)

Surface diameter: 4-1/4 in. (108 mm)

EXPLODED VIEW OF FLOOR BOX

Figure 9-11   A floor box with leveling screws.

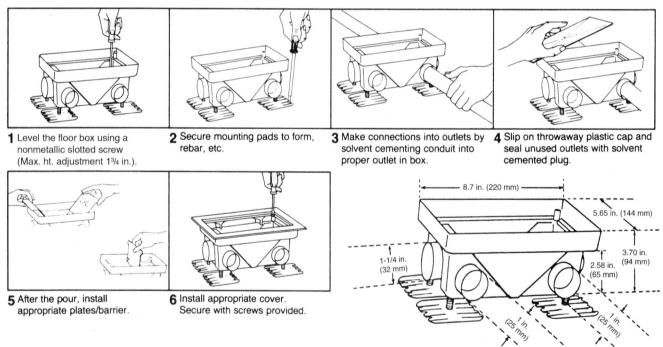

1 Level the floor box using a nonmetallic slotted screw (Max. ht. adjustment 1¾ in.).

2 Secure mounting pads to form, rebar, etc.

3 Make connections into outlets by solvent cementing conduit into proper outlet in box.

4 Slip on throwaway plastic cap and seal unused outlets with solvent cemented plug.

5 After the pour, install appropriate plates/barrier.

6 Install appropriate cover. Secure with screws provided.

Figure 9-12   Installation of multifunction nonmetallic floor box.

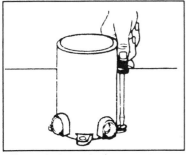

1 Fasten the box to the form or set on level surface.

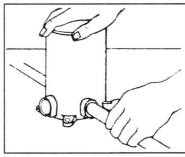

2 Make connections into outlets by solvent cementing conduit into the proper outlet in box.

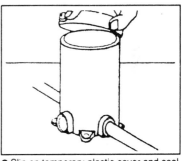

3 Slip on temporary plastic cover and seal unused outlets with solvent cemented plugs.

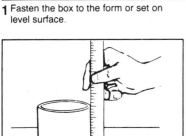

4 Remove temporary plastic cover and determine thickness of flooring to be used. Scribe a line around box at this distance from the floor.

5 Using a handsaw, cut off box at scribed line.

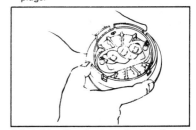

6 Install leveling ring to underside of cover, so that four circular posts extend through slots in outer ring.

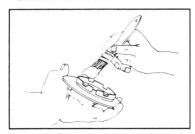

7 Apply PVC cement to leveling ring and to top inside edge of box.

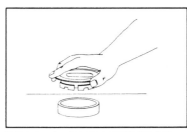

8 Press outer ring cover assembly into floor box for perfect flush fit.

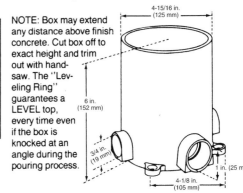

NOTE: Box may extend any distance above finish concrete. Cut box off to exact height and trim out with handsaw. The "Leveling Ring" guarantees a LEVEL top, every time even if the box is knocked at an angle during the pouring process.

4-15/16 in. (125 mm)
6 in. (152 mm)
3/4 in. (19 mm)
1 in. (25 mm)
4-1/8 in. (105 mm)

Figure 9-13   Installation and trim-out of nonmetallic floor box.

REVIEW QUESTIONS

Refer to the *National Electrical Code®* or the working drawings when necessary. Where applicable, responses should be written in complete sentences.

The responses to Questions 1 through 3 are to cite the applicable *NEC®* references.

1. Surface raceways may be extended through dry partitions if _____
_____
_____ .

2. Where power and communications circuits are to be installed in a combination raceway, the different types of circuits must be installed in _____
_____
_____ .

3. If it is necessary to make splices in a multioutlet assembly, the conductors plus the splices shall not fill more than _____ percent of the cross-sectional area at the point where the splices are made.
_____ .
_____ .

Respond to Questions 4 and 5 by showing required calculations and citing *NEC® Sections*.

4. Using a maximum conductor fill of 40 percent for a surface metal surface raceway that measures ½ in. × ¾ in., the maximum allowable cross-sectional area fill is _____ sq. in.

5. Each compartment of the multioutlet assembly installed in the insurance office has a cross-sectional area of 3.7 sq. in. (2390 sq. mm). The maximum number of 10 AWG, Type THWN conductors that may be installed in a compartment of the multioutlet assembly is _____
_____ .

6. Several computer circuits are to be installed in a building where nonmetallic cable and metal boxes are being used. Specify the number of conductors required for a cable serving a computer outlet similar to those in the insurance office. Note any special actions that must be taken to use that cable, and cite the *NEC®* references.
_____
_____
_____

UNIT 10

Working Drawings – Upper Level

OBJECTIVES

After studying this unit, the student should be able to

- tabulate materials required to install an electrical rough-in.
- select the components to install an electric water heater.
- discuss the advantages/disadvantages between single- and three-phase supply systems.

Before examining the three offices on the upper level of the commercial building, the working drawings, the loading schedules, and the panelboard worksheets should be reviewed. The schedules and worksheets are located in the Appendix.

INSURANCE OFFICE

Several special wiring systems are shown in the insurance office, most of which have been discussed in earlier units.

- In-the-floor receptacles are installed in the center of the office area.
- A special raceway is installed in the two exterior walls. This raceway provides for both electrical power and communications.
- Special circuits for personal computers are installed to the computer room and special luminaires (fixtures) are installed in that room.
- A reception area is provided with a combination lighting system to provide a different "look" for the clients.

Computer Room Circuits

A room in the insurance office is especially designed to be a computer room, but it is not a computer room by the definitions set forth in *NEC®* *645.1*. In *NEC®* language, it is a room with some computers in it, and the special requirements for a computer room are not applicable. These actions have been taken to comply with good practice:

- A special surge protection receptacle is specified.
- A special grounding system is provided.
- A separate neutral is provided with each of the receptacle circuits.
- A luminaire (fixture) lens that lessens the glare on the computer screens has been specified.

The special receptacles are discussed in Unit 5, "Switches and Receptacles." The grounding terminal of each of these receptacles is connected to an insulated grounding conductor and not to the box. These grounding conductors are installed in the conduit with the circuit conductors. The grounding conductors are connected to a special isolated grounding terminal in the panelboard; see *NEC®* *250.146(D)*. A separate, insulated-grounding conductor is installed with the feeder circuit and is connected to a grounding electrode at the switchgear. This grounding system reduces, if not eliminates, the electromagnetic interference that often is present in the conventional grounding system. The surge protection feature of the receptacle provides protection from lightning and other severe electrical surges that may occur in the electrical system. These surges are often severe enough to cause a loss of data in the computer memory and, in some cases, damage to the computer. These surge protectors are available in separate units, which can be used to provide protection on any type of sensitive equipment, such as radios and televisions. Three separate neutrals are installed to reduce the possibility of a conductor overheating from nonlinear currents. There are six current-carrying conductors in the raceway, three neutrals and three phase conductors; thus their ampacity must be adjusted.

Material Take-off

A journeyman electrician should be able to look at an electrical drawing and prepare a list of the materials required to install the wiring system. A great deal of time can be lost if the proper materials are not on the job when needed. It is essential that the electrician prepare in advance for the installation so that the correct variety of material is available in sufficient quantities to complete the job.

Experience is by far the best teacher for learning to tabulate materials. Guidance can be given in preparing for the experience:

- Be certain that there is a clear understanding of how the system is to be installed. Decide from where home runs (raceway directly to panelboard) will be made, and establish the sequence of connecting the outlet boxes. The electrician has considerable freedom in making these decisions, even when a scheme is shown on the working drawings.

- Take special note of long runs of raceway and determine whether voltage drop is a problem. This can mean an increase in conductor and/or raceway size.

- Be organized and meticulous. Have several sheets of paper (ruled preferred) and a sharp pencil with a good eraser. Start with the home run, and complete that system before starting another.

BEAUTY SALON

Of special interest in the beauty salon are the connection to the water heater, a small laundry, and the lighting for the stations where customers are given special attention. A reduced image is presented in Figure 10-1. For a larger image refer to working drawing E3.

Water Heater Circuits

A beauty salon (Figure 10-1) uses a large amount of hot water. To accommodate this need, the specifications indicate that a circuit in the beauty salon is to supply an electric water heater. The water heater is not furnished by the electrical contractor, but the circuit is to be installed and connected by the electrical contractor, Figure 10-2. The water heater is rated for 3800 watts at 208 volts single-phase. In most cases the rating of the overcurrent protective device will also be indicated, but if not, the following method can be used to determine the correct circuit rating. The water heater is connected to an individual branch-circuit, see *NEC® Article 100*.

▶*NEC® 422.13* requires that the branch-circuit for a storage-type electric water heater with a capacity of 120 gallons (450 L) or less is considered to be a continuous load. This means that the conductor and overcurrent protection sizing is subject to the 125 percent factor as found in *NEC® 210.19(A)(1)*, *210.20(A)*, *215.2(A)(1)*, *215.3*, and *230.42*.◀

Load current:

$$\frac{3800 \text{ W}}{208 \text{ V}} = 18.3 \text{ A}$$

Minimum BC rating:

$$18.3 \text{ A} \times 1.25 = 22.9 \text{ A}$$

NEC® 422.11(E)(3) establishes the OCPD rating as 150 percent of the current or the next higher standard OCPD rating.

Maximum OCPD rating:

$$18.3 \text{ A} \times 1.50 = 27.45 \text{ A}$$

Next higher standard OCPD rating:

30-ampere

Washer-Dryer Combination

One of the uses of the hot water is to supply a washer-dryer combination used to launder the many towels and other items commonly used in a beauty salon. The unit is rated for 4000 VA at 208Y/120 volts single-phase.

- The branch-circuit supplying this unit must have a rating of at least 125 percent of the appliance load, *NEC® 422.10(A)*.

- The branch-circuit ampacity must be at least 20 amperes (4000 VA/208 V = 19.23 A).

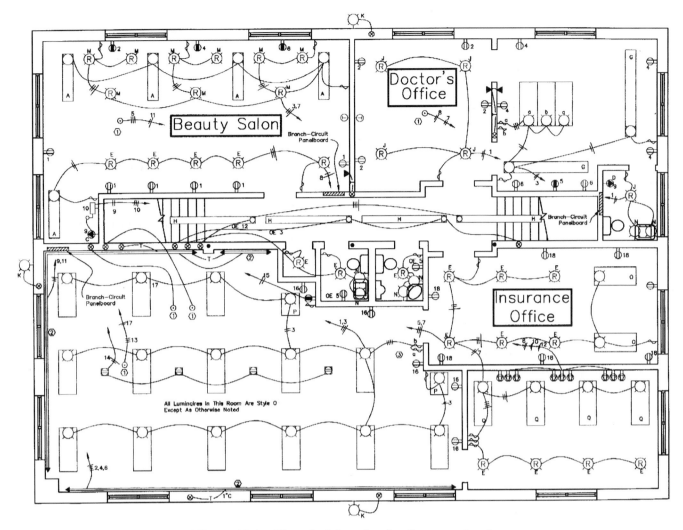

Figure 10-1 Electrical drawing for the upper level.

- The overcurrent protective device must be rated at least 25 amperes (load in amperes multiplied by 1.25 and then raised to the next standard size overcurrent device).

- The conductor ampacity must be at least 21 amperes to allow the use of an overcurrent device with a rating of 25-ampere, *NEC® 240.6*.

- The use of a 25-ampere OCPD requires the installation of a size 10 AWG conductor or larger, see *NEC® 240.4(D)*.

Conductor Selection

The two conductors to the water heater outlet and the three conductors to the washer-dryer combination receptacle are to be installed in a single raceway.

- A reference to *NEC® 240.4* is made at the bottom of *NEC® Table 310.16*. This section limits the OCPD rating for the small conductors 14, 12, and 10 AWG. The limit for copper conductors of 14 AWG is 15-ampere; for 12 AWG it is 20-ampere; and for 10 AWG, it is 30-ampere.

- For four or more current-carrying conductors in a raceway, the conductor ampacity must be adjusted according to *NEC® 310.15(B)(2)*.

- The ampacity of a size 10 AWG Type THHN conductor will be 32 amperes (the allowable ampacity of 40 amperes multiplied by the adjustment factor of 0.8). This will permit the use of a 30-ampere overcurrent protective device.

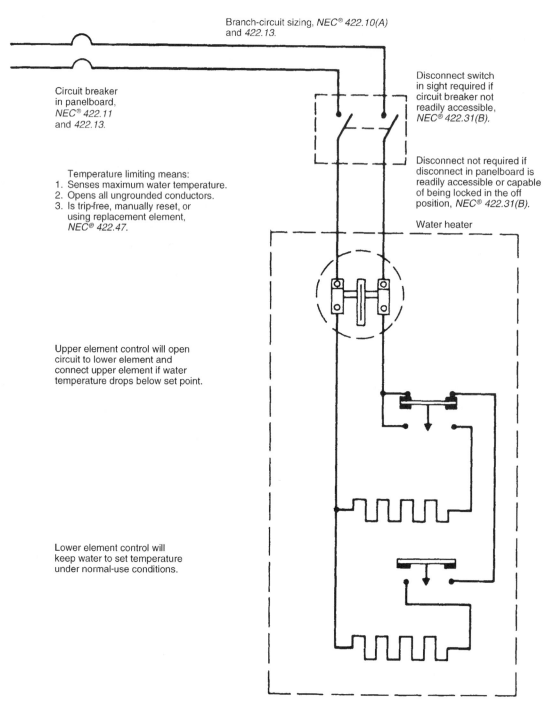

Branch-circuit sizing, *NEC® 422.10(A)*
and *422.13.*

Circuit breaker
in panelboard,
NEC® 422.11
and *422.13.*

Disconnect switch
in sight required if
circuit breaker not
readily accessible,
NEC® 422.31(B).

Disconnect not required if
disconnect in panelboard is
readily accessible or capable
of being locked in the off
position, *NEC® 422.31(B).*

Water heater

Temperature limiting means:
1. Senses maximum water temperature.
2. Opens all ungrounded conductors.
3. Is trip-free, manually reset, or
   using replacement element,
   NEC® 422.47.

Upper element control will open
circuit to lower element and
connect upper element if water
temperature drops below set point.

Lower element control will
keep water to set temperature
under normal-use conditions.

Figure 10-2   Water heater installation.

DOCTOR'S OFFICE

The special feature of the electrical layout for the doctor's office is the single-phase, three-wire feeder to the panelboard. In comparing the loading schedules of the beauty salon and the doctor's office, it would be noted that although the beauty salon has a 50 percent higher load, the feeder conductors are smaller.

When a three-phase, four-wire system is available in a building, all panelboards are usually supplied by a four-wire feeder. In the commercial building this occupancy is supplied by a single-phase system to allow for a comparison.

These are the smallest occupancies in the commercial building, and their areas are approximately the same. The total electrical loading in the doctor's office is 18,618 volt-amperes, and in the beauty salon, it is 27,809. The beauty salon has a 50 percent higher load, but the feeder for the beauty salon (3 AWG) is smaller than the feeder for the doctor's office (2 AWG). This reversal is because the beauty salon has a three-phase feeder, and the doctor's office has a single-phase feeder. There are several differences in these two systems.

Electrical Power Systems

Three-phase systems are preferred where there is a large number of motors. Three-phase motors do not require starting windings and thus are less expensive.

In commercial-type buildings the common three-phase systems are 208Y/120, 480Y/277, and 240Δ/120. The 480Y provides a higher voltage for motor loads, and lighting is available that operates on 277 volts. This system allows the use of smaller (less costly) circuit conductors, raceways, and equipment. It is necessary to install transformers to service 120-volt loads. The 240-volt delta system provides a slightly higher voltage than 208Y/120-volt systems for motor operation but limits the single-phase loading. Only two phases provide 120 volts; the third phase is referred to as the "high-leg." (The conductor color must be, or be tagged with, orange.)

It is important that all motors have a rating compatible with the system voltage.

Single-phase systems may be taken from a three-phase system or from an independent source. If the system has a voltage rating of 208 or 277, it is a part of a three-phase system. If the higher voltage is 240, then it may be from a 240-volt delta system or a 240/120, single-phase system.

Number of Conductors

- A three-phase circuit requires three or four conductors depending on whether a neutral is available. A 208Y/120 and a 480Y/277 require four conductors; 240-volt delta and 480-volt delta circuits require only three.

- Single-phase circuits are either two- or three-wire. A two-wire connection may provide either the higher or lower voltage (120 or 240 from a 120/240-volt source).

Special Problems

- When a single-phase circuit is taken from a three-phase, wye-connected system, the neutral is considered a current-carrying conductor and shall not be reduced in size.

Special Notes

- The beauty salon requires an OCPD rated at 90-ampere, which places the minimum size conductor selection in the 140°F (60°C) column. If it were certain that all the terminations and equipment were rated for 167°F (75°C) then the minimum conductor size would be 3 AWG, and the raceway size would be trade size 1, assuming that the neutral were reduced.

- The neutral of the feeder to the doctor's office [see *NEC® 220.61(C)*] shall not be reduced in size.

- In this comparison, shown in Table 10-1, the best choice is clouded by special rules. Usually, the three-phase system will provide about 50 percent more power at the same installation cost. In this case, the single-phase system was a reasonable choice. The message is clear; each circuit should be evaluated based on its situation.

Load and Feeder Comparison		
Item	Doctor's Office	Beauty Salon
Total Load	18,618 VA	27,809 VA
Feeder OCPD Rating	110-ampere	90-ampere
Minimum Conductor Size	2 AWG	2 AWG
Neutral Size	2 AWG	6 AWG
Number of Conductors	3	4
Raceway Trade Size	1	1¼

Table 10-1

REVIEW QUESTIONS

Refer to the *National Electrical Code®* or the working drawings when necessary. Where applicable, responses should be written in complete sentences.

Questions 1 through 3 refer to the water heater circuit to be installed in the beauty salon.

1. Is it permissible to install a receptable for the water heater in the salon? What section of the *Code* permits or prohibits this installation?

   _____

   _____

2. The water heater has two elements. Under what conditions are both elements heating?

   _____

   _____

3. If the water heater is not marked with the maximum overcurrent protection rating, the maximum overcurrent protection is to be sized not to exceed _____ percent of the water heater's ampere rating; see *NEC® 422.11(E)*.

The following questions refer to the branch-circuit panelboard in the doctor's office. The panelboard schedule is shown on drawing E3 and the panelboard summary on E4.

4. How many single-pole branch-circuits can be added to the panelboard?

   _____

   _____

5. How many general-purpose branch-circuits are to be installed?

   _____

6. If an appliance that will operate for long periods of time is marked with the rating of the overcurrent device, is it permissible to install a device with the next higher rating?

Refer to the working drawings for the insurance office for Questions 7 and 8.

7. Tabulate the following materials:

   a. Tabulate the luminaires (fixtures) by style, and count the number of each.

   _____

   _____

   b. Tabulate the lamps by type/rating, and count the number of each.

   _____

   _____

   c. Tabulate the receptacles by type/rating, and count the number of each.

   _____

   _____

d. Tabulate the switches by type/rating, and count the number of each.

_____

_____

8. Determine the conductor count for each ceiling box and the minimum box size. Then tabulate the number of each size required for the installation. Boxes should not be used to support the luminaires (fixtures) and conductors that do not need to be spliced are to be considered as being pulled through.

Suggestions: Examine each box and mark the plans with the required capacity. Check the luminaire (fixture) styles carefully before counting conductors. The following forms are provided to help and guide the tabulation of information. It is suggested that after the number of conductors in each box has been determined, each unique value be entered in the first column of the first table. Then the required boxes can be identified and a count made.

After all the boxes have been identified and the count established, the second table can be used to summarize by box type and size.

Form for tabulation of conductors and boxes; all conductors are 12 AWG:

Number of Conductors	Box Type and Size	Box Count

Form for summary by box dimension and type:

Box Type and Size	Box Count

UNIT 11

Special Circuits (Owner's Circuits)

OBJECTIVES

After studying this unit, the student should be able to

- describe typical connection schemes for photocells and timers.
- list and describe the main parts of an electric boiler control.
- describe the connections necessary to control a sump pump.

The occupants of the commercial building are each responsible for the electric energy used within their areas. It is the owner's responsibility to provide illumination in the public areas and to provide heat for the entire building. The circuits supplying those areas and the devices for which the building is responsible will be called the owner's circuits.

LOADING SCHEDULE

The loading schedule for the owner's circuits is printed in the Appendix. These circuits serve:

- the boiler that has the greatest electrical demand.
- the circulating pumps that have an emergency power source that will keep them running even if the utility power is interrupted.
- selected lighting that can be supplied by the special power source.

The loading schedule can be divided into three parts:

1. the lighting and receptacles that do not require emergency backup.
2. the lighting, pumps, and receptacles that require emergency backup.
3. the boiler feed, which has a separate 600-ampere main switch in the service equipment.

LIGHTING CIRCUITS

Those lighting circuits that are considered to be the owner's responsibility can be divided into five groups according to the method used to control the circuit:

- continuous operation.
- manual control.
- automatic control.
- timed control.
- photocell control.

Continuous Operation

The luminaires (fixtures) installed at the top of each of the stairways to the second floor of the commercial building are in continuous operation. Because the stairways have no windows, it is necessary to provide lighting even in the daytime. The power to these luminaires (fixtures) is supplied directly from the owner's panelboard.

Manual Control

A conventional manual lighting control system has been selected to operate the remaining second-floor corridor lighting as well as the lighting in the utility and toilet rooms.

Automatic Control

The sump pump is an example of equipment that is operated by an automatic control system. No human intervention is necessary—the control system starts the pump when the liquid level in the sump rises above the maximum allowable level, and then it is pumped down to the minimum level.

Timed Control

The luminaire (fixture) at the entrance to the commercial building is controlled by an astronomical clock located near the owner's panel in the boiler room. The circuit is connected as shown in Figure 11-1. When the clock is first installed, it must be adjusted to the correct time, date, and operational period. Thereafter, it automatically adjusts the period for the changes in nighttime hours. The clock motor is connected to the emergency panel so that the correct time is maintained in the event that the utility power fails.

Photocell Control

The luminaires (fixtures) located on the exterior of the building are controlled by individual photocells, Figure 11-2. The photocell control consists of a light-sensitive cell and an amplifier that increases the signal until it is sufficient to operate a relay that controls the light. A circuit is connected to the photocell to provide power for the amplifier and relay.

SUMP PUMP CONTROL

The sump pump is used to remove water entering the building because of sewage line backups, water main breakage, minor flooding due to natural causes, or plumbing system damage within the building. Because the sump pump is a critical item, it is connected to the emergency panel. The pump motor is protected by a manual motor starter, Figure 11-3, and is controlled by a float switch, Figure 11-4. When the water in the sump rises, a float is lifted, which mechanically completes the circuit to the motor and starts the pump, Figure 11-5. When the water level falls, the pump shuts off.

Another type of sump pump that is available commercially has a start-stop operation due to water pressure against a neoprene gasket. This gasket, in turn, pushes against a small integral switch within the submersible pump. This type of sump pump does not use a float.

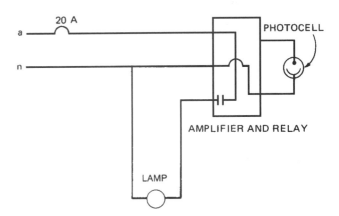

Figure 11-2 Photocell control of lighting.

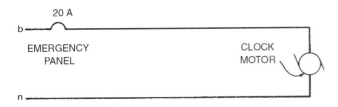

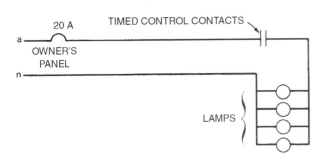

Figure 11-1 Timed control of lighting.

Figure 11-3 Manual motor controller for sump pump.

Figure 11-4   Float switch.

BOILER CONTROL

An electrically heated boiler supplies heat to all areas of the commercial building. The boiler has a full-load rating of 200 kW. As purchased, the boiler is completely wired except for the external control wiring, Figure 11-6. A heat sensor plus a remote bulb, Figure 11-7, is mounted so that the bulb is on the outside of the building. The bulb should be mounted where it will not receive direct sunlight and shall be spaced at least ½ in. from the brick wall. If these mounting instructions are not followed, the bulb will give inaccurate readings. The heat sensor is adjusted so that it closes when the outside temperature falls below a set point, usually 65°F (18°C). The sensor remains closed until the temperature increases to a value above the differential setting, approximately 70°F (21°C). When the outside temperature causes the heat sensor to close, the boiler

automatically maintains a constant water temperature and is always ready to deliver heat to any zone in the building that requires heat.

A typical heating control circuit is shown in Figure 11-8.

Figure 11-9 is a schematic drawing of a typical connection scheme for the boiler heating elements.

The following points summarize the wiring requirements for a boiler:

- A disconnect means must be installed to disconnect the ungrounded supply conductors.

- The disconnect shall be in sight or capable of being locked in the open position.

- The branch-circuit conductor sizes are to be based on 125 percent of the rated load.

- The overcurrent protective devices are to be sized at 125 percent of the rated load.

NEC® Article 424 establishes the requirements for the installation of fixed electric space heating equipment.

It should be noted that the boiler for the commercial building is not sized for a particular heat loss; its selection is based on certain electrical requirements. The building's occupants each have a heating thermostat located within their particular area. The thermostat operates a relay, Figure 11-10, which controls a circulating pump, Figure 11-11, in the hot-water piping system serving the area. The circuit to the circulating pump is supplied from the emergency panel so that the water will continue to circulate if the utility power fails, Figure 11-12. In subfreezing weather, the continuous circulation prevents the freezing of the boiler water; such freezing could damage the boiler and the piping system.

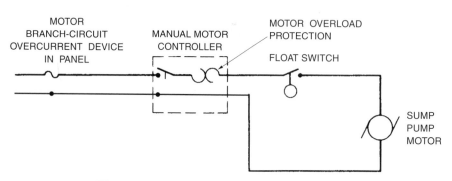

Figure 11-5   Sump pump control diagram.

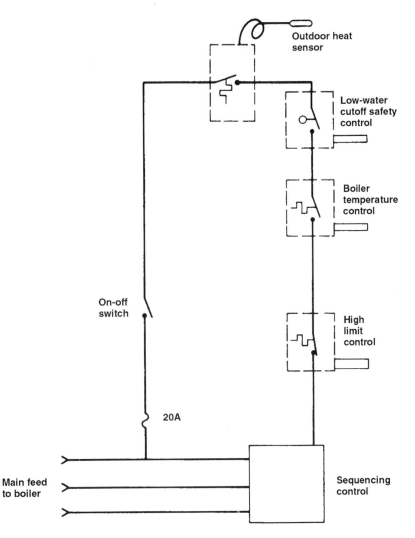

Figure 11-6   Boiler control diagram.

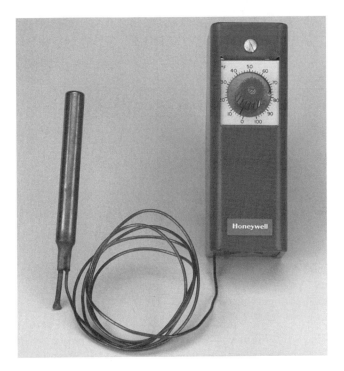

Figure 11-7   Heat sensor with remote mounting bulb.

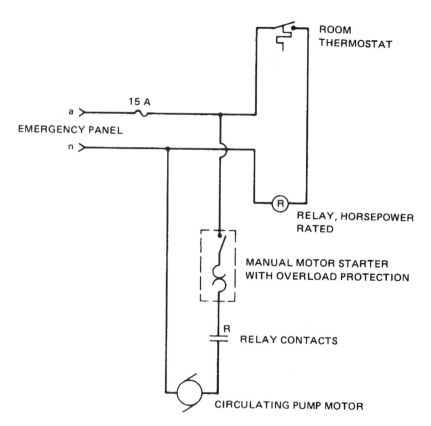

Figure 11-8   Heating control circuit.

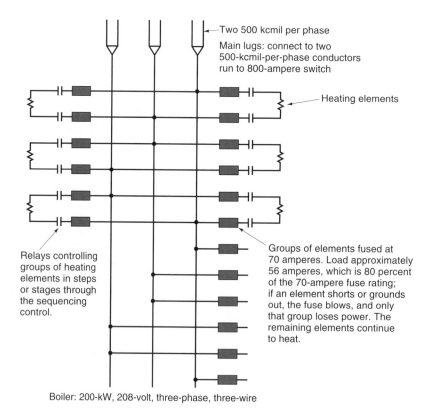

Two 500 kcmil per phase

Main lugs: connect to two 500-kcmil-per-phase conductors run to 800-ampere switch

Heating elements

Relays controlling groups of heating elements in steps or stages through the sequencing control.

Groups of elements fused at 70 amperes. Load approximately 56 amperes, which is 80 percent of the 70-ampere fuse rating; if an element shorts or grounds out, the fuse blows, and only that group loses power. The remaining elements continue to heat.

Boiler: 200-kW, 208-volt, three-phase, three-wire

Figure 11-9   Boiler heating elements.

Figure 11-10   Relay in NEMA 1 enclosure.

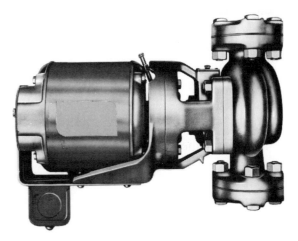

Figure 11-11   Water-circulating pump.

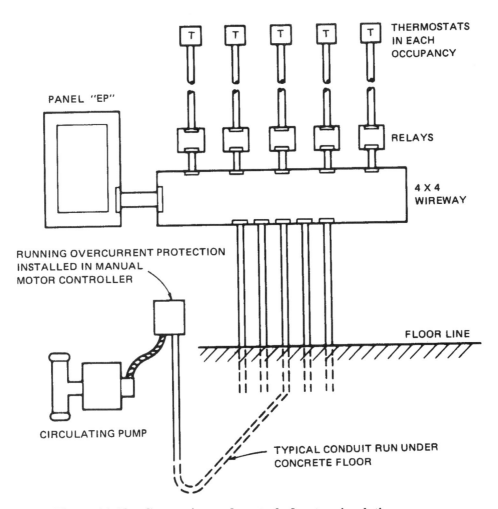

Figure 11-12   Connection and control of water-circulating pumps.

REVIEW QUESTIONS

Refer to the *National Electrical Code*® or the working drawings when necessary. Where applicable, responses should be written in complete sentences.

The first four questions refer to the installation of the electrical supply to the boiler.

1. What is the proper name of the electrical supply to the boiler?

   _____

   _____

2. Describe, in detail, the electrical supply to the boiler, indicating options available and any special precautions.

   _____

   _____

   _____

   _____

   _____

   _____

3. Show the calculations to determine the minimum size Type THHN conductors and the rating of the overcurrent protective device (fuses) for the circuit supplying the boiler. Use two conductors per phase in two raceways. Also determine the size of the equipment grounding conductors to be installed in each raceway. The expected maximum temperature in the room where the boiler is located is 212°F (100°C). The boiler electrical panel and the distribution panel are rated 167°F (75°C). All conductors are copper.

4. Verify, or disprove, the correctness of the sizing of the boiler branch-circuit.

The following three questions refer to the installation of the various items of equipment in the equipment room.

5. What is the proper rating of the overload protection to be installed for the sump pump?

_____

_____

_____

_____

_____

_____

6. Detail the installation of the outside heat sensor.

_____

_____

_____

_____

_____

_____

7. What is the difference between a timer and a clock?

_____

_____

_____

_____

_____

_____

_____

UNIT 12

Panelboard Selection and Installation

OBJECTIVES

After studying this unit, the student should be able to

- identify the criteria for selecting a panelboard.
- correctly place and number circuits in a panelboard.
- compute the correct feeder size for a panelboard.
- determine the correct overcurrent protection for a panelboard.
- prepare a panelboard directory.

The installation of panelboards is always a part of new building construction, such as the commercial building. After a building is occupied, it is common for the electrical needs to exceed the capacity of the installed system. In this case the electrician is often expected not only to install a new panelboard, but to select a panelboard that will satisfy the needs of the occupant.

PANELBOARDS

Separate feeders are to be run from the main service equipment to each of the areas of the commercial building. Each feeder will terminate in a panelboard, which is to be installed in the area to be served. Figure 12-1 shows two types of typical panelboards. The first panel has a main breaker plus many branch-circuit breakers. The second panel has branch-circuit breakers only.

A panelboard is defined by the *NEC®* as *a single panel or group of panel units designed for assembly in the form of a single panel, including buses and automatic overcurrent devices, and equipped with or without switches for the control of light, heat, or power circuits; designed to be placed in a cabinet or cutout box placed in or against a wall, partition,* or other support; and accessible only from the front.*

A panelboard is designed to be installed in a cabinet or cutout box placed in or against a wall or partition. This panelboard is to be accessible only from the front. Panelboards shall be dead front (see *NEC® 408.18*); thus, no current-carrying parts are exposed to a person operating the devices.

Although a panelboard is accessible only from the front, a switchboard may be accessible from the rear as well. *NEC® 408.34* defines a lighting and appliance branch-circuit panelboard *as a panelboard having more than 10 percent of its overcurrent devices rated at 30 amperes or less with neutral connections provided for these devices. NEC® 408.35* sets the maximum number of overcurrent devices in a lighting and appliance panelboard at 42. *NEC® 408.36(C)* stipulates that if the panelboard contains snap switches with a rating of 30 amperes or less, the panelboard overcurrent protection shall not exceed 200 amperes. Snap switches are switches that are snapped in place in a panelboard or other similar enclosure. They are additional to the overcurrent devices and shall be rated for the load they are to control.

---

*Reprinted with permission from NFPA 70-2005.

Lighting and appliance branch-circuit panelboards generally are used on single-phase, three-wire, and three-phase, four-wire electrical systems. This type of installation provides individual branch-circuits for lighting and receptacle outlets and permits the connection of appliances such as the equipment in the bakery.

Panelboard Construction

In general, panelboards are constructed so that the main feed busbars run the height of the panelboard. The buses to the branch-circuit protective devices are connected to the alternate main buses as shown in Figure 12-2A and Figure 12-2B. The busing arrangement shown in these figures is typical.

Figure 12-1   Panelboards.

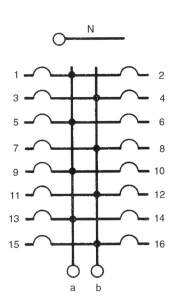

Figure 12-2A   Lighting and appliance branch-circuit panelboard—single-phase, three-wire.

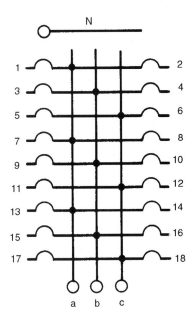

Figure 12-2B   Lighting and appliance branch-circuit panelboard—three-phase, four-wire.

Always check the wiring diagram furnished with each panelboard to verify the busing arrangement. In an arrangement of this type, the connections directly across from each other are on the same phase, and the adjacent connections on each side are on different phases. As a result, multiple protective devices can be installed to serve the 208-volt equipment. The numbering sequence shown in Figure 12-2 and Figure 12-3 is the system used for most panelboards. Figure 12-4 and Figure 12-5 show the phase arrangement requirements of *NEC® 408.3*.

Numbering of Circuits

The number of overcurrent devices in a panelboard is determined by the needs of the area being served. Using the bakery as an example, there are thirteen single-pole circuits and five three-pole circuits. See Figure 12-3 for a panelboard circuit

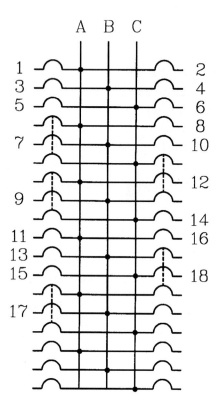

Figure 12-3   Panelboard circuit numbering scheme.

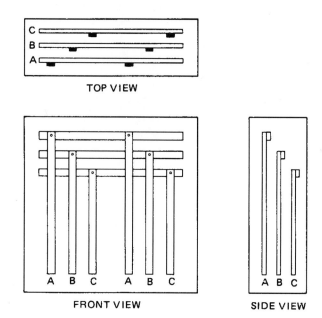

Figure 12-4   Phase arrangement requirements for switchboards and panelboards.

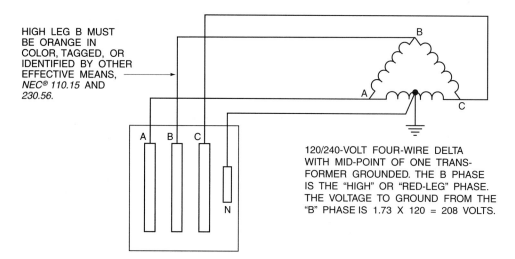

HIGH LEG B MUST BE ORANGE IN COLOR, TAGGED, OR IDENTIFIED BY OTHER EFFECTIVE MEANS, *NEC® 110.15* AND *230.56*.

120/240-VOLT FOUR-WIRE DELTA WITH MID-POINT OF ONE TRANS-FORMER GROUNDED. THE B PHASE IS THE "HIGH" OR "RED-LEG" PHASE. THE VOLTAGE TO GROUND FROM THE "B" PHASE IS 1.73 X 120 = 208 VOLTS.

Figure 12-5   Panelboard or switchboard supplied by four-wire, delta-connected system.

numbering scheme example. This is a total of 28 poles. When using a three-phase supply, the incremental number is six (a pole for each of the three phases on both sides of the panelboard). The minimum number of poles that could be specified for the bakery is 30. This would limit the power available for growth, and it would not permit the addition of a three-pole load. The reasonable choice is to go to 36 poles, which provides considerable flexibility for growth, Figure 12-3.

Panelboard Ratings

Panelboards are available in various ratings including 100-, 225-, 400-, and 600-ampere. These ratings are determined by the current-carrying capacity of the main bus. After the feeder capacity is determined, a panelboard rating is selected in compliance with *NEC® 408.30*, which has a rating not less than the load calculated in accordance with *NEC® Article 220*.

Panelboard Overcurrent Protection

In many installations, a single feeder may be sized to serve more than one panelboard, see Figure 12-6. For this situation, it is necessary to install an overcurrent device with a trip rating not greater than the lowest rating of the panelboards. In the case shown in Figure 12-7, a main device is not required because the panelboard is protected by the feeder protective device. Other examples of methods of providing panelboard protection are shown in Figure 12-7, Figure 12-8, and Figure 12-9.

Panelboard Cabinet Sizing

If there are special requirements for using the wiring gutters, the physical size of the cabinet shall be sufficient to accommodate those requirements. Often a feeder serves two or more panelboards, in which case either a junction box is installed or the feeder connects to the first panelboard and then is extended to the others. This requires the installation

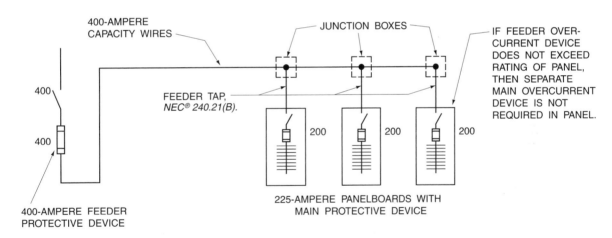

Figure 12-6   Panelboards with main overcurrent protection in each panel.

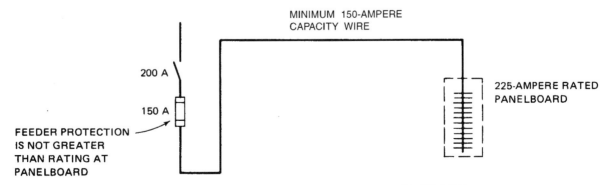

Figure 12-7   Panelboard without main, *NEC® 408.36(A)*.

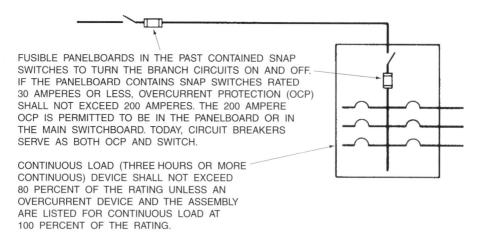

FUSIBLE PANELBOARDS IN THE PAST CONTAINED SNAP
SWITCHES TO TURN THE BRANCH CIRCUITS ON AND OFF.
IF THE PANELBOARD CONTAINS SNAP SWITCHES RATED
30 AMPERES OR LESS, OVERCURRENT PROTECTION (OCP)
SHALL NOT EXCEED 200 AMPERES. THE 200 AMPERE
OCP IS PERMITTED TO BE IN THE PANELBOARD OR IN
THE MAIN SWITCHBOARD. TODAY, CIRCUIT BREAKERS
SERVE AS BOTH OCP AND SWITCH.

CONTINUOUS LOAD (THREE HOURS OR MORE
CONTINUOUS) DEVICE SHALL NOT EXCEED
80 PERCENT OF THE RATING UNLESS AN
OVERCURRENT DEVICE AND THE ASSEMBLY
ARE LISTED FOR CONTINUOUS LOAD AT
100 PERCENT OF THE RATING.

Figure 12-8   Panelboard with snap switches, *NEC® 408.36(C)*.

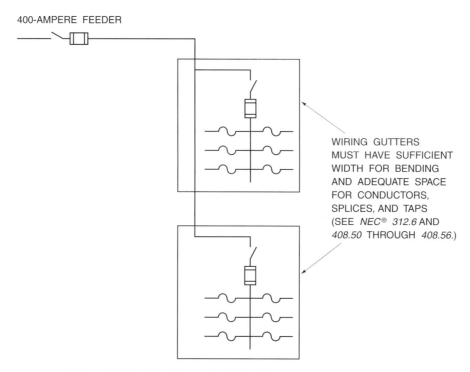

400-AMPERE FEEDER

WIRING GUTTERS
MUST HAVE SUFFICIENT
WIDTH FOR BENDING
AND ADEQUATE SPACE
FOR CONDUCTORS,
SPLICES, AND TAPS
(SEE *NEC® 312.6* AND
408.50 THROUGH *408.56*.)

Figure 12-9   Panelboards with adequate space for wiring.

of additional feeder conductors. The cabinet must have sufficient gutter width for bending these conductors and for the required splices. These requirements are illustrated in Figure 12-9 and are given in *NEC® 312.6, 312.7, 312.8* and *312.9*.

Panelboard Directory

One of the final actions in installing an electrical circuit or system is to update or create the panelboard directory. A holder will be located on the inside of the panelboard door. An examination of Figure 12-1 reveals that the size and shape of these holders will vary. The single-column schedule shown on the left has the advantage that the numbers appear in numerical order (1, 2, . . . *x*). The advantage of the holder on the right is that the schedule mimics the layout of the overcurrent devices and their numbering (odd numbers on the left, even on the right).

A circuit directory is required by *NEC® 408.4*. There is the option of creating a directory in your choice of size and shape and securing it to the inside of the panelboard door. If this is done, a sheet of clear plastic covering the directory will lengthen its life. Securing the plastic sheet only at the top will permit removal to make modifications.

Circuit Directory or Circuit Identification

▶*NEC® 408.4* is pretty clear as to its requirements on this subject. It requires that *Every circuit and circuit modification shall be legibly identified as to its clear, evident, and specific purpose or use. The identification shall include sufficient detail to allow each circuit to be distinguished from all others. The identification shall be included in a circuit directory that is located on the face of inside of the panel door in the case of a panelboard, and located at each switch on a switchboard.*∗◀

Table 12-1 exhibits a directory prepared for the drugstore; all directories for the commercial building are recorded on the working drawings. It is common for them to appear in the specification or on the working drawings. The directory gives the "load/area" in a wording that would be clear for "non-electrical" people to decipher. In the next column is the size of the overcurrent protective device, then the circuit number, and in the center column the phase connection. The center column would not be considered as required, but it helps the schedule in mimicking the panelboard. The remaining columns follow the same sequence in reverse.

Close Unused Openings!

It seems obvious that there should be no unused openings in panelboards and switches. Sometimes a

---

∗Reprinted with permission from NFPA 70-2005.

circuit breaker "knockout" in the cover of a panelboard, or a conduit knockout is inadvertently removed, leaving an opening for someone or something to enter the panelboard and come in contact with live parts. Very hazardous! It is now a requirement that all unused openings be closed with an "identified" or approved means. See *NEC® 408.7*.

WORKING SPACE AROUND ELECTRICAL EQUIPMENT

Personnel working on or needing access to electrical equipment shall have enough room and adequate lighting to safely move around and work on the equipment. This will enable them to perform the required function such as examining, adjusting, servicing, or maintaining the equipment. Many serious injuries and deaths have occurred as a result of arc blasts and electrical shock. Should this occur, the personnel must have enough space to get away safely and quickly from the faulted electrical equipment.

NEC® 110.26 covers the minimum working space, access, headroom, and lighting requirements for electrical equipment such as switchboards, panelboards, and motor control centers operating at 600 volts or less.

Working Space Considerations

When discussing working space, the major concern is ACCESS TO and ESCAPE FROM the working space.

Drugstore – Panelboard Directory						
Load/Area Served	OCPD	BC #	Phase	BC #	OCPD	Load/Area Served
Sales Area Lighting	20	1	A	2	20	Sales Receptacles, North
Sales Area Lighting	20	3	B	4	20	Toilet Receptacles
Toilet Lighting	20	5	C	6	20	Sales Receptacles, South
Pharmacy Receptacles	20	7	A	8	20	Show Window Receptacles
Basement Lighting	20	9	B	10	20	Pharmacy Lighting
Show Window Track	20	11	C	12	20	Basement Receptacles South
Show Window Track	20	13	A	14	20	Basement Receptacles North
Exterior Sign	20	15	B			
Roof Receptacle	20	17	C			
			A	16	50	Compressor
			B			
			C			

Table 12-1

NEC® 110.26 addresses "working space," not the room itself. Do not confuse the term *working space* with *room*. The concern for safety requires adequate working space around the equipment, which may or may not be the entire room.

The diagrams on the following pages show the required working space in front of electrical equipment. If access is needed on the sides and/or rear of the electrical equipment, the required space shall be provided on the sides and/or rear of the equipment.

Working space requirements are classified by Conditions. In these Conditions, metal water or steam pipes; electrical conduits; electrical equipment; metal ducts; all metal grounded surfaces or objects; and all concrete, brick, tile, or similar conductive building materials are considered as "grounded." Refer to *NEC® Table 110.26(A)(1)*.

Condition 1: Exposed live parts on one side of the working space and no live or grounded parts on the other side, or exposed live parts on both sides of the working space if effectively guarded by suitable insulating material as illustrated in Figure 12-10.

Condition 2: Exposed live parts on one side of the working space and grounded parts on the other side of the working space as illustrated in Figure 12-11.

Condition 3: Exposed live parts on both sides of the working space as illustrated in Figure 12-12.

Working Space "Depth"

Working space is that space required to have safe access to the live parts of electrical equipment. Generally, this is in front of the equipment, but it could be on the sides or back, depending upon the design of the equipment. If there is a need to have to access live parts from the sides and/or rear of the equipment, minimum working space shall be provided for those needs as illustrated in Figure 12-13.

All working spaces shall allow for the opening of equipment doors and hinged panels to at least a 90° position as is shown in Figure 12-14.

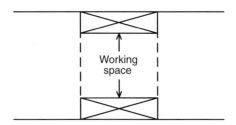

Figure 12-12   Condition 3.

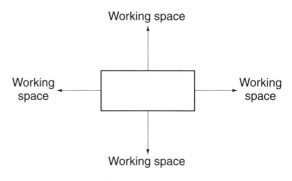

Figure 12-13   Working space depth.

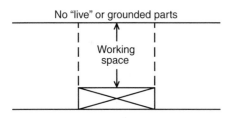

Figure 12-10   Condition 1.

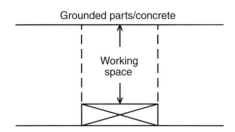

Figure 12-11   Condition 2.

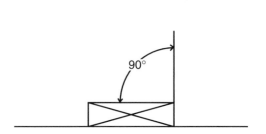

Figure 12-14   Hinged door and panels.

Working spaces shall be kept clear and not be used for storage as is illustrated in Figure 12-15.

Working space distances are measured from exposed live parts, or from the enclosed "dead front" of switchboards. When this enclosed equipment is "opened," live parts will be exposed, see Figure 12-16.

Different Voltages to Ground . . . Different Working Space Requirements

For system voltages of 0 to 150 volts to ground, the minimum working space requirement is 3 ft (900 mm). An example of this is the 208Y/120-volt, wye-connected system in the commercial building discussed in this text, where the voltage to ground is 120 volts. See Figure 12-17.

For system voltages of 151 to 600 volts to ground, the minimum working space requirement varies from 3 ft to 4 ft (900 mm to 1.2 m), depending upon the Condition. An example of this would be a

480/277-volt, wye-connected system in which the voltage to ground is 277 volts. See Figure 12-18, Figure 12-19, and Figure 12-20.

Working Space "Width"

For equipment with a width equal to or less than 30 in. (750 mm), as shown in Figure 12-21, provide a working space with a width of not less than 30 in.

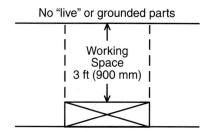

Figure 12-18   151–600 volts to ground—Condition 1.

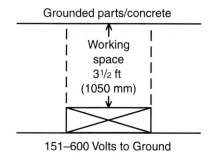

Figure 12-19   151–600 volts to ground—Condition 2.

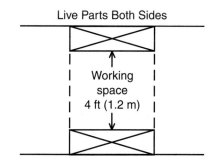

Figure 12-20   151–600 volts to ground—Condition 3.

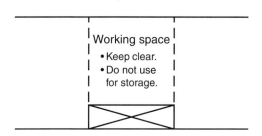

Figure 12-15   Clear working spaces.

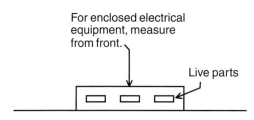

Figure 12-16   Measured from live parts.

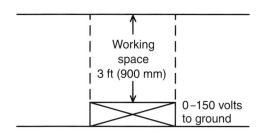

Figure 12-17   0–150 volts to ground.

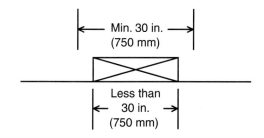

Figure 12-21   Equipment 30 in. (750 mm) or less wide.

(750 mm). For equipment with a width greater than 30 in. (750 mm), as shown in Figure 12-22, provide a working space width of not less than the width of the equipment. This working space width will enable a person to stand in front of the equipment and not be in contact with a grounded object.

Working Space "Height"

Working space height shall not be less than 6.5 ft (2.0 m). For equipment more than 6.5 ft (2.0 m) high, working space height shall be at least as high as the equipment. This working space height will enable most people to stand in front of the equipment. See Figure 12-23 and *NEC® 110.26(E)*.

Access and Entrance to Working Space

Access and entrance refer to the actual working space requirements, not to the room itself. However, in instances where the electrical equipment is located in a small room, the working space could, in fact, be the entire room.

For electrical equipment, at least one entrance, of stipulated size (as illustrated in Figure 12-24), is required.

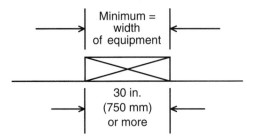

Figure 12-22   Equipment more than 30 in. (750 mm) wide.

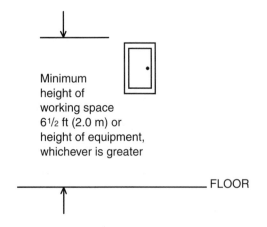

Figure 12-23   Working space height.

For electrical equipment rated 1200 volts (see Figure 12-25) or higher and greater than 6 ft (1.8 m) wide, two entrances to the working space are required. The entrances to this space shall have a minimum width of 2 ft (600 mm) and a minimum height of 6.5 ft (2 m). An entrance shall be located at each end of the equipment.

Exception: When there is an unobstructed, continuous way of exit travel from the working space, one means of egress is permitted, as illustrated in Figure 12-26.

Exception: If the minimum working space requirements listed in *NEC® Table 110.26(A)(1)* are

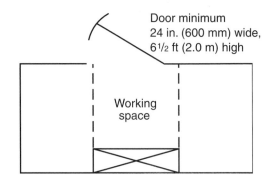

Figure 12-24   Access to working space.

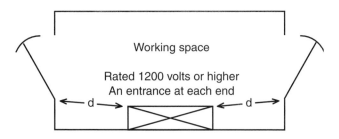

Figure 12-25   Space for electrical equipment.

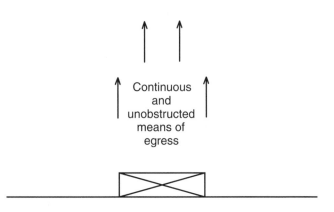

Figure 12-26   Unobstructed exit travel.

doubled, only one entrance to the working space is required, as shown in Figure 12-27.

Dedicated Electrical Space

A "zone" equal to the width and depth of the electrical equipment and extending from the floor to a height of 6 ft (1.8 m) above the equipment or to the structural ceiling, whichever is lower, shall be dedicated to the electrical installation (see Figure 12-28). This zone shall be kept clear of all piping, ducts, or other equipment that is not related to the electrical installation. See *NEC® 110.26(F)(1)(a)*.

Above the dedicated space, foreign systems are permitted, but only if adequate protection is provided so water leaks, condensation, breaks, ruptures, etc. will not damage the electrical equipment. See *NEC® 110.26(F)(1)(b)*.

The minimum headroom (shown in Figure 12-28) for the working space around electrical equipment is 6.5 ft (2.0 m), or the height of the electrical equipment, whichever height is greater.

Illumination

For safety reasons, *NEC® 110.26(D)* requires that illumination be provided for all working spaces around electrical equipment. This could be the general illumination in the room where the equipment is located, or it might be dedicated illumination (see Figure 12-29). Control for this illumination shall not be by automatic systems only; there shall be a manual means to turn on the illumination. In the commercial building the equipment room illumination is controlled by a single-pole switch located inside the equipment room entry door.

SUMMARY

- The working space width shall be not less than 30 in. (750 mm) for each panelboard.

- The working space depth shall be not less than 3 ft (900 mm).

- The working space height shall be not less than 6.5 ft (2.0 m).

- A dedicated space, not less than the depth and width of the panelboard, shall extend from the panelboard to the structural ceiling.

- The illumination in the space shall have a manual control.

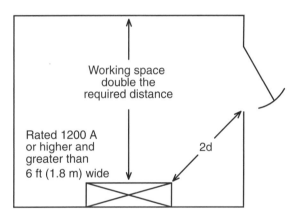

Figure 12-27   Single exit.

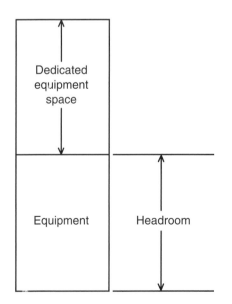

Figure 12-28   Dedicated space and headroom.

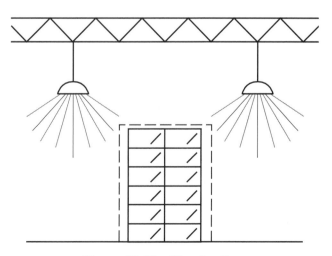

Figure 12-29   Illumination.

REVIEW QUESTIONS

Refer to the *National Electrical Code®* or the working drawings when necessary. Where applicable, responses should be written in complete sentences.

The *NEC®* refers to switchboards, distribution boards, and lighting and appliance panelboards. For Questions 1, 2, and 3, write the distinguishing characteristics of each. By observing these characteristics, a person would know which type of board was being viewed.

1. Switchboard:

    _____

    _____

    _____

2. Panelboard:

    _____

    _____

    _____

3. Lighting and appliance panelboard:

    _____

    _____

    _____

4. There are several requirements that are the same for distribution boards and panelboards. List at least two that an electrician should be aware of.

    _____

    _____

    _____

    _____

Refer to the drugstore panelboard directory (Table 12-1) and the drugstore feeder selection summary in the Appendix for the information necessary to answer Questions 5 through 9.

5. The panelboard must be rated for at least _____ ampere.

6. The panelboard has a total of _____ poles.

7. There is space for the addition of _____ three-pole breaker(s).

8. There is space for the addition of _____ two-pole breaker(s).

9. There is space for the addition of _____ single-pole breaker(s).

10. How much working space must be provided in front of a 480/277-volt dead front switchboard according to *Table 110.26(A)(1)*? The switchboard faces a concrete block wall.   3 ft (900 mm)   3½ ft (1.0 m)   4 ft (1.2 m).   Circle the correct answer.

11. A wall-mounted panel has a number of electrical conduits running out of the top and bottom of the panel. A sheet metal worker started to run a cold-air return duct through the wall approximately 8 ft (2.5 m) high directly above the panel. The electrician stated that this is not permitted, and that this space is only to be used for electrical equipment. He cited *NEC*® _____ of the *National Electrical Code*® to support his argument.

12. For the main switchboard in the commercial building, what "condition" describes the installation relative to working space? Refer to *NEC*® *110.26.*

UNIT 13

The Electric Service

OBJECTIVES

After studying this unit, the student should be able to

- install power transformers to meet the *NEC®* requirements.
- draw the basic transformer connection diagrams.
- recognize different service types.
- connect metering equipment.
- apply ground-fault requirements to an installation.
- install a grounding system.

The installation of the electric service to a building requires the cooperation of the electrician and the local power company and, in some cases, the electrical inspector. The availability of high voltage and the power company requirements determine the type of service to be installed. This unit will investigate several common variations in electrical service installations together with the applicable *NEC®* rules.

TRANSFORMERS

The principal reason for installing a transformer is to either increase or decrease the voltage. Because it is more economical to transport electricity at a high voltage, transformers are installed, by either the utility company or the owner to step the voltage down to a level that can be used by the equipment in the building. Transformers are available in two basic types, liquid-filled or dry.

Liquid-Filled Transformers

Many transformers are immersed in a liquid that may be askarel or oil. This liquid performs important functions:

- It is part of the required insulation dielectric.
- It acts as a coolant by conducting heat away from the core and the winding of the transformer to the surface of the enclosing tank, which is then cooled by radiation or fan cooled.

Dry-Type Transformers

Dry-type transformers are widely used because they are lighter in weight than comparably rated liquid-filled transformers. Installation is simpler because there is no need to take precautions against liquid leaks.

Dry-type transformers are constructed so that the core and coil are open to allow for cooling by the free movement of air. Fans may be installed to increase the cooling effect. In this case, the transformer can be used at a greater load level. A typical dry-type transformer installation is shown in Figure 13-1. An installation of this type is known as a unit substation and consists of three main components:

1. the high-voltage switch.
2. the dry-type transformer.
3. the secondary distribution section.

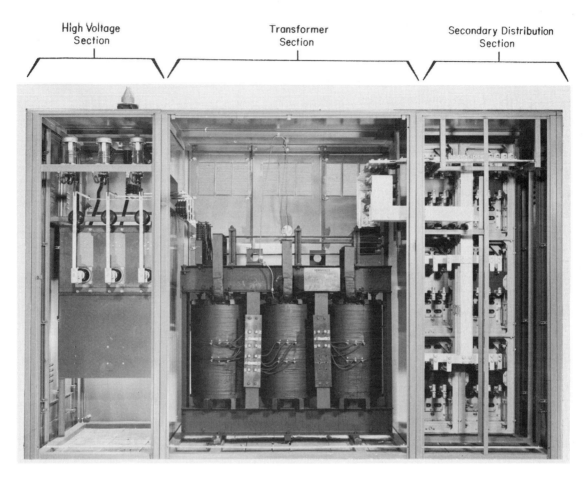

Figure 13-1   A unit substation.

TRANSFORMER OVERCURRENT PROTECTION

NEC® Article 450 addresses transformer installations and groups transformers into two voltage levels:

1. Over 600 volts, see *NEC® Table 450.3(A)*.
2. 600 volts or less, see *NEC® Table 450.3(B)*.

NEC® Table 450.3(A) is divided horizontally into two sections. The upper section applies to transformers in any location. The lower section allows some reduction in the device sizes if the location is supervised as defined in *NEC® Table 450.3(A), Note 3*. The table is also divided vertically into two sections; the left section addresses primary protection, the right, secondary protection.

Figure 13-2 illustrates five of the common situations found in commercial building transformer installations. As specified in *NEC® Article 450*, the overcurrent devices protect the transformer only.

The conductors supplying or leaving the transformer may require additional overcurrent protection according to *NEC® Articles 240* and *310*. (See Units 17, 18, and 19 for information concerning fuses and circuit breakers.)

Figure 13-3 shows one method of installing a dry-type transformer in a commercial or industrial building.

TRANSFORMER CONNECTIONS

A transformer is used in a commercial building primarily to change the transmission line high voltage to the value specified for the building, such as 480Y/277 or 208Y/120 volts. A number of connection methods can be used to accomplish the changing of the voltage. The connection used depends upon the requirements of the building. The following paragraphs describe several of the more commonly used secondary connection methods.

Transformers over 600 volts with primary and secondary protection.		I. Primary over 600 volts	II. Secondary over 600 volts	III. Secondary 600 volts or less
	For transformers with impedance ▶ not over 6%:	• Maximum fuse–300%* • Maximum breaker–600%*	• Maximum fuse–250% • Maximum breaker–300%*	• Maximum fuse–125%* • Maximum breaker–125%*
	For transformers with impedance ▶ over 6%, but not over 10%:	• Maximum fuse–300%* • Maximum breaker–400%*	• Maximum fuse–225%* • Maximum breaker–250%*	• Maximum fuse–125%* • Maximum breaker–125%*

A

* Where percentages marked with asterisks do not correspond to a standard rating or setting, the next higher standard rating or setting may be used.

Transformers over 600 volts with primary protection only, Supervised installation only.	Primary over 600 volts	Secondary over 600 volts or under 600 volts
	• Maximum fuse–250%* • Maximum breaker–300%*	• No overcurrent protection required

B

* Where percentages marked with asterisks do not correspond to a standard rating or setting, the next higher standard rating or setting may be used.

** Individual primary overcurrent protection is not required if the primary feeder overcurrent device is sized as shown to the left. This may allow more than one transformer to be connected to one feeder.

Transformers over 600 volts with primary and secondary protection. Supervised installations only.		I. Primary over 600 volts	II. Secondary over 600 volts	III. Secondary 600 volts or less
	For transformers with impedance ▶ not over 6%:	• Maximum fuse–300%* • Maximum breaker–600%*	• Maximum fuse–250%* • Maximum breaker–300%*	• Maximum fuse–250%* • Maximum breaker–250%*
	For transformers with impedance ▶ over 6%, but not over 10%:	• Maximum fuse–300%* • Maximum breaker–400%*	• Maximum fuse–225%* • Maximum breaker–250%*	• Maximum fuse–250%* • Maximum breaker–250%*

C

* Where percentages marked with asterisks do not correspond to a standard rating or setting, the next higher standard rating or setting may be used.

** Individual primary overcurrent protection is not required if protection on secondary conforms to the values in columns II and III, or if the transformer is equipped by the manufacturer with coordinated thermal overload protection, AND if primary feeder overcurrent protection does not exceed the values in Column I. This may permit more than one transformer to be connected to one feeder.

Transformers 600 volts and less with primary protection only.	Primary 600 volts or less	Secondary 600 volts or less
	Maximum fuse size is not to exceed 125% of the transformer's rated primary current if secondary protection is not provided. For transformers having rated primary current of more than 9 amperes and if the 125% sizing does not correspond to a standard size, then the next higher standard size of fuse or nonadjustable breaker may be used. For transformers having rated primary current of less than 9 amperes, if the 125% sizing does not correspond to a standard size, then a fuse or nonadjustable trip circuit breaker, not to exceed 167% may be used.	Transformer secondary protection is not required when the primary fuse does not exceed 125% of the rated primary current.

D

Transformers 600 volts and less with primary and secondary protection.	Primary 600 volts or less	Secondary 600 volts or less
	Maximum fuse or breaker is not to exceed 250% of the transformer's rated primary current when transformer secondary protection does not exceed 125%. If transformer is equipped with coordinated overload protection furnished by the manufacturer, individual primary overcurrent protection is not required if the primary feeder has overcurrent protection: • Set not over 6 times primary current for transformers with not more than 6% impedance. • Set not over 4 times primary current for transformers with more than 6% but not over 10% impedance. This may allow more than one transformer to be connected to one feeder.	Maximum fuse or breaker is not to exceed 125% of the transformer's rated secondary current when the transformer primary protection is not over 250% of the rated primary full-load current. For transformers having rated secondary current of 9 amperes or more, if the 125% sizing does not correspond to a standard size, then the next higher standard size fuse or nonadjustable breaker may be used. For transformers having rated secondary current of less than 9 amperes, if the 125% sizing does not correspond to a standard size, then a fuse or breaker not to exceed 167% may be used.

E

Figure 13-2   Typical transformer overcurrent protection requirements. See *Table 450.3(A)* and *Table 450.3(B)*.

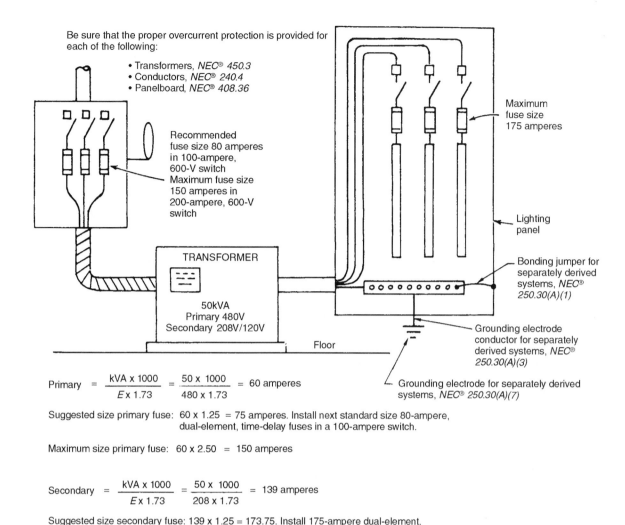

Be sure that the proper overcurrent protection is provided for each of the following:

- Transformers, *NEC® 450.3*
- Conductors, *NEC® 240.4*
- Panelboard, *NEC® 408.36*

Recommended fuse size 80 amperes in 100-ampere, 600-V switch
Maximum fuse size 150 amperes in 200-ampere, 600-V switch

Maximum fuse size 175 amperes

Lighting panel

Bonding jumper for separately derived systems, *NEC® 250.30(A)(1)*

Grounding electrode conductor for separately derived systems, *NEC® 250.30(A)(3)*

Grounding electrode for separately derived systems, *NEC® 250.30(A)(7)*

TRANSFORMER

50kVA
Primary 480V
Secondary 208V/120V

Floor

$$\text{Primary} = \frac{kVA \times 1000}{E \times 1.73} = \frac{50 \times 1000}{480 \times 1.73} = 60 \text{ amperes}$$

Suggested size primary fuse: 60 x 1.25 = 75 amperes. Install next standard size 80-ampere, dual-element, time-delay fuses in a 100-ampere switch.

Maximum size primary fuse: 60 x 2.50 = 150 amperes

$$\text{Secondary} = \frac{kVA \times 1000}{E \times 1.73} = \frac{50 \times 1000}{208 \times 1.73} = 139 \text{ amperes}$$

Suggested size secondary fuse: 139 x 1.25 = 173.75. Install 175-ampere dual-element, time-delay fuses, *NEC® 450.3(A)*

Figure 13-3   Diagram of dry-type transformer installation.

Single-Phase System

Single-phase systems usually provide 120 and/or 240 volts with a two- or three-wire connection, Figure 13-4. The center tap of the transformer secondary shall be grounded in accordance with *NEC® Article 250, Part II*, as will be discussed later. Grounding is a safety measure and should be installed with great care.

Open Delta System

This connection scheme has the advantage of being able to provide either three-phase or three-phase and single-phase power using only two transformers. It is usually installed where there is a strong probability that the power requirement will increase, at which time a third transformer can be added. The open delta connection is illustrated in Figure 13-5.

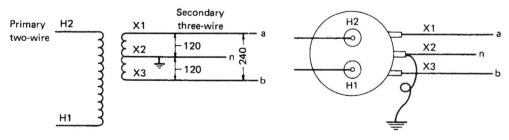

Figure 13-4   Single-phase transformer connection.

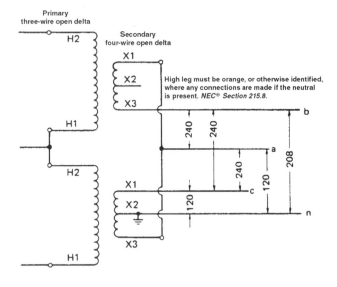

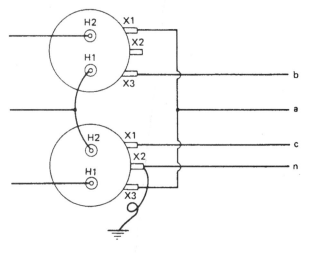

Figure 13-5   Three-phase open delta connection.

In an open delta transformer bank, 86.6 percent of the capacity of the transformers is available. For example, if each transformer in Figure 13-5 has a 100-kilovolt-ampere rating, then the capacity of the bank is

$$100 + 100 = 200 \text{ kVA}$$

$$200 \text{ kVA} \times 0.866 = 173 \text{ kVA}$$

Another way of determining the capacity of an open delta bank is to use 57.7 percent of the capacity of a full delta bank. Thus, the capacity of three 100-kilovolt-ampere transformers connected in full delta is 300 kilovolt-amperes. Two 100-kilovolt-ampere transformers connected in open delta have a capacity of

$$300 \text{ kVA} \times 0.577 = 173 \text{ kVA}$$

When an open delta transformer bank is to serve three-phase power loads only, the center tap is not connected.

Four-Wire Delta System

This connection, illustrated in Figure 13-5, has the advantage of providing both three-phase and single-phase power from either an open (two transformers) or a closed (three transformers) delta system as shown in Figure 13-6. A center tap is brought out of one of the transformers, which is grounded and becomes the neutral conductor to a single-phase, three-wire power system. The voltage to ground between phases "A" and "C," which are connected to the transformer that has been tapped, will be equal to ground and additive when measured phase to phase. For example, 120 volts to ground and 240 volts between phases. The voltage measured between the grounded tap and the "B" phase will be higher than 120 and lower than 240. This phase is called the high leg and cannot be used for lighting purposes. This high leg shall be identified by using orange-colored conductors or by effectively identifying it as the "B" phase conductor whenever it is present in a box or cabinet with the neutral of the system. See *NEC® 110.15, 230.56,* and *408.3(E)*. This "B" phase is to be connected to the center bus bar in panelboards and switchboards.

Three-Wire Delta System

This connection, illustrated in Figure 13-6, provides only three-phase power. One phase of the system may be grounded, in which case it is often referred to as a corner-grounded delta. The power delivered and the voltages measured between phases remains unchanged. Overcurrent devices are not to be installed in the grounded phase. See Unit 17, Figure 17-22.

Three-Phase, Four-Wire Wye System

The most commonly used system for modern commercial buildings is the three-phase, four-wire wye, Figure 13-7. This system has the advantage of being able to provide three-phase power and permitting lighting to be connected between any of the three phases and the neutral. Typical voltages available with this type of system are 120Y/208, 265Y/460, and 277Y/480. In each case, the transformer connections are the same.

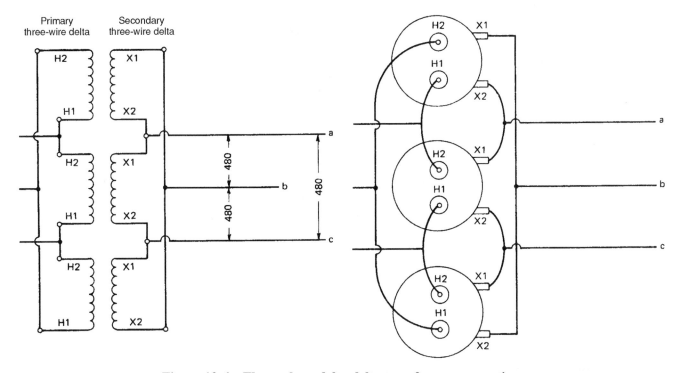

Figure 13-6   Three-phase delta-delta transformer connection.

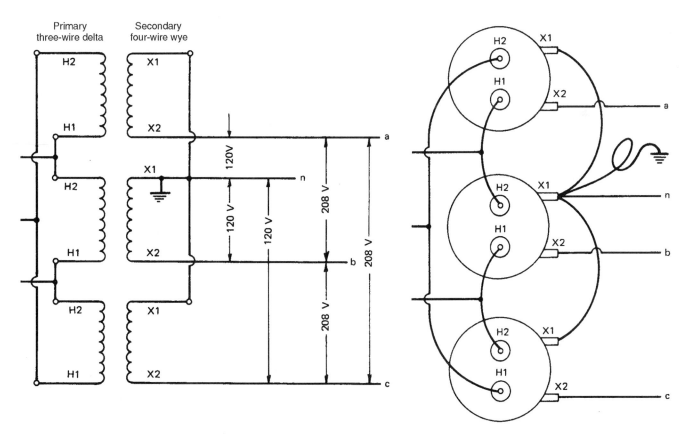

Figure 13-7   Three-phase delta-wye transformer connection.

Notes to All Connection Diagrams

All three-phase connections are shown with the primary connected in delta. For other connections, it is recommended that qualified engineering assistance be obtained.

THE SERVICE ENTRANCE

The regulations governing the method of bringing the electric power into a building are established by the local utility company. These regulations vary considerably between utility companies. Several of the more common methods of installing the service entrance are shown in Figure 13-8, Figure 13-9, Figure 13-10, and Figure 13-11.

Except when it has been legally rejected or amended or is specifically not covered, all wiring shall conform to the requirements of the *NEC*, *NFPA 70*. As stated in *NEC* *90.2(B)(5)*, installations under the exclusive control of electric utilities are exempt from the *NEC* and are governed by the requirements of the *National Electrical Safety Code*. These regulations are considerably different than the *NEC* and are intended to govern the installation of electrical equipment installed by an electric utility for the purpose of generating, metering, distributing, transforming, etc., as they function as a utility.

A common example is that of parking lot lighting around shopping centers and industrial and commercial complexes. As the *Code* is presently written, no matter who makes the installation—an electrical contractor, a building maintenance electrician, or an electrical utility—this type of lighting installation clearly comes under the jurisdiction and requirements of the *NEC*, *NFPA 70*.

Pad-Mounted Transformers

Liquid-insulated transformers as well as dry-type transformers are used for this type of installation. These transformers may be fed by an underground or overhead service, Figure 13-8. The

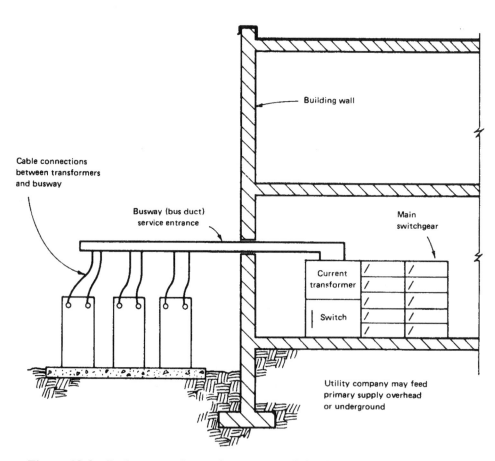

Figure 13-8   Pad-mounted transformers supplying bus duct service entrance.

secondary normally enters the building through a bus duct or large cable.

Unit Substation

For this installation, the primary runs directly to the unit substation where all of the necessary equipment is located, Figure 13-9. The utility company will require the building owner to buy the equipment for a unit substation installation or be charged a higher rate for the energy used.

Pad-Mounted Enclosure

Figure 13-10 illustrates an attractive arrangement in which the transformer and metering equipment are enclosed in a weatherproof cabinet. This type of installation (also known as a *transclosure*) is particularly adapted to smaller service-entrance requirements.

Underground Vault

This type of service is used when available space is an important factor and an attractive site is desired. The metering may be at the utility pole or in the building, Figure 13-11.

METERING

Electricians working on commercial installations seldom make metering connections. However, electricians should be familiar with the following two basic methods of metering.

High-Voltage Metering

When a commercial building is occupied by a single tenant, the utility company may elect to meter the high-voltage side of the transformer. To accomplish this, a potential transformer and two current transformers are installed on the high-voltage lines, and the leads are brought to the meter as shown in Figure 13-12.

The left-hand meter socket in the illustration is connected to receive a standard socket-type watt-hour meter; the right-hand meter socket will receive a var-hour meter (volt-ampere reactive meter). The two meters are provided with 15-minute demand attachments, which register kilowatt (kW) and kilovolt-ampere reactive (kVAR) values, respectively. These demand attachments will indicate the maximum usage of electrical energy for a 15-minute period during the interval between the readings made by

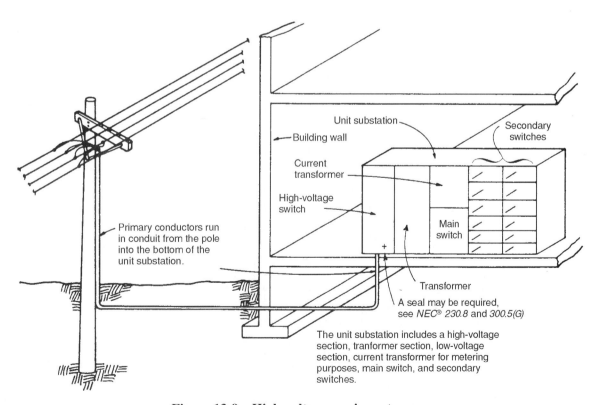

Figure 13-9   High-voltage service entrance.

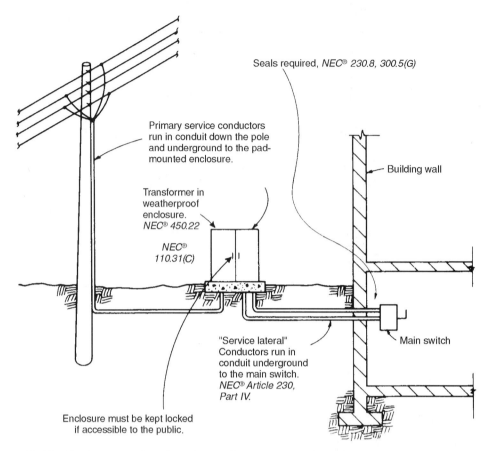

Seals required, *NEC® 230.8, 300.5(G)*

Primary service conductors run in conduit down the pole and underground to the pad-mounted enclosure.

Transformer in weatherproof enclosure. *NEC® 450.22*

NEC® 110.31(C)

Building wall

Main switch

"Service lateral" Conductors run in conduit underground to the main switch. *NEC® Article 230, Part IV.*

Enclosure must be kept locked if accessible to the public.

Figure 13-10   Pad-mounted enclosed transformer supplying underground service entrance.

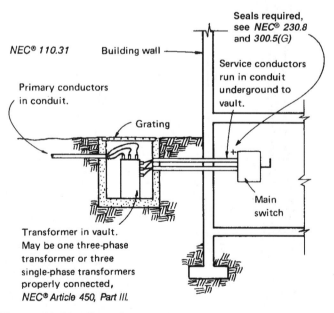

NEC® 110.31

Building wall

Seals required, see *NEC® 230.8* and *300.5(G)*

Service conductors run in conduit underground to vault.

Primary conductors in conduit.

Grating

Main switch

Transformer in vault. May be one three-phase transformer or three single-phase transformers properly connected, *NEC® Article 450, Part III.*

Figure 13-11   Transformer in underground vault supplying underground service entrance.

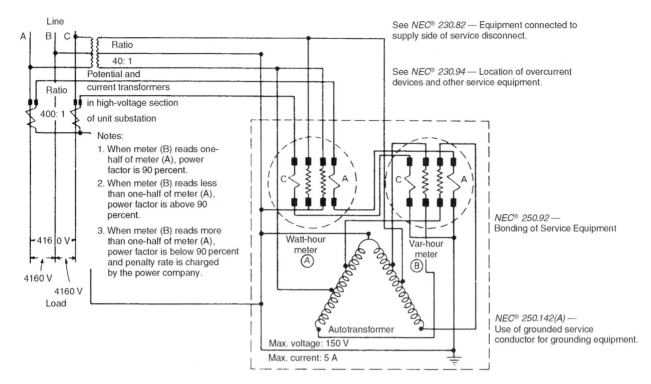

Figure 13-12   Connections for high-voltage watt-hour and var-hour meters.

the utility company. The rates charged by the utility company for electrical energy are based on the maximum demand and the power factor as determined from the two meters. A high demand or a low power factor will result in higher rates.

Low-Voltage Metering

Low-voltage metering of loads greater than 200 amperes is accomplished in the same manner as high-voltage metering. In other words, potential and current transformers are used. For loads of 200 amperes or less, the feed wires from the primary supply are run directly to the meter socket, Figure 13-13. For multiple occupancy buildings, such as the commercial building discussed in this text, the meters are usually installed as a part of the service-entrance equipment, called a *switchboard*.

Figure 13-13   Meter socket.

SERVICE-ENTRANCE EQUIPMENT

When the transformer is installed at a location far from the building, the service-entrance equipment consists of the service-entrance conductors, the main switch or switches, the metering equipment, and the secondary distribution switches, Figure 13-14. The commercial building shown in the plans is equipped in this manner and will be used as an example for the following paragraphs.

The Service

The service for the commercial building is similar to the service shown in Figure 13-10. A pad-mounted, three-phase transformer is located outside the building, and rigid conduit running underground serves as the service raceway.

The service-entrance equipment is located in the basement of the commercial building. This equipment and the service and feeder raceways are shown in Figure 13-14. There is a switch and meter for each of the panelboards. In addition, the building main disconnect and meter and a boiler disconnect and meter are shown. The exact configuration of this equipment would be developed by the manufacturer.

Figure 13-15 provides a diagram of the connections made within the equipment. For each occupant, there is an overcurrent protection device (a set of fuses), a feeder disconnect switch, a meter, and terminals for connection of the feeder. In addition there are two main switches, which together would disconnect the utility electrical power from the building. As many as six disconnects may be installed and still be in compliance with *NEC® 230.71(A)*.

It is possible that the sum of the feeder overcurrent devices is greater than that of the main service overcurrent device. That is because each feeder is calculated for the load on that feeder. The service-entrance conductors and overcurrent devices are calculated according to the requirements for service-entrance calculations.

Adding the feeder fuse sizes in the commercial building will yield a sum of 695 amperes. The main disconnect is fused at 600 amperes. This is because the requirements for the main are somewhat different from the feeders as is illustrated in the following calculations. It is the responsibility of the consulting electrical engineer or electrical contractor to provide the manufacturer of the electrical equipment with the sizes and types of feeder disconnecting means (fusible or circuit breakers) and main disconnecting means (fusible or circuit breakers), and the desired metering information.

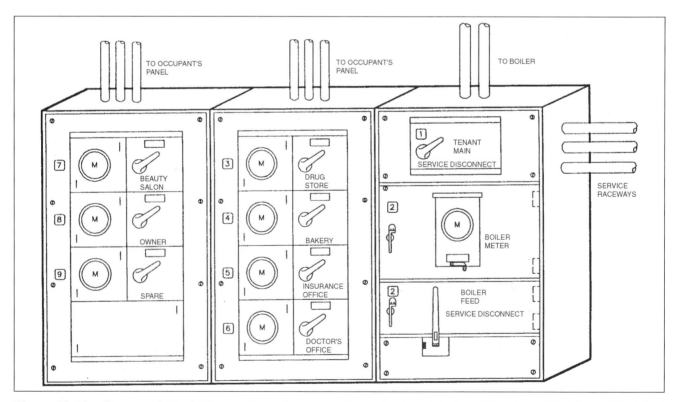

Figure 13-14 **Commercial building service-entrance equipment, pictorial view. (*Courtesy of* Erickson Electrical Equipment Co.)**

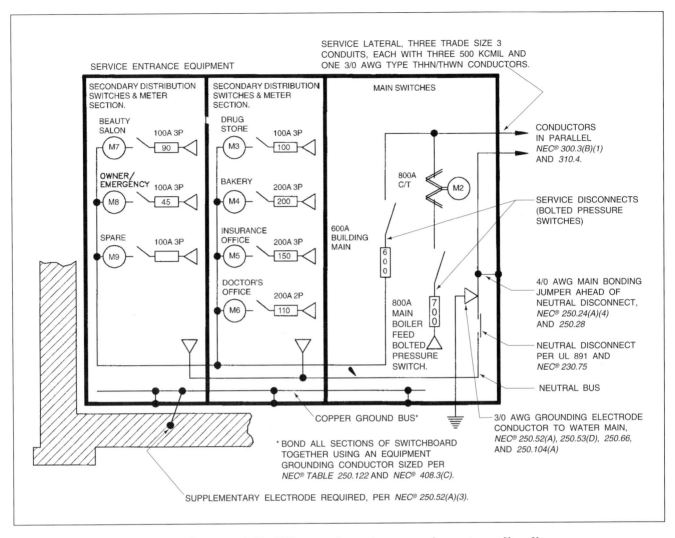

Figure 13-15   Commercial building service-entrance equipment, one-line diagram.

Commercial Building — Load Summary by Type						
Occupancies Load Types	Drugstore	Bakery	Insurance Office	Doctor's Office	Beauty Salon	Owner
Continuous	8838	22,566	8974	5330	5605	4096
Noncontinuous + Receptacle	5280	6380	15,780	5300	7080	2556
Highest Motor	8604	8604	8604	4264	5762	1128
Total Motor	8604	17,333	8604	4264	9562	3768
Noncoincident	8604	0	8604	4264	5762	0
Balanced	8604	34,985	8604	8064	13,562	0
Nonlinear	4962	4446	21,864	660	1155	2022

Table 13-1

In Unit 8, "Feeders," the process for sizing the feeder to the drugstore panelboard was explained in detail. Similar information for the other feeders is recorded in the Appendix. A tabulation of these values is recorded in Table 13-1 and Table 13-2. In Table 13-1, the sums of the various load types are shown. There is no total in the noncoincident row. This value only applies after the boiler is included. The boiler branch-circuit load is added, and the noncoincident load is subtracted, to obtain the service requirement.

Commercial Building—Load Summary by Phase							
Occupancies Load Connections	Drugstore	Bakery	Insurance Office	Doctor's Office	Beauty Salon	Owner	Totals
Phase A	4503	4213	7886	0	3387	3420	23,409
Phase B	5343	2893	8998	3950	1950	3560	26,694
Phase C	4272	4188	7870	2880	3348	3440	25,998

Table 13-2

In Table 13-2 the loads are listed by the phase connection. It is required that these be reasonably equal as is the case in the commercial building. It is unnecessary to arrange the loads to be exactly equal because there will always be a variation in actual loads either over or under the calculated load.

Building Main

The internal connections between the feeder switches and the building main will be determined by the manufacturer. These connections are usually made with copper or aluminum bars sized for the load. The calculations for the overcurrent protection are shown in Table 13-3 and Table 13-5. The calculated load values are taken from Table 13-1. They are adjusted to determine the OCPD selection load just as was done for the feeders. There is one important difference, the OCPD is a bolted pressure switch and is listed for operation at 100 percent of rating, thus the conductors shall be adjusted for the continuous load, but the OCPD may be sized for the calculated load. In Table 13-3 the internal connections of the building main are shown as needing an ampacity of 550 amperes. The minimum OCPD rating is 600-ampere.

Boiler Branch-Circuit

The operation of the boiler is discussed in Unit 11, "Special Circuits (Owner's Circuits)." The supply is three-phase, and no neutral is required. There is a requirement for an equipment grounding conductor to be installed in the feeder raceway because of the necessity of installing a short section of flexible metal conduit to reduce vibration transmissions. The sizing of this equipment grounding conductor is discussed later in this unit. See Figure 13-24. The branch-circuit conductors to the boiler shall have an ampacity of not less than 125 percent of the load to be in compliance with *NEC® 424.82. Annex D* of the *NEC®* simplifies calculations when it states:

Commercial Building—Load Summary for Building Main		
Load Summary, 208Y/120 Three-Phase, Four-Wire	Computed Load	Conductor Sizing
Continuous Load	55,409	69,261
Noncontinuous + Receptacle Load	42,376	42,376
Highest Motor Load Allowance	8604	10,755
Total Motor Load	52,135	52,135
Growth	39,570	49,463
Total Loads (Calculated) (OCPD Selection)	198,155	224,066
Component Selection	**Input**	**Output**
Selection Amperes	550	622
Ampere Rating	600	

Table 13-3

Fractions of an Ampere. Except where the calculations result in a major fraction of an ampere (0.5 or larger), such fractions are permitted to be dropped.

As shown in Table 13-4 and in the specifications, the electric boiler has a rating of 200 kilowatts at 208 volts, 3-phase. In amperes, this equates to:

$$I = \frac{kW \times 1000}{E \times 1.73} = \frac{200 \times 1000}{208 \times 1.73} = 556 \text{ amperes}$$

Apply the 1.25 multiplier as required by *NEC® 424.82:*

$$556 \times 1.25 = 695 \text{ amperes}$$

Therefore, the conductors and the overcurrent protective device for the boiler must have an ampacity of not less than 695 amperes.

To derate, correct, and adjust the ampacity of the conductor to be used, *NEC® 110.3(C)* permits the use of the allowable ampacity in the column that indicates the temperature rating of the type of conductor to be used. This permission is also found in *NEC® 210.19(A)(1)* for branch-circuits, *NEC®*

Commercial Building Boiler Branch-Circuit Calculation		
Load Summary, 208Y Three-Phase, Four-Wire	Input	Output
Continuous Load	200,000	250,000
OCPD Selection	**Input**	**Output**
OCPD Selection Amperes	556	695
OCPD Ampere Rating		700
Minimum Conductor Sets and Size	2	500 kcmil
Minimum Ampacity		347
Phase Conductor Selection	**Input**	**Output**
Ambient Temperature and Correction	38	0.91
Current Carrying Conductors and Adjustment	3	1
Reduction Factor		0.91
Number of Parallel Sets	2	
Minimum Ampacity		347
Minimum Allowable Ampacity		382
Conductor Size		500 kcmil
Conductor Type and Allowable Ampacity	THHN/ THWN	430
Derated Ampacity		391
Raceway Size Determination		
Circuit Conductors Size and Area	500 kcmil	2.1219 in.² (1369 mm²)
Equipment Grounding Conductor Size and Area	1/0 AWG	0.1855 in.² (119.7 mm²)
Total Conductor Area		2.3074 in.² (1489 mm²)
Minimum Raceway Type and Trade Size	RMC	Trade Size 3

Table 13-4

215.19(A)(1) for feeders, and *NEC® 230.42(A)* for services. In this case, THHN/THWN conductors with a 194°F (90°C) rating are being used. *NEC® Table 310.16* indicates that a 500-kcmil copper THHN/THWN conductor has an allowable ampacity of 430 amperes.

Two conductors are run per phase. The total allowable ampacity for two of these conductors in parallel, in separate raceways, is:

$$430 \times 2 = 860 \text{ amperes}$$

Because the boiler room is a relatively hot location, to be safe and conservative, an arbitrary correction factor of 0.91 found at the bottom of *NEC® Table 310.16* is used:

$$860 \times 0.91 = 783 \text{ amperes}$$

Therefore, two 500-kcmil copper THHN/THWN conductors in parallel are more than adequate for the boiler load requirement of 695 amperes.

When paralleling conductors, all of the phase conductors shall be identical in length, type, size, material, and terminations. It is customary for electricians to lay all the conductors, in this case six, on a floor where a perfect match in lengths can be ensured. See *NEC® 310.4* and Unit 8.

The main disconnect for the boiler is an 800-ampere bolted pressure switch (illustrated in Figure 13-16A and Figure 13-16B). It is listed for operation at 100 percent of rating. The fuses in this switch are 700-ampere Class L fuses.

Fixed electric heating is considered to be a continuous load and is subject to the 125 percent multiplier as indicated in *NEC® 210.19(A)(1), 210.20(A), 215.2(A)(1), 215.3,* and *230.42.* By *Code* definition, a *continuous load* is a load that is expected to continue for three hours or more.

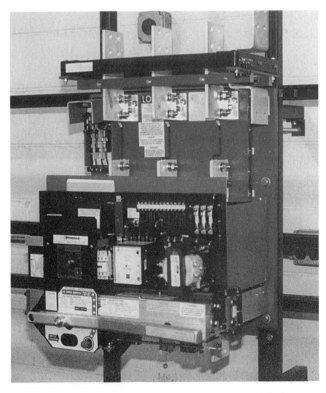

Figure 13-16A   Bolted pressure contact switch for use with high-capacity fuses. (*Courtesy* Square D Company)

Figure 13-16B   A bolted pressure contact switch with shunt trip operator, integral ground-fault protection, auxiliary contacts, and key interlock. (*Courtesy Pringle Electrical.*)

Service-Entrance Conductor Size

As previously stated, the two main overcurrent protective devices are bolted pressure switches listed for operation at 100 percent of rating. This allows the service-entrance conductors to be sized to the calculated load as shown in Table 13-5. This sum is reduced by the noncoincident load as the boiler load greatly exceeds that of the air conditioning, resulting in a total load of 1031 amperes. Three parallel sets of 500-kcmil, Type THHN/THWN conductors will be installed for the service. The total derated ampacity will be 1290 amperes. The same procedure is followed to size the neutral. As with a feeder neutral, the balanced load is subtracted, and the nonlinear load is added. Two additional *NEC®* requirements shall be applied when sizing neutrals:

- *NEC® 250.24(C)(1)* requires that the grounded conductor, i.e., the neutral, in a service shall not be smaller than the required grounding electrode conductor. Grounding electrode conductor sizes are given in *NEC® Table 250.66*.

- In *NEC® 250.24(C)(2)* the grounded conductor shall not be smaller than 1/0 AWG when part of a parallel set.

Commercial Building Service Calculations		
Load Summary—Service	**Calculated**	
Building Main	198,155	
Boiler	200,000	
Total Load	398,155	
Service Conductor Selection	**Input**	**Output**
Calculated Load Total	398,155	
Noncoincident Load *AC load*	−27,234	
Adjusted Load VA and A	370,921	1031
Number of Parallel Sets and Amperes per Set	3	344
Minimum Conductor Size		500 kcmil
Minimum Ampacity		344
Phase Conductor Sizing	**Input**	**Output**
Ambient Temperature, Correction	82°F (28°C)	1
Current-Carrying Conductors (#) (Adjustment)	3	1
Reduction Factor		1
Minimum Allowable Ampacity		344
Conductor Size		500 kcmil
Conductor Type and Ampacity	THHN/ THWN	430
Total Derated Ampacity (Three Sets)		1290
Neutral Sizing	**Input**	**Output**
Calculated Load Total	398,115	
Balanced Load *Motor 3ϕ*	−31,092	
Boiler *3ϕ*	−200,000	
Nonlinear Load *Lights/*	35,109	
Neutral Load VA and A	202,132	561
Neutral Sets and Load per Set	3	187
Minimum Neutral Size		3/0 AWG
Minimum Allowable Ampacity		187
Conductor Type and Size	THHN/ THWN	3/0 AWG
Allowable and Derated Ampacity	225	225
Raceway Size Determination		
Feeder Conductor Size and Area	500 kcmil	2.1219 in.² (1369 mm²)
Neutral Conductor Size and Area	3/0 AWG	0.2679 in.² (518.4 mm²)
Equipment Grounding Conductor Size and Area		0
Total Conductor Area		2.3898 in.² (1887 mm²)
Raceway Type and Trade Size	RMC	Trade Size 3

Table 13-5

When applied to the commercial building service, both of these stipulate a minimum neutral size of 1/0 AWG. The minimum size required by the load is larger, so these requirements are fulfilled.

GROUNDING

The principal reason that electrical systems, equipment, and conductive materials enclosing the preceding are grounded is to facilitate, i.e., enable, the immediate response of overcurrent protective devices to ground faults. This action removes the hazard, to people and animals, that would otherwise exist.

Electrical systems, such as a three-phase, four-wire wye, and circuits, such as those supplied by a transformer connected to a system that exceeds 150 volts to ground, are grounded to stabilize and limit the voltage to ground. This is the grounding of current-carrying conductors. *NEC® Article 250, Part II* identifies the systems and circuits to be grounded.

Electrical equipment and conductive materials enclosing this equipment are bonded to limit the voltage to ground on these items. See *NEC® Article 250, Parts IV and VII*, which detail the items to be bonded to ground. Ideally, except in special cases where isolation is necessary, comprehensive bonding of all conductive materials should be the goal.

This includes all metallic piping, duct work, framing, partitions, siding, and all other items that could come in contact with electrical system wiring or circuits and with a person or animal.

The system/circuit grounding and the equipment/materials grounding are to be joined at only one location on a premise. Failure to maintain the separation will result in extraneous currents from the electrical conductors, i.e., through the pipes, ducts, etc. These currents would increase the level of electromagnetic radiation, which, many have alleged, has a harmful effect on the human body.

It is important that the ground connections and the grounding electrode system be properly installed. To achieve the best possible ground system, the electrician must use the recommended procedures and equipment when making the installation. Figure 13-17 illustrates some of the terminology used in *NEC® Article 250*.

If the grounding is installed in accordance with *NEC® Article 250*, there will be a good current path should a ground fault occur. The reason for this is that the lower the impedance of the grounding path, the greater is the ground-fault current. This increased ground-fault current causes the overcurrent device protecting the circuit to respond faster. For example,

$$I = \frac{E}{Z} = \frac{277}{0.1} = 2770 \text{ amperes}$$

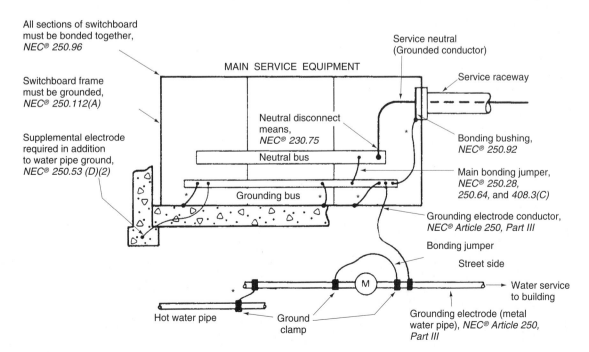

Figure 13-17   Terminology of service grounding and bonding.

but with a lower impedance,

$$I = \frac{E}{Z} = \frac{277}{0.01} = 27,700 \text{ amperes}$$

The amount of ground-fault damage to electrical equipment is related to (1) the response time of the overcurrent device and (2) the amount of current. One common term used to relate the time and current to the ground-fault damage is *ampere squared seconds* (I^2t):

Amperes × Amperes × Time in seconds = I^2t

It can be seen in the equation for I^2t that when the current *(I)* and time *(t)* in seconds are kept to a minimum, a low value of I^2t results. Lower values mean that less ground-fault damage will occur. Units 17, 18, and 19 provide detailed coverage of overcurrent protective devices, fuses, and circuit breakers.

Comprehensive (System) Grounding

The concept of comprehensive (system) grounding dictates that all conductive material such as water pipes, gas pipes, ductwork, siding, and framing be bonded to all electrical equipment and connected by a main bonding jumper to the system ground. The interconnection of the equipment/material makes it highly unlikely that any of the items could become isolated, for even if one connection failed, the integrity of the grounding would be maintained through other connections. The value of this concept is illustrated in Figure 13-18 and described in the following example:

1. A live wire contacts the gas pipe. The bonding jumper A is not installed originally.

2. The gas pipe is now energized at 120 volts.

3. The insulating joint in the gas pipe results in a poor path to ground; assume the resistance is 8 ohms.

4. The 20-ampere fuse will not open.

$$\frac{E \text{ volts}}{R \text{ ohms}} = I \text{ amperes}$$

$$\frac{120 \text{ volts}}{8 \text{ ohms}} = 15 \text{ amperes}$$

5. If a person touches the hot gas pipe and the water pipe at the same time, a current passes through the person's body. If the body's resistance is 12,000 ohms, then the current is

$$\frac{120 \text{ volts}}{12,000 \text{ ohms}} = 0.01 \text{ ampere}$$

This value of current passing through a human body can cause death.

6. The fuse is now seeing 15 + 0.01 = 15.01 amperes; however, it still does not blow.

7. If comprehensive grounding had been achieved, bonding jumper A would have kept the voltage difference between the water pipe and the gas pipe at virtually zero, and the fuse would have opened. If 10 ft (3.0 m) of 4 AWG copper wire were used as the jumper, then the resistance of the jumper would be 0.00308 ohm per *NEC® Chapter 9, Table 8*. The current is

$$\frac{120 \text{ volts}}{0.00308 \text{ ohms}} = 38,961 \text{ amperes}$$

In an actual system, the impedance of all of the parts of the circuit would be much higher.

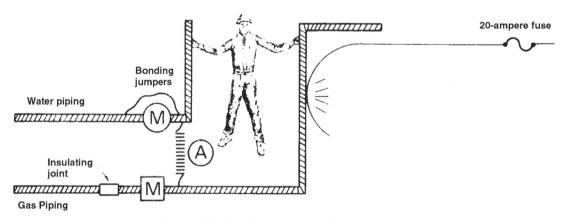

Figure 13-18   System grounding.

Thus, a much lower current would result. The value of the current, however, would be enough to cause the fuse to open.

The advantages of comprehensive system grounding are:

- the potential voltage differentials between the different parts of the system are kept at a minimum, thereby reducing shock hazard.

- impedance of a ground path is kept at a minimum, which results in higher current flow in the event of a ground fault. The lower the impedance, the greater is the current flow, and the faster the overcurrent device opens.

Grounding Electrode Systems

In Figure 13-15 and Figure 13-17, the main service equipment, the service raceways, the neutral bus, the grounding bus, and the hot and cold water pipes have been bonded together.

For discussion purposes of the commercial building, at least 20 ft (6.0 m) of 3/0 AWG bare copper conductor is installed in the footing to serve as the additional supplemental electrode that is required by *NEC® 250.52(A)(3)*. The minimum size permitted is 4 AWG copper. This conductor, buried in the concrete footing, is supplemental to the water pipe electrode. *NEC® 250.52(A)* permits the supplemental electrode to be bonded to the grounding electrode conductor, the grounded service-entrance conductor, the grounded service raceway, or the grounded service enclosure.

For commercial or industrial installations, the supplemental electrode may be bonded to the interior metal water piping only where the entire length of water piping that will serve as the conductor is exposed and will be under qualified maintenance conditions. See *NEC® 250.52(A)(1), Exception.*

Had a 3/0 AWG bare copper conductor not been selected for installation in the concrete footing, the supplemental grounding electrode could have been:

- the metal frame of the building where effectively grounded.

- at least 20 ft (6.0 m) of steel reinforcing bars (re-bars) ½-inch (12.7 mm) minimum diameter, encased in concrete at least 2 in. (50 mm) thick, in direct contact with the earth, near the bottom of the foundation or footing.

- at least 20 ft (6.0 m) of bare copper wire, encircling the building, minimum size 2 AWG, buried directly in the earth at least 2.5 ft (750 mm) deep.

- ground rods.

- ground plates.

All of these factors are discussed in *NEC® Article 250, Parts III, V,* and *VI.*

Many *Code* interpretations are probable as a result of the grounding electrode systems concept. Therefore, the local *Code* authority should be consulted. For instance, some electrical inspectors may not require the bonding jumper between the hot and cold water pipes, as shown in Figure 13-17. They may determine that an adequate bond is provided through the water heater itself of either type, electric or gas. Other electrical inspectors may require that the hot and cold water pipes be bonded together because some water heaters contain insulating fittings that are intended to reduce corrosion inside the tank caused by electrolysis. According to their reasoning, even though the originally installed water heater contains no fittings of this type, such fittings may be included in a future replacement heater, thereby necessitating a bond between the cold and hot water pipes. The local electrical inspector should be consulted for the proper requirements. However, bonding the pipes together, as shown in Figure 13-17, is the recommended procedure.

Grounding Requirements

When grounding service-entrance equipment, the following *Code* rules shall be observed:

- The electrical system shall be grounded when maximum voltage to ground does not exceed 150 volts; see *NEC® 250.20(B)(1).*

- The electrical system shall be grounded when the neutral is used as a circuit conductor (for example, a 480/277-volt, wye-connected, three-phase, four-wire system), *NEC® 250.20(B)(2).* This is similar to the connection of the system shown in Figure 13-7.

- The electrical system shall be grounded where the midpoint of one transformer is used to establish the grounded neutral circuit conductor, as on a 240/120-volt, three-phase, four-wire delta system; see *NEC® 250.20(B)* and refer to Figure 13-5.

- ▶The earth shall not be considered as an effective ground-fault current path. See *NEC® 250.4(A)(5)*.◀

- ▶*NEC® 250.4(A)(5)* requires that an effective ground-fault current path be established for all electrical equipment, wiring, and other electrically conductive material likely to become energized. This path shall be installed:

  1. permanent, and
  2. be of low impedance, capable of safely carrying the maximum ground-fault current likely to be imposed on it.◀

  Details of this subject are discussed in Units 17, 18, and 19.

- All grounding schemes shall be installed so that no objectionable currents will exist in the grounding conductors and other grounding paths, see *NEC® 250.6*.

- The grounding electrode conductor shall be connected to the supply side of the service disconnecting means, see *NEC® 250.24(A)*.

- The identified neutral conductor is the conductor that shall be grounded, see *NEC® 250.26*.

- Tie (bond) everything together, see *NEC® 250.50*.

- Do not use interior metal water piping to serve as a means of interconnecting steel framing members, concrete-encased electrodes, and the ground ring for the purposes of establishing the grounding electrode system as required by *NEC® 250.104*. Should any PVC nonmetallic water piping be interposed in the metal piping runs, the continuity of the bonding-grounding system is broken.

  The proper location to connect the grounding electrode conductor, the bonding conductors associated with the metal framing members, concrete-encased electrodes, and ground ring is anywhere on the first 5 ft (1.5 m) of metal water piping after it enters the building. The first five ft (1.5 m) may include a water meter. See *NEC® 250.52(A)(1)*.

  The only exception to this is for industrial and commercial buildings, where the entire length of the water piping being used as the conductor to interconnect the various electrodes

is exposed and where only qualified people will be doing maintenance on the installation; see *NEC® 250.52(A)(1), Exception*.

- The grounding electrode conductor is to be sized as given in *NEC® Table 250.66*.

- The metal frame of the building shall be bonded to the grounding electrode system where it is effectively grounded, *NEC® 250.50(B)*.

- The hot and cold water metal piping system shall be bonded to the service-equipment enclosure, to the grounded conductor at the service, and to the grounding electrode conductor, *NEC® 250.104*.

- *NEC® 250.64(C)* states that grounding electrode conductors shall not be spliced. There are exceptions for commercial and industrial installations where the grounding electrode conductor may be spliced by either exothermic welding or irreversible compression-type connectors that are listed for that purpose.

- The grounding electrode conductor shall be connected to the metal underground water pipe when 10 ft (3.0 m) long or longer and in direct contact with the earth. The 10 ft (3.0 m) includes the metal well casing. See *NEC® 250.52(A)(1)*.

- In addition to grounding the service equipment to the underground water pipe, one or more additional electrodes are required, such as a bare conductor in the footing, a grounding ring, rod or pipe electrodes, or plate electrodes. All of these items shall be bonded if they are available on the premises. See *NEC® 250.50* and *250.104*.

- *NEC® 250.52(B)* prohibits using a metal underground gas piping system as the grounding electrode. However, where metal gas piping comes into a building, and if the gas piping is likely to become energized, it shall be bonded to the service equipment, the grounded conductor at the service, the grounding electrode conductor, or to one or more of the accepted grounding electrodes used. See *NEC® 250.104(B)*.

- The grounding electrode conductor shall be copper, aluminum, or copper-clad aluminum, *NEC® 250.62*.

- The grounding electrode conductor may be solid or stranded, uninsulated, covered or bare, and shall not be spliced, see *NEC® 250.62* and *250.64(C)*.

- Bonding is required around all insulating joints or sections of the metal piping system that might be disconnected, *NEC® 250.68(B)*.

- The connection to the grounding electrode shall be accessible, *NEC® 250.68(A)*. A connection to a concrete-encased, drive, or buried grounding electrode does not have to be accessible.

- Make sure that the length of bonding conductor is long enough so that if the equipment is removed, the bonding will remain in place, *NEC® 250.68(B)*. The bonding jumper around the water meter is a good example where this requirement applies.

- The grounding electrode conductor shall be tightly connected by using proper lugs, connectors, clamps, or other approved means, *NEC® 250.70*. One type of grounding clamp is shown in Figure 13-19.

Sizing the Grounding Electrode Conductors

The grounding electrode conductor connects the grounding electrode, to the equipment grounding conductor, and to the system grounded (neutral) conductor. In the commercial building, this means that the grounding electrode conductor connects the main water pipe (grounding electrode) to the grounding bus (equipment grounding conductor) and to the neutral bus [system grounded (neutral) conductor]. See Figure 13-17.

NEC® Table 250.66 (Table 13-6) is referred to when selecting grounding electrode conductors for services where there is no overcurrent protection ahead of the service-entrance conductors other than the utility companies' primary or secondary overcurrent protection.

In *NEC® Table 250.66*, the service conductor sizes are given in the wire size and not the ampacity values. To size a grounding electrode conductor for

Figure 13-19   Grounding clamp.

Table 250.66 Grounding Electrode Conductor for Alternating-Current Systems

| Size of Largest Ungrounded Service-Entrance Conductor or Equivalent Area for Parallel Conductors[a] (AWG/kcmil) | | Size of Grounding Electrode Conductor (AWG/kcmil) | |
Copper	Aluminum or Copper-Clad Aluminum	Copper	Aluminum or Copper-Clad Aluminum[b]
2 or smaller	1/0 or smaller	8	6
1 or 1/0	2/0 or 3/0	6	4
2/0 or 3/0	4/0 or 250	4	2
Over 3/0 through 350	Over 250 through 500	2	1/0
Over 350 through 600	Over 500 through 900	1/0	3/0
Over 600 through 1100	Over 900 through 1750	2/0	4/0
Over 1100	Over 1750	3/0	250

Notes:
1. Where multiple sets of service-entrance conductors are used as permitted in 230.40, Exception No. 2, the equivalent size of the largest service-entrance conductor shall be determined by the largest sum of the areas of the corresponding conductors of each set.
2. Where there are no service-entrance conductors, the grounding electrode conductor size shall be determined by the equivalent size of the largest service-entrance conductor required for the load to be served.
[a]This table also applies to the derived conductors of separately derived ac systems.
[b]See installation restrictions in 250.64(A).

Table 13-6
Reprinted with permission from NFPA 70-2005

a service with three parallel 500-kcmil service conductors, it is necessary to total the wire size and select a grounding electrode conductor for an equivalent 1500-kcmil service conductor. In this case, the grounding electrode conductor is 3/0 AWG copper (see Figure 13-14). Figure 13-20 shows another example.

The main bonding jumper, see Figure 13-21 and Figure 13-22, connects the neutral busbar to the equipment grounding busbar. The jumper is sized according to *NEC® 250.28(D)*. The service to the commercial building consists of three 500-kcmil conductors in parallel. This adds up to a total conductor area of 1500 kcmil. Multiplying by 0.125 yields a minimum size for the main bonding jumper of 187.5 kcmil. Referring to *NEC® Chapter 9, Table 8*, this requires a 4/0 AWG conductor.

The calculated neutral load is 560 amperes, 187 amperes per conductor. The 167°F (75°C) column of *NEC® Table 310.16* indicates that 3/0 AWG conductors have an ampacity of 200 amperes. This would

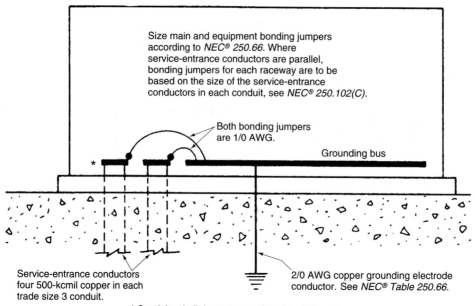

Size main and equipment bonding jumpers according to NEC® 250.66. Where service-entrance conductors are parallel, bonding jumpers for each raceway are to be based on the size of the service-entrance conductors in each conduit, see NEC® 250.102(C).

Both bonding jumpers are 1/0 AWG.

Grounding bus

*

Service-entrance conductors four 500-kcmil copper in each trade size 3 conduit.

2/0 AWG copper grounding electrode conductor. See NEC® Table 250.66.

* Conduits shall rise not more than 3 in. (75 mm) above bottom of enclosure, see NEC® 408.5.

Figure 13-20   Typical service entrance (grounding and bonding).

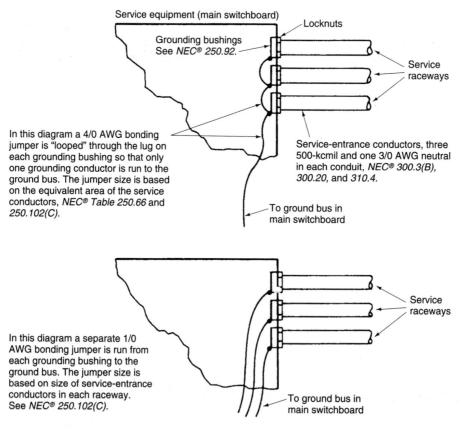

Service equipment (main switchboard)

Locknuts

Grounding bushings See NEC® 250.92.

Service raceways

In this diagram a 4/0 AWG bonding jumper is "looped" through the lug on each grounding bushing so that only one grounding conductor is run to the ground bus. The jumper size is based on the equivalent area of the service conductors, NEC® Table 250.66 and 250.102(C).

Service-entrance conductors, three 500-kcmil and one 3/0 AWG neutral in each conduit, NEC® 300.3(B), 300.20, and 310.4.

To ground bus in main switchboard

Service raceways

In this diagram a separate 1/0 AWG bonding jumper is run from each grounding bushing to the ground bus. The jumper size is based on size of service-entrance conductors in each raceway. See NEC® 250.102(C).

To ground bus in main switchboard

Figure 13-21   Sizing of main bonding jumper for the service to the commercial building.

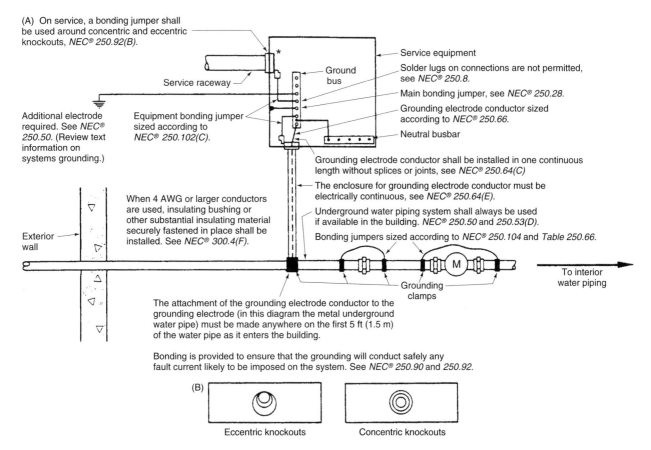

(A) On service, a bonding jumper shall be used around concentric and eccentric knockouts, *NEC® 250.92(B)*.

Service raceway

Additional electrode required. See *NEC® 250.50*. (Review text information on systems grounding.)

Ground bus

Service equipment

Solder lugs on connections are not permitted, see *NEC® 250.8*.

Main bonding jumper, see *NEC® 250.28*.

Grounding electrode conductor sized according to *NEC® 250.66*.

Neutral busbar

Equipment bonding jumper sized according to *NEC® 250.102(C)*.

Grounding electrode conductor shall be installed in one continuous length without splices or joints, see *NEC® 250.64(C)*

The enclosure for grounding electrode conductor must be electrically continuous, see *NEC® 250.64(E)*.

Underground water piping system shall always be used if available in the building. *NEC® 250.50* and *250.53(D)*.

Bonding jumpers sized according to *NEC® 250.104* and *Table 250.66*.

When 4 AWG or larger conductors are used, insulating bushing or other substantial insulating material securely fastened in place shall be installed. See *NEC® 300.4(F)*.

Exterior wall

To interior water piping

Grounding clamps

The attachment of the grounding electrode conductor to the grounding electrode (in this diagram the metal underground water pipe) must be made anywhere on the first 5 ft (1.5 m) of the water pipe as it enters the building.

Bonding is provided to ensure that the grounding will conduct safely any fault current likely to be imposed on the system. See *NEC® 250.90* and *250.92*.

(B)

Eccentric knockouts   Concentric knockouts

Figure 13-22   Bonding and grounding of service equipment.

seem to be adequate. However, there are other *Code* sections that must be checked.

Referring to *NEC® 250.24(B)*, the requirement is set forth that the minimum size for the neutral is determined by the requirement for a grounding electrode conductor. When the service conductors are paralleled, the neutral shall be routed with the phase conductors and shall be sized no smaller than the required grounding electrode conductor, *NEC® Table 250.66*. The three 500-kcmil service-entrance conductors in each service raceway will require that the neutral conductor be no smaller than 1/0 AWG, see *NEC® 310.4*. Although the calculated neutral load may allow a smaller conductor, a 1/0 AWG is the minimum size because this conductor would carry high-level fault currents, if a ground fault occurs.

Remember that *NEC® 250.4(A)(5)*, *250.90*, and *250.96* require that the grounding and bonding conductors be capable of carrying any fault current that they might be called upon to carry under fault conditions. This subject is covered in Units 17, 18, and 19.

Nonconductive paint, enamel, or similar coating shall be removed at contact points when sections of electrical equipment are bolted together to ensure that the sections are effectively bonded. Tightly driven metal bushings and locknuts generally will bite through paints and enamels, thus making the removal of the paint unnecessary. See *NEC® 250.96(A)* and refer to Figure 13-21, Figure 13-22, and Figure 13-23.

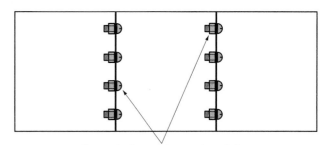

Nonconductive paint, enamel, or similar coating must be removed at contact points.

Figure 13-23   Ensuring bonding when electrical equipment is bolted together.

NEC® Table 250.122 is referred to when selecting equipment grounding conductors when there is overcurrent protection ahead of the conductor supplying the equipment.

NEC® Table 250.122 (Table 13-7) is based on the current setting or rating, in amperes, of the overcurrent device installed ahead of the equipment (other than service-entrance equipment) being supplied.

The electric boiler branch-circuit in the commercial building consists of two 500-kcmil conductors per phase protected by 700-ampere fuses in the main switchboard. The use of flexible metal conduit at the boiler means that a bonding jumper shall be installed in each conduit as shown in Figure 13-24. An illustration of a typical application for the electric boiler feeder in the commercial building is shown in Figure 13-24.

GROUND-FAULT PROTECTION

The *NEC®* requires the use of ground-fault protection for equipment (GFPE) devices on services that meet the conditions outlined in *NEC® 230.95.* Thus, ground-fault protection devices are installed:

- on solidly grounded wye services above 150 volts to ground, but not greater than 600 volts between phases on service disconnects rated at 1000 amperes or more (for example, on 480Y/277-volt systems).

Table 250.122 Minimum Size Equipment Grounding Conductors for Grounding Raceway and Equipment

Rating or Setting of Automatic Overcurrent Device in Circuit Ahead of Equipment, Conduit, etc., Not Exceeding (Amperes)	Size (AWG or kcmil)	
	Copper	Aluminum or Copper-Clad Aluminum*
15	14	12
20	12	10
30	10	8
40	10	8
60	10	8
100	8	6
200	6	4
300	4	2
400	3	1
500	2	1/0
600	1	2/0
800	1/0	3/0
1000	2/0	4/0
1200	3/0	250
1600	4/0	350
2000	250	400
2500	350	600
3000	400	600
4000	500	800
5000	700	1200
6000	800	1200

Note: Where necessary to comply with 250.4(A)(5) or (B)(4), the equipment grounding conductor shall be sized larger than given in this table.

*See installation restrictions in 250.120.

Table 13-7

Reprinted with permission from NFPA 70-2005

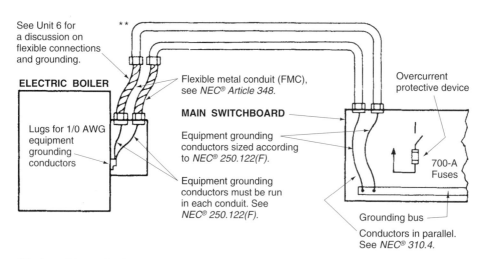

Figure 13-24   Sizing equipment grounding conductors.

**Each conduit contains three 500-kcmil conductors and one 1/0 AWG equipment grounding conductor, which is solidly connected to the grounding bus in the switchboard and to the terminal box on the boiler. The grounding conductor may be insulated or bare. See *NEC® 250.118.*

- to operate at 1200 amperes or less.

- so that the maximum time of opening the service switch or circuit breaker does not exceed one second for ground-fault currents of 3000 amperes or more.

- to limit damage to equipment and conductors on the load side of the service disconnecting means. GFPE will not protect against damage caused by faults occurring on the line side of the service disconnect.

These ground-fault protection requirements do not apply on services for a continuous industrial process where a nonorderly shutdown will introduce additional or increased hazards, *NEC® 230.95*, *Exception 1*, and *240.12*.

When a fuse-switch combination serves as the service disconnect, the fuses shall have adequate interrupting capacity to interrupt the available fault current *(NEC® 110.9)* and shall be capable of opening any fault current that exceeds the interrupting rating of the switch during any time when the ground-fault protective system will not cause the switch to open, see *NEC® 230.95(B)*.

Ground-fault protection is not required on

- delta-connected, three-phase systems.

- ungrounded wye-connected, three-phase systems.

- single-phase systems.

- 120/240-volt, single-phase systems.

- 208Y/120-volt, three-phase, four-wire systems.

- systems greater than 600 volts; for example, 2400/4160 volts.

- service disconnecting means rated at less than 1000 amperes.

- systems in which the service is subdivided; for example, a 1600-ampere service may be divided between two 800-ampere switches.

The time of operation of the device as well as the ampere setting of the GFPE device must be considered carefully to ensure that the continuity of the electrical service is maintained. The time of operation of the device includes:

1. the sensing of a ground fault by the GFPE monitor.

2. the monitor signaling the disconnect switch to open.

3. the actual opening of the contacts of the disconnect device (either a switch or a circuit breaker). The total time of operation may result in a time lapse of several cycles or more (Units 17 and 18).

GFPE circuit devices were developed to overcome a major problem in circuit protection: the low-value phase-to-ground arcing fault, Figure 13-25. The amount of current in an arcing phase-to-ground fault can be low when compared to the rating or setting of the overcurrent device. For example, an arcing fault can generate a current of 600 amperes. A main breaker rated at 1600 amperes will allow this current without tripping, because the 600-ampere current appears to be just another load current. The operation of the GFPE device assumes that under normal conditions the total instantaneous current in all of the conductors of a circuit will exactly balance, Figure 13-26. Thus, if a current coil is installed so that all of the circuit conductors run through it, the normal current measured by the coil will be zero. If a ground fault occurs, some current will return through the grounding system, and an unbalance will result in the conductors. This unbalance is then detected by the GFPE device, Figure 13-27.

The purpose of ground-fault protection devices is to sense and protect equipment against *low-level ground faults*. GFPE monitors do not sense phase-to-phase faults, three-phase faults, or phase-to-neutral faults. These monitors are designed to sense phase-to-ground faults only.

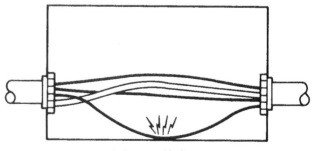

Arcing ground fault can occur when a phase wire and the conduit grounding system contact each other.

Figure 13-25 Ground fault.

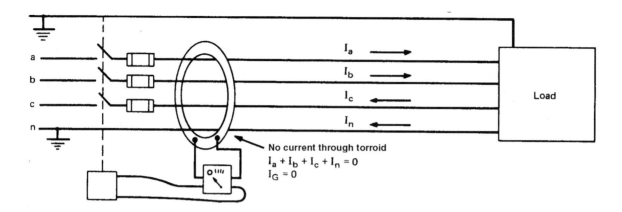

Figure 13-26   Normal condition.

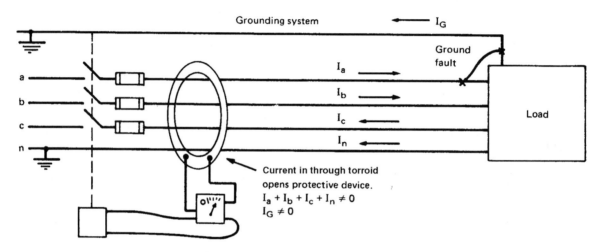

Figure 13-27   Abnormal condition.

High-magnitude ground-fault currents can cause destructive damage even though a GFPE is installed. The amount of arcing damage depends upon:

1. how much current flows.

2. the length of time that the current exists.

For example, if a GFPE device is set for a ground fault of 500 amperes and the time setting is six cycles, then the device will need six cycles to signal the switch or circuit breaker to open the circuit whether the ground fault is slightly more than 500 amperes or as large as 20,000 amperes. The six cycles needed to signal the circuit breaker plus the operation time of the switch or breaker may be long enough to result in damage to the switchgear.

The damaging effects of high-magnitude ground faults, phase-to-phase faults, three-phase faults, and phase-to-neutral faults can be reduced substantially by the use of current-limiting overcurrent devices,

NEC® 240.2. These devices reduce both the peak let-through current and the time of opening once the current is sensed. For example, a ground fault of 20,000 amperes will open a current-limiting fuse in less than one-half cycle. In addition, the peak let-through current is reduced to a value much less than 20,000 amperes (see Units 17 and 18).

►Ground-fault protection for equipment (GFPE) is not to be confused with personal ground-fault protection (GFCI). In commercial buildings, GFCI devices are required in bathrooms, in kitchens, on rooftops, and outdoors where readily accessible to the public. This is found in *NEC® 210.8(B).*◄

The GFPE is connected to the normal fused switch or circuit breaker, which serves as the circuit protective device. For feeders, the GFPE requirement is found in *NEC® 215.10.* For services, the GFPE requirement is found in *NEC® 230.95.* GFPE is required on services and feeders of solidly grounded systems of 1000 amperes or more where the voltage to ground is more than 150 volts but not more than 600 volts. The GFPE device is adjusted so that it will signal the protective device to open under ground-fault conditions. The maximum setting of the GFPE is 1200 amperes.

In the commercial building in the plans, the service voltage is 208Y/120 volts. The voltage to ground on this system is 120 volts. This value is not large enough to sustain an electrical arc. Therefore, it is not required (according to the *NEC®*) that ground-fault protection for the service disconnecting means be installed for the commercial building. The electrician can follow a number of procedures to minimize the possibility of an arcing fault. Examples of these procedures follow:

- Be sure that conductor insulation is not damaged when the conductors are pulled into raceway.

- Be sure that the electrical installation is properly grounded and bonded.

- Tighten locknuts and bushings.

- Tighten all electrical connections.

- Tightly connect bonding jumpers around concentric and/or eccentric knockouts.

- Be sure that conduit couplings and other fittings are installed properly.

- Check insulators for minute cracks.

- Install insulating bushings on all raceways.

- Insulate all bare busbars in switchboards when possible.

- Be sure that conductors do not rest on bolts or other sharp metal edges.

- Be sure that electrical equipment does not become damp or wet either during or after construction.

- Be sure that all overcurrent devices shall have an adequate interrupting capacity.

- Do not work on hot panels.

- Be careful when using fish tape because the loose end can become tangled with the electrical panelboard.

- Be careful when working with live parts; do not drop tools or other metal objects on top of such parts.

- Avoid large services; for example, it is usually preferable to install two 800-ampere service disconnecting means rather than to install one 1600-ampere service disconnecting means.

SAFETY IN THE WORKPLACE

Many injuries have occurred by individuals working electrical equipment "hot." A fault, whether line-to-line, line-to-line-to-line, or line-to-ground, can develop tremendous energy. Splattering of melted copper and steel, the heat of the arc blast, the blinding light of the arc, and electric shock are hazards that are present. The temperatures of an electrical arc, such as might occur on a 480/277-volt solidly grounded, wye-connected system, are hotter than the surface of the sun. The enormous pressures of an "arc blast" can blow a person clear across the room. These awesome pressures will vent (discharge) through openings, such as the open cover of a panel, just where the person is standing.

Federal laws are in place to protect workers from injury while on the job. When working on or near electrical equipment, certain safety practices must be followed.

In the Occupational Safety and Health Act (OSHA), Sections 1910.331 through 1910.360 are devoted entirely to safety-related work practices. Proper training in work practices, safety procedures, and other personnel safety requirements are described. Such things as turning off the power, locking the switch off, or properly tagging the switch are discussed in the OSHA Standards.

The fundamental rule is **NEVER WORK ON EQUIPMENT WITH THE POWER TURNED ON!**

In the few cases where the equipment absolutely must be left on, proper training regarding electrical installations, proper training in established safety practices, proper training in first aid, properly insulated tools, safety glasses, hard hats, protective insulating shields (rubber blankets to cover live parts), rubber gloves, nonconductive and nonflammable clothing, and working with more than one person are required.

Proper training is required to enable a person to become qualified. The *National Electrical Code®* describes a qualified person as *One familiar with the construction and operation of the equipment and the hazards involved.*

The National Fire Protection Association Standard NFPA 70B, "*Electrical Equipment Maintenance*," discusses safety issues and safety procedures and mirrors the OSHA regulations in that **NO ELECTRICAL EQUIPMENT SHOULD BE WORKED ON WHILE IT IS ENERGIZED.**

The National Electrical Manufacturers Association (NEMA) in Publication PB 2.1–1996, "General Instructions for Proper Handling, Installation, Operation, and Maintenance of Deadfront Distribution Switchboards Rated 600 Volts or Less," repeats the safety rules just discussed. This standard states that "The installation, operation, and maintenance of switchboards should be conducted only by qualified personnel." This standard discusses the **"Lock-Out, Tag-Out"** procedures. It clearly states that:

WARNING: HAZARDOUS VOLTAGES IN ELECTRICAL EQUIPMENT CAN CAUSE SEVERE PERSONAL INJURY OR DEATH. UNLESS OTHERWISE SPECIFIED, INSPECTION AND PREVENTATIVE MAINTENANCE SHOULD ONLY BE PERFORMED ON SWITCHBOARDS, AND EQUIPMENT TO WHICH POWER HAS BEEN TURNED OFF, DISCONNECTED AND ELECTRICALLY ISOLATED SO THAT NO ACCIDENTAL CONTACT CAN BE MADE WITH ENERGIZED PARTS.

There have been lawsuits where serious injuries have occurred on equipment that had ground-fault protection (GFP). The claims were that the ground-fault protection for equipment also provides protection for personnel. **This is not true!** Throughout the *NEC,* references are made to *Ground Fault Protection For Personnel (GFCI)* and *Ground Fault Protection For Equipment (GFPE).*

Many texts have been written about the hazards of electricity. All of these texts say **TURN THE POWER OFF!**

IT'S THE LAW!

NEC® 110.16 has added a new requirement that addresses the serious hazard of arc blasts. This section states that:

▶ *Flash protection.   Switchboards, panelboards, industrial control panels, meter socket enclosures, and motor control centers that are in other than dwelling occupancies and are likely to require examination, adjustment, servicing, or maintenance while energized shall be field marked to warn qualified persons of potential electric arc flash hazards. The marking shall be located so as to be clearly visible to qualified persons before examination, adjustment, servicing, or maintenance of the equipment.* *◀

The *Fine Print Notes* to *110.16* reference NFPA 70E *Standard for Electrical Safety in the Workplace*, for "*assistance in determining severity of potential exposure, planning safe work practices, and selecting personal protective equipment,*" and to ANSI Z535.4, *Product Safety Signs and Labels*, for "*guidelines for the design of safety signs and labels for application to products.*"*

Flash protection is a very complicated subject, and will be discussed over and over again as to how the electrical contractor and electrician will comply.

If you want to learn more, visit Bussmann's Web site at http://www.bussmann.com. There you will find an easy-to-use computer program for making arc-flash and fault-current calculations.

---

*Reprinted with permission from NFPA 70-2005.

REVIEW QUESTIONS

Refer to the *National Electrical Code®* or the working drawings when necessary. Where applicable, responses should be written in complete sentences.

A 300-kilovolt-ampere transformer bank consisting of three 100-kilovolt-ampere transformers has a three-phase, 480-volt delta primary and a three-phase, 208Y/120 secondary.

For Questions 1 through 4 show calculations and/or source of information:

1. What is the full-load current in the primary?

2. What is the full-load current in the secondary?

3. What is the proper rating of the fuses to install in the secondary?

4. What is the kilovolt-ampere of the bank if one of the transformers is removed?

5. Sketch the secondary connections, and indicate where a 120-volt load could be connected.

6. A cabinet with panelboard, shown on the right in the following figure, is added to an existing panelboard installation. Two knockouts, one near the top of the panelboard and the other near the bottom, are cut in the adjoining sides of the cabinets. List any parts to be checked or added and indicate on the drawing the proper way of extending the phase and neutral conductors to the new panelboard.

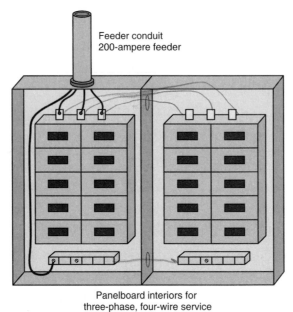

Feeder conduit
200-ampere feeder

Panelboard interiors for
three-phase, four-wire service

The following questions refer to grounding requirements. Give the source of your responses.

7. What is the proper size copper grounding electrode conductor for a 100-ampere service that consists of 3/0 AWG phase conductors? _____

_____

_____

8. What is the minimum size copper grounding electrode conductor for a service with three sets of parallel 350-kcmil conductors (three conductors per phase)? _____

_____

_____

9. If they are available, what six items shall be bonded together to form a grounding electrode system? _____

_____

_____

10. The engineering calculations for an 800-ampere service entrance call for two 500-kcmil copper conductors per phase connected in parallel. The neutral calculations show that the neutral conductors need only be 3 AWG copper conductors. The riser diagram shows two rigid metal conduits, each containing three 500-kcmil phase conductors and one 1/0 AWG neutral conductor. Why is the neutral conductor sized as a 1/0 AWG, or is this a mistake?

_____

_____

_____

_____

_____

11. Electric utility installations and equipment under the exclusive control of the utility do not have to conform to the *National Electrical Code.* However, they do have to conform to the _____.

12. You are an electrical apprentice. The electrical foreman on the job tells you to connect some equipment grounding conductors to the ground bus that runs along the front bottom of a dead-front switchboard. The system is 480/277-volt solidly grounded wye. The building is still under construction, and a number of other trades are working in the building. The switchboard has already been energized, and it is totally enclosed except for two access covers near the front bottom where the ground bus runs. The ground bus already has the proper number of and size of lugs for the equipment grounding conductors. You ask permission to turn off the main power so that you can work on connecting the equipment grounding conductors to the ground bus without having to worry about accidentally touching the phase buses with your hands or with a screwdriver or wrench. The foreman says, "No." What would you do?

_____

_____

_____

_____

_____

_____

13. Electrical explosions directly into the face of a person working on or near energized electrical equipment can and will cause serious injury or death. OSHA laws and *NFPA 70E* have strict safety requirements for working on electrical equipment. The basic rules are to turn off the power before working on the equipment, and to wear the proper type of nonflammable clothing. To further provide safety for the worker, the *National Electrical Code*® has added a new requirement pertaining to **Flash Protection**. This requirement is found in _____ of the *NEC.*®

UNIT 14

Lamps for Lighting

OBJECTIVES

After studying this unit, the student should be able to

- define the technical terms associated with lamp selection.
- list the lamps scheduled to be used in the commercial building.
- explain the application of lamps used in the commercial building.
- identify the parts of the three most popular types of lamps.
- order the lamp types according to certain characteristics.
- define lamps by their designations.

For most construction projects, the electrical contractor is required to purchase and install lamps in the luminaires (fixtures). Thomas Edison provided the talent and perseverance that led to the development of the incandescent lamp in 1879 and the fluorescent lamp in 1896. Peter Cooper Hewitt produced the mercury lamp in 1901. All three of these lamp types have been refined and greatly improved since they were first developed.

NEC® Article 410 contains the provisions for the wiring and installation of luminaires (fixtures), lampholders, lamps, receptacles, and rosettes.

Several types of lamps are used in the commercial building. In the *NEC®* these lamps are referred to as either incandescent or electric discharge lamps. In the industry, the incandescent lamp is also referred to as a filament lamp because the light is produced by a heated wire filament. The electric discharge lamps include a variety of types, but all require a ballast. The most common types of electric discharge lamps are fluorescent, mercury, metal halide, high-pressure sodium, and low-pressure sodium.

Mercury, metal halide, and high-pressure sodium lamps are also classified as high-intensity discharge (HID) lamps.

LIGHTING TERMINOLOGY

Candela (cd)

The luminous intensity of a source, when expressed in candelas, is the candlepower (cp) rating of the source.

Lumen (lm)

Lumen (lm) is the amount of light received in a unit of time on a unit area at a unit distance from a point source of one candela, Figure 14-1. The surface area of a sphere is 12.57 times the radius. When the measurement is in customary units, the unit area is one square foot and the unit distance is one foot, thus a one-candela source produces one lumen on each square foot (one footcandle) on the sphere for a total of 12.57 lumens. If the units are in SI, then the unit area is one square meter, and the unit distance is one meter, thus a one-candela source produces one lumen on each square meter (one lux) for a total of 12.57 lumen.

Illuminance

The measure of illuminance on a surface is the lumen per unit area expressed in footcandles (fc) or lux (lx). The recommended illuminance levels vary

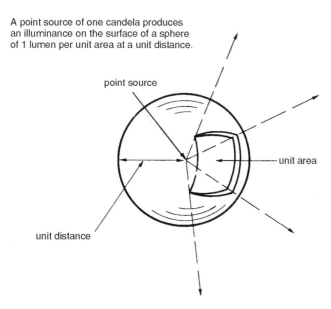

A point source of one candela produces an illuminance on the surface of a sphere of 1 lumen per unit area at a unit distance.

point source

unit area

unit distance

Figure 14-1   Pictorial presentation of unit sphere.

greatly, depending on the task to be performed and the ambient lighting conditions. For example, while 5 footcandles (fc), or 54 lux (lx), is often accepted as adequate illumination for a dance hall, 200 fc (2152 lx) may be necessary on a drafting table for detailed work.

Lumens per Watt (lm/W)

Lumens per watt is a measure of the effectiveness (efficacy) of a light source in producing light from electrical energy. A 100-watt incandescent lamp producing 1670 lumens has an efficacy of 16.7 lumens per watt. (An efficiency is when the input and output have the same units of measure; an efficacy is when the units are different. A transformer has an efficiency, watts in–watts out. A lamp has an efficacy, watts in–light out.)

Kelvin (K)

The Kelvin (sometimes incorrectly called degree Kelvin) is measured from absolute zero; it is equivalent to a degree value in the Celsius scale plus 273.16. The color temperature of lamps is given in Kelvin. The lower the number, the warmer appearing is the light (more red content). The higher the number, the cooler appearing is the light (more blue content).

Color Rendering Index (CRI)

This value is often given for lamps so the user can have an idea of the color rendering probability. The CRI uses filament light as a base for 100 and the warm white fluorescent for 50. The CRI can be used only to compare lamps that have the same color temperature. (It is our recommendation that this system be avoided. The only sure way to determine whether a lamp will provide good color rendition is to see the material in the light produced by that lamp.)

INCANDESCENT LAMPS

The incandescent lamp has the lowest efficacy of the types listed in Table 14-1. However, incandescent lamps are popular and account for more than 50 percent of the lamps sold in the United States. This popularity is due largely to the low cost of incandescent lamps and luminaires (fixtures).

Federal energy legislation that became effective in 1994 removed several of the more commonly used lamps from the marketplace. No longer can the standard reflector lamps, like the R40, or standard PAR lamps, like the PAR38, be manufactured. The

Characteristics of Electric Lamps					
	Filament	**Fluorescent**	**Mercury**	**Metal Halide**	**HPS**	**LPS**
Lumens per watt	6 to 23	25 to 100	30 to 65	65 to 120	75 to 140	130 to 180
Wattage range	40 to 1500	4 to 215	40 to 1000	175 to 1500	35 to 1000	35 to 180
Life	750 to 8k	9k to 20k	16k to 24k	5k to 15k	20k to 24k	18k
Color temperature	2400 to 3100	2700 to 7500	3000 to 6000	3000 to 5000	2000	1700
Color Rendition Index	90 to 100	50 to 110	25 to 55	60 to 70	20 to 25	0
Potential for good color rendition	high	highest	fair	good	color discrimination	no color discrimination
Lamp cost	low	moderate	moderate	high	high	moderate
Operational cost	high	good	moderate	moderate	low	low

The values given above are generic for general service lamps. A survey of lamp manufacturer's catalogs should be made before specifying or purchasing any lamp.

Table 14-1

use of halogen and krypton-filled elliptical reflector (ER), and low-voltage lamps is encouraged. The ER lamp is designed primarily for use as a replacement and is not recommended for new installations. Typical of the recommended substitutions are a 50-watt PAR30 halogen lamp with a 65° spread to replace the 75-watt R30 reflector lamp and a 60- or 90-watt PAR halogen flood with a 30° spread to replace the 150-watt PAR38. A full list of substitutions is available at major hardware stores and electrical distributors.

Construction

The light-producing element in the incandescent lamp is a tungsten wire called the filament, Figure 14-2. This filament is supported in a glass envelope (bulb). The air is evacuated from the bulb and is replaced with an inert gas, such as argon. The filament is connected to the base by the lead-in wires. The base of the incandescent lamp supports the lamp and provides the connection means to the power source. The lamp base may be any one of the base styles shown in Figure 14-3.

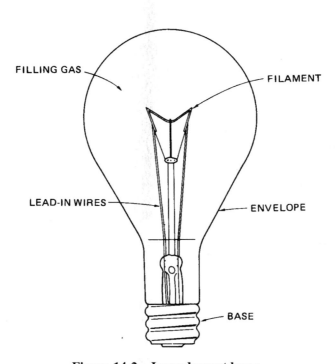

Figure 14-2   Incandescent lamp.

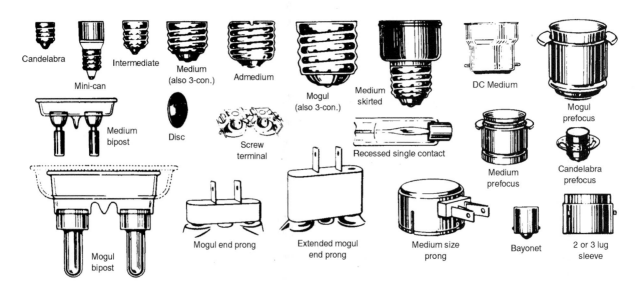

Figure 14-3   Incandescent lamp bases.

Characteristics

Incandescent lamps are classified according to the following characteristics:

Voltage Rating.   Incandescent lamps are available with many different voltage ratings. When installing lamps, the electrician should be sure that a lamp with the correct rating is selected because a small difference between the rating and the actual voltage has a great effect on the lamp life and lumen output, Figure 14-4.

Wattage.   Lamps are usually selected according to their wattage rating. This rating is an indication of the consumption of electrical energy but is not a true measure of light output. For example, at the rated voltage, a 60-watt lamp produces 840 lumens and a 300-watt lamp produces 6000 lumens; therefore, one 300-watt lamp produces more light than seven 60-watt lamps.

Shapes.   Figure 14-5 illustrates the common lamp configurations and their letter designations.

Size.   Lamp size is usually indicated in eighths of an inch and is the diameter of the lamp at the widest place. Thus, the lamp designation A19 means that the lamp has an arbitrary shape and is $1\frac{9}{8}$ in. or $2\frac{3}{8}$ in. (60.3 mm) in diameter, Figure 14-6.

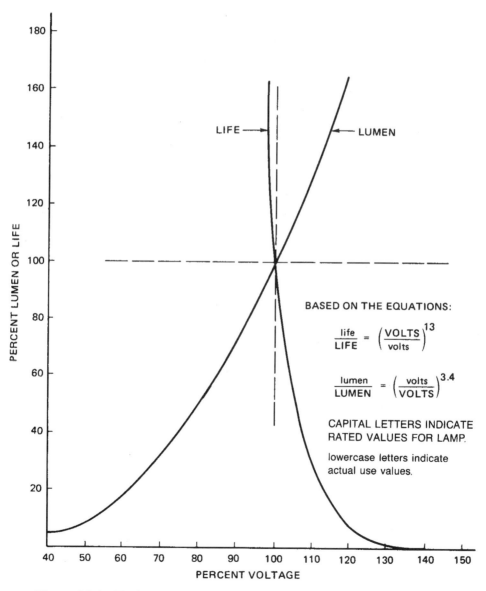

BASED ON THE EQUATIONS:

$$\frac{life}{LIFE} = \left(\frac{VOLTS}{volts}\right)^{13}$$

$$\frac{lumen}{LUMEN} = \left(\frac{volts}{VOLTS}\right)^{3.4}$$

CAPITAL LETTERS INDICATE RATED VALUES FOR LAMP.

lowercase letters indicate actual use values.

Figure 14-4   Typical operating characteristics of an incandescent lamp.

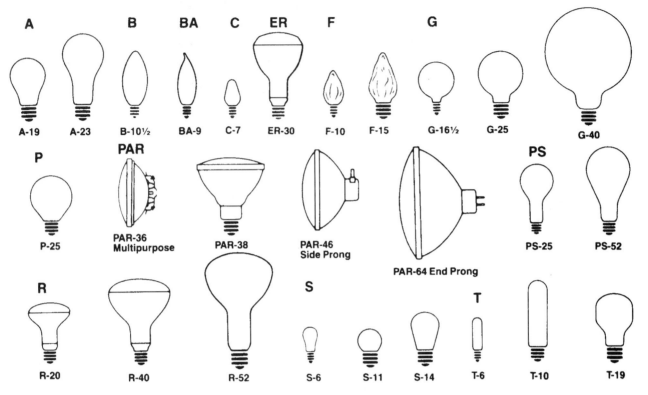

Figure 14-5   Incandescent lamps.

Operation. The light-producing filament in an incandescent lamp is a resistance element that is heated to a high temperature by an electric current. This filament is usually made of tungsten, which has a melting point of 3655 K. At this temperature, the tungsten filament produces light with an efficacy of 53 lumens per watt. However, to increase the life of the lamp, the operating temperature is lowered, which also means a lower efficacy. For example, if a 500-watt lamp filament is heated to a temperature of 3000 K, the resulting efficacy is 21 lumens per watt.

Catalog Designations. Catalog designations for incandescent lamps usually consist of the lamp wattage followed by the shape and ending with the diameter and other special designations as appropriate. Common examples are

60A19	60 watt, arbitrary shape, $^{19}/_8$ in. diameter
60PAR/HIR/WFL55°	60 watt, parabolic reflector, halogen, wide flood with 55° spread

LOW-VOLTAGE INCANDESCENT LAMPS

In recent years, low-voltage (usually 12-volt) incandescent lamps have become very popular for accent lighting. Many of these lamps are tungsten halogen lamps. They have a small source size, as shown in Figure 14-7. This feature allows precise control of the light beam. A popular size is the MR16, which has a reflector diameter of just 2 in.

— 19/8 in. (60.3 mm) —

Figure 14-6   An A19 lamp.

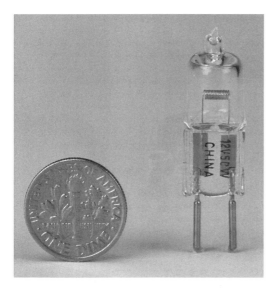

Figure 14-7   A 100-watt, 12-volt tungsten halogen lamp.

These lamps provide a whiter light than regular incandescent lamps. When dimming tungsten halogen lamps, a special dimmer is required because of the transformer that is installed to reduce the voltage. The dimmer is installed in the line voltage circuit supplying the transformer. Dimmed lamps will darken if they are not occasionally operated at full voltage.

Catalog Designations

Catalog designations for low-voltage incandescent lamps are similar to other incandescent lamps except for some special cases, such as:

MR16   mirrored reflector, 2 in. diameter

FLUORESCENT LAMPS

Luminaires (fixtures) using fluorescent lamps are considered to be electric-discharge lighting by the *NEC.*® Fluorescent lighting has the advantages of a high efficacy and long life.

Construction

A fluorescent lamp consists of a glass tube with an electrode and a base at each end, Figure 14-8. The inside of the tube is coated with a phosphor (a fluorescing material), the air is evacuated, and an inert gas plus a small quantity of mercury is released into the tube. The base styles for fluorescent lamps are shown in Figure 14-9.

Characteristics

Fluorescent lamps are classified according to type, length or wattage, shape, and color. See Figure 14-10.

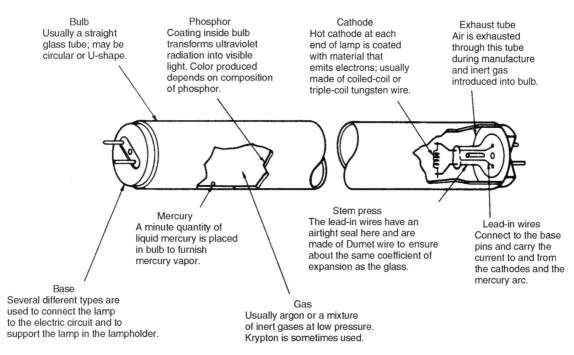

Bulb
Usually a straight glass tube; may be circular or U-shape.

Phosphor
Coating inside bulb transforms ultraviolet radiation into visible light. Color produced depends on composition of phosphor.

Cathode
Hot cathode at each end of lamp is coated with material that emits electrons; usually made of coiled-coil or triple-coil tungsten wire.

Exhaust tube
Air is exhausted through this tube during manufacture and inert gas introduced into bulb.

Mercury
A minute quantity of liquid mercury is placed in bulb to furnish mercury vapor.

Stem press
The lead-in wires have an airtight seal here and are made of Dumet wire to ensure about the same coefficient of expansion as the glass.

Lead-in wires
Connect to the base pins and carry the current to and from the cathodes and the mercury arc.

Base
Several different types are used to connect the lamp to the electric circuit and to support the lamp in the lampholder.

Gas
Usually argon or a mixture of inert gases at low pressure. Krypton is sometimes used.

Figure 14-8   Basic parts of a hot cathode fluorescent lamp.

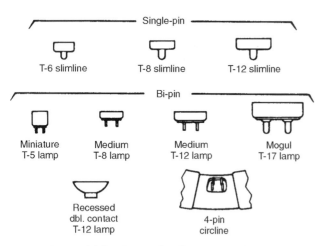

Figure 14-9   Bases for fluorescent lamps.

Type. The lamps may be preheat, rapid start, or instant start depending upon the ballast circuit.

Length or Wattage. Depending on the lamp type, either the length or the wattage is designated. For example, both the F40 preheat and the F48 instant start are 40-watt lamps, 48 in. (1.22 m) long. The bases of these two lamps are different, however.

Shapes. The fluorescent lamp usually has a straight, tubular shape. Exceptions are the circline lamp, which forms a complete circle, and the U-shaped lamp, which is an F40T12 lamp having a 180° bend in the center to fit a 2-ft (610 mm) long luminaire (fixture), and the PL lamp, which has two parallel tubes with a short connecting bridge at the ends opposite the base.

Catalog Designation and Color. An earlier reference was made to federal energy legislation; this legislation also impacted the manufacturing of fluorescent lamps. The popular cool white and warm white lamps are no longer available. Instead, a series of lamps with superior color are available as a substitution. A lamp with a 30 in its nomenclature indicates it produces light with a color temperature of 3000 K, which would provide good rendition to warm colors. Lamps with a 50 or 65 would provide good rendition to cool colors.

Common examples of recommended lamps are:

F40/41U/RS   A substitute for the
                  F40CW, it is a T12,
                  34-watt lamp producing
                  2900 initial lumen with
                  a CRI of 85 at a color
                  temperature of 4100 K.

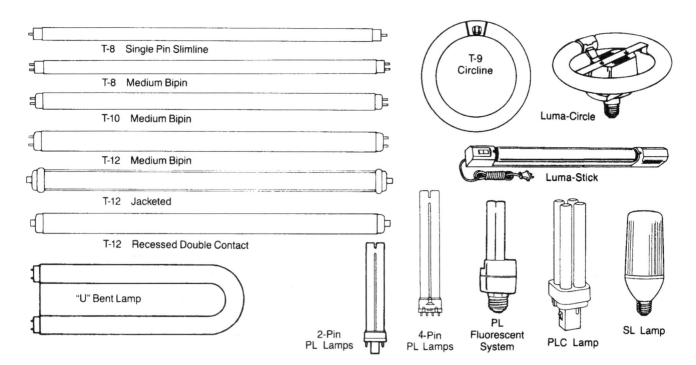

Figure 14-10   Fluorescent lamps.

F96T12/SPEC30/HO   A substitute for the F96T12/HO/CW, is a 95-watt lamp, producing 8350 initial lumen with a CRI of 70 at a color temperature of 4100 K.

PL*18/27   Compact fluorescent, 18 watts, 2700 K color temperature.

Operation

If a substance is exposed to such rays as ultraviolet and X-rays and emits light as a result, then the substance is said to be fluorescing. The inside of the fluorescent lamp is coated with a phosphor material, which serves as the light-emitting substance. When sufficient voltage is applied to the lamp electrodes, electrons are released. Some of these electrons travel between the electrodes to establish an electric discharge or arc through the mercury vapor in the lamp. As the electrons strike the mercury atoms, radiation is emitted by the atoms. This radiation is converted into visible light by the phosphor coating on the tube, Figure 14-11.

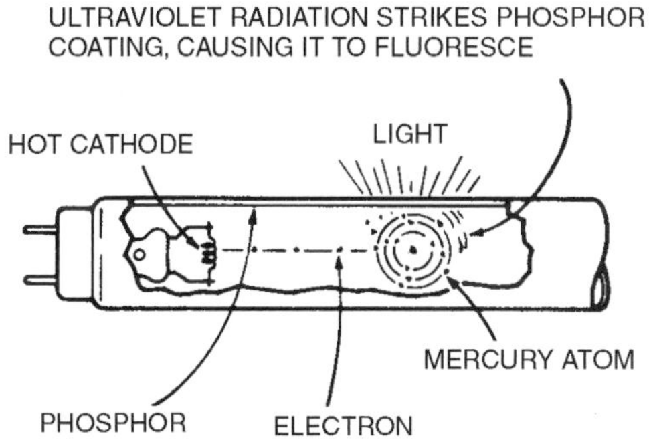

As the mercury atoms are ionized, the resistance of the gas is lowered. The resulting increase in current ionizes more atoms. If allowed to continue, this process will cause the lamp to destroy itself. As a result, the arc current must be limited. The standard method of limiting the arc current is to connect a reactance (ballast) in series with the lamp.

Ballasts and Ballast Circuits

Today, the focus is on energy conservation. This resulted in a major change in the design of ballasts.

The most popular type of ballast today is the electronic ballast. Figure 14-12 is a photo of a modern electronic fluorescent ballast. Electronic ballasts are quieter, are 25 to 40 percent more efficient, and weigh considerably less than older core-and-coil ballasts. They also increase the nonlinear load component of a circuit, so care must be exercised when sizing branch-circuit and feeder conductors.

The older magnetic ballasts consisted of an assembly of a core and coil, a capacitor, and a thermal protector installed in a metal case. When the assembled parts are placed in the case, it is filled with a potting compound to improve the heat dissipation and reduce ballast noise.

Ballasts are required for high-intensity discharge lamps.

Ballasts serve two basic functions:

1. to provide the proper voltage for starting.

2. to control the current during operation.

The installation requirements for ballasts are enumerated in *NEC® Article 410, Part XIII*.

Figure 14-11 How light is produced in a hot cathode fluorescent lamp.

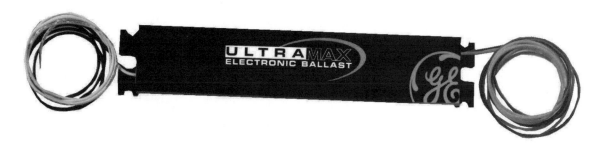

Figure 14-12 An electronic Class P ballast. (*Courtesy of* General Electric Corporation)

Preheat Circuit. The first fluorescent lamps developed were of the preheat type and required a starter in the circuit. This type of lamp is now obsolete and is seldom found except in smaller sizes, which may be used for items such as desk lamps. The starter serves as a switch and closes the circuit until the cathodes are hot enough. The starter then opens the circuit, and the lamp lights. The cathode temperature is maintained by the heat of the arc after the starter opens. Note in Figure 14-13 that the ballast is in series with the lamp and acts as a choke to limit the current through the lamp.

Rapid-Start Circuit. In the rapid-start circuit, the cathodes are heated continuously by a separate winding in the ballast, Figure 14-14, with the result that almost instantaneous starting is possible. This type of fluorescent lamp requires the installation of a continuous grounded metal strip within an inch of the lamp. The metal wiring channel or the reflector of the luminaire (fixture) can serve as this grounded strip. The standard rapid-start circuit operates with a lamp current of 430 mA. Two variations of the basic

circuit are available; the high-output (HO) circuit operates with a lamp current of 800 mA, and the very high-output (VHO) circuit has 1500 mA of current. Although high-current lamps are not as efficacious as the standard lamp, they do provide a greater concentration of light, thus reducing the required number of luminaires (fixtures).

Instant-Start Circuit. The lamp cathodes in the instant-start circuit are not preheated. Sufficient voltage is applied across the cathodes to create an instantaneous arc, Figure 14-15. As in the preheat circuit, the cathodes are heated during lamp operation by the arc. The instant-start lamps require single-pin bases, Figure 14-16, and are generally called *slimline lamps*. Bipin base fluorescent lamps are available, such as the 40-watt F40T12/CW/IS lamp. For this style of lamp, the pins are shorted together so that the lamp will not operate if it is mistakenly installed in a rapid-start circuit.

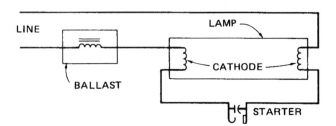

Figure 14-13   Preheat circuit.

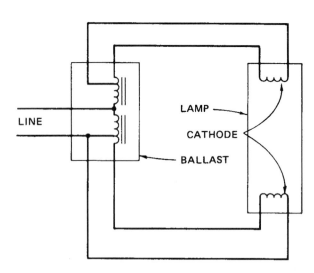

Figure 14-14   Rapid-start circuit.

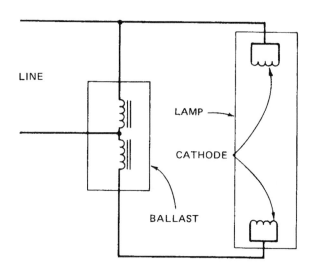

Figure 14-15   Instant-start circuit.

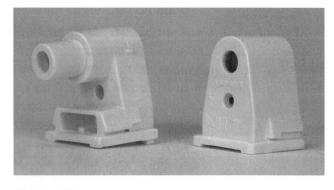

Figure 14-16   Single-pin base for instant-start fluorescent lamp.

Special Circuits

Most fluorescent lamps are operated by one of the circuits just covered: the preheat, rapid-start, or instant-start circuits. Variations of these circuits, however, are available for special applications.

Dimming Circuit.   The light output of a fluorescent lamp can be adjusted by maintaining a constant voltage on the cathodes and controlling the current passing through the lamp. Devices such as thyratrons, silicon-controlled rectifiers, and autotransformers provide this type of control. The manufacturer of the ballast should be consulted about the installation instructions for dimming circuits.

Flashing Circuit.   The burning life of a fluorescent lamp is greatly reduced if the lamp is turned on and off frequently. Special ballasts are available that maintain a constant voltage on the cathodes and interrupt the arc current to provide flashing.

High-Frequency Circuit.   Fluorescent lamps operate more efficiently at frequencies above 60 hertz. The gain in efficacy varies according to the lamp size and type. However, the gain in efficacy and the lower ballast cost generally are offset by the initial cost and maintenance of the equipment necessary to generate the higher frequency.

Direct-Current Circuit.   Fluorescent lamps can be operated on a dc power system if the proper ballasts are used. A ballast for this type of system contains a current-limiting resistor that provides an inductive kick to start the lamp.

Special Ballast Designation

Special ballasts are required for installations in cold areas such as outdoors. Generally, these ballasts are necessary for installations in temperatures lower than 50°F (10°C). These ballasts have a higher open-circuit voltage and are marked with the minimum temperature at which they will operate properly.

Class P Ballast.   The National Fire Protection Association reported that the second most frequent cause of electrical fires in the United States was the overheating of fluorescent ballasts. To lessen this hazard, Underwriters Laboratories, Inc., established a standard for a thermally protected ballast, which was designated as a Class P ballast. This type of ballast has an internal protective device that is sensitive to the ballast temperature. This device opens the circuit to the ballast if the average case temperature exceeds 194°F (90°C) when operated in a 77°F (25°C) ambient temperature. The requirements for the thermal protection of ballasts are established in *NEC® 410.73(E)*. After the ballast cools, the protective device is automatically reset. As a result, a fluorescent lamp with a Class P ballast is subject to intermittent off–on cycling when the ballast overheats.

It is possible for the internal thermal protector of a Class P ballast to fail. The failure can be in the welded-shut mode or it can be in the open mode. Because the welded-shut mode can result in overheating and possible fire, it is recommended that the ballast be protected with in-line fuses as indicated in Figure 14-17.

External fuses can be added for each ballast so that a faulty ballast can be isolated to prevent the shutdown of the entire circuit because of a single failure, Figure 14-17. The ballast manufacturer normally provides information on the fuse type and its ampere rating. The specifications for the commercial building require that all ballasts shall

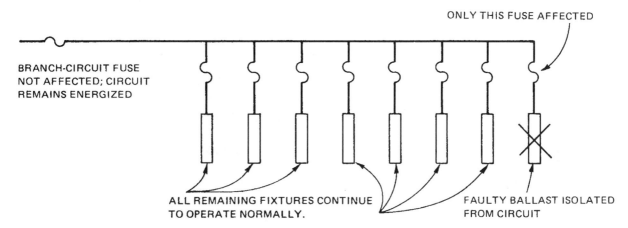

Figure 14-17   Ballast protected by fuse.

be individually fused; the fuse size and type are selected according to the ballast manufacturer's recommendations, Figure 14-18.

Sound Rating.   All ballasts emit a hum that is caused by magnetic vibrations in the ballast core. Ballasts are given a sound rating (from A to F) to indicate the severity of the hum. The quietest ballast has a rating of A. The need for a quiet ballast is determined by the ambient noise level of the location where the ballast is to be installed. For example, the additional cost of the A ballast is justified when it is to be installed in a doctor's waiting room. In the bakery work area, however, a ballast with a C rating is acceptable; in a factory, the noise of an F ballast probably will not be noticed.

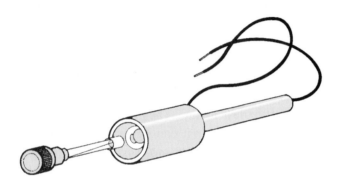

Figure 14-18   In-line fuseholder for ballast protection.

Power Factor.   The ballast limits the current through the lamp by providing a coil with a high reactance in series with the lamp. An inductive circuit of this type has an uncorrected power factor of from 40 percent to 60 percent; however, the power factor can be corrected to within 5 percent of unity by the addition of a capacitor. In any installation where there are to be a large number of ballasts, it is advisable to install ballasts with a high power factor.

Compact Fluorescent Lamps

One of the newest items is the compact fluorescent lamp. These lamps usually consist of a twin tube arrangement, which has a connecting bridge at the end of the tubes, Figure 14-19. Lamps are available that have two sets of tubes; these are called double twin tube or quad tube lamps, Figure 14-20.

This type of lamp has a rated life ten times that of an incandescent lamp and provides about three times the light per watt of power. A typical socket, along with a low-power-factor ballast, is shown in Figure 14-21. Some lamps have a medium screw

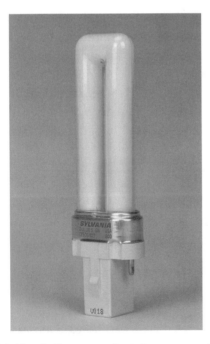

Figure 14-19   A 5-watt, twin-tube, compact fluorescent lamp.

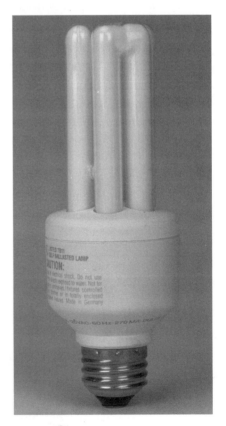

Figure 14-20   A 10-watt, quad-tube, compact fluorescent lamp.

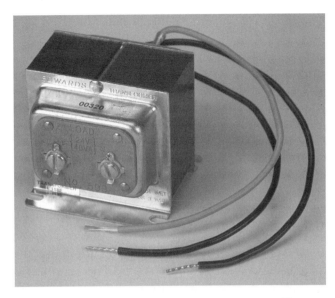

Figure 14-21   Socket and ballast for compact fluorescent lamp.

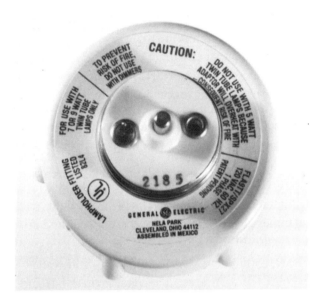

Figure 14-23   Medium-base, screw-in socket with retractable pin for positioning twin-tube compact fluorescent lamp.

base, with the ballast in the base, Figure 14-22. This type of lamp can directly replace an incandescent lamp. The bases for these lamps often have a retractable pin for the base connection, Figure 14-23. This allows the positioning of the lamp.

Class E Ballast.   Manufacturers of certain fluorescent ballasts are required to meet a Ballast Efficiency Factor (BEF). Initially, this applied only to F40T12 one- and two-lamp ballasts and to two-lamp ballasts for F96T12 and F96T12HO lamps. The BEF for the two-lamp F40T12, as used in the commercial building, requires that the lamp produce a minimum 84.8 lumens per watt efficacy.

HIGH-INTENSITY DISCHARGE (HID) LAMPS

Two of the lamps in this category, mercury and metal halide, are similar in that they use mercury as an element in the light-producing process. The other

HID lamp, high-pressure sodium, uses sodium in the light-producing process. In all three lamps, the light is produced in an arc tube that is enclosed in an outer glass bulb. This bulb serves to protect the arc tube from the elements and to protect the elements from the arc tube. An HID lamp will continue to give light after the bulb is broken, but it should be promptly removed from service. When the outer bulb is broken, people can be exposed to harmful ultraviolet radiation. Mercury and metal halide lamps produce a light with a strong blue content. The light from sodium lamps is orange in color.

Mercury Lamps

Many people consider the mercury lamp to be obsolete. They have the lowest efficacy of the HID family, which ranges from 30 lumens per watt for the smaller-wattage lamps to 65 lumens per watt for

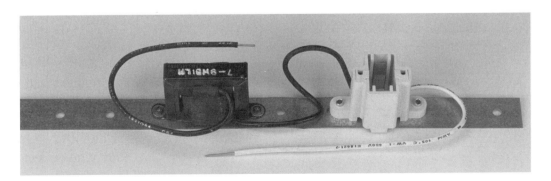

Figure 14-22   Medium-base socket with ballast for compact fluorescent lamp.

the larger-wattage lamps. Some of the positive features of mercury lamps are that they have a long life, with many lamps still functioning at 24,000 hours. With a clear bulb, they give a greenish light that makes them popular for landscape lighting.

Catalog Designations

Catalog designations vary considerably for HID lamps depending on the manufacturer. In general the designation for a mercury lamp will begin with an *H*, for metal halide lamps it will begin with an *M*, and either an *L* or *C* for high-pressure sodium lamps. See Figure 14-24.

MH250/C/U	metal halide, 250-watt, phosphor coated, base up (Philips)
MVR250/C/U	same as preceding (General Electric)
M250/C/U	same as preceding (Sylvania)
H38MP-100/DX	mercury, type 38 ballast, 100-watt, deluxe white (Philips)
HR100DX38/A23	same as preceding (General Electric)
H38AV-100/DX	same as preceding (Sylvania)

Metal Halide Lamps

The metal halide lamp has the disadvantage of a relatively short life and a rapid drop-off in light output as the lamp ages. Rated lamp life varies from 5000 hours to 15,000 hours. During this period the light output can be expected to drop by 30 percent or more. These lamps are considered to have good color-rendering characteristics and are often used in retail clothing and furniture stores. The lamp has a high efficacy rating, which ranges from 65 to 120 lumens per watt. Only a few styles are available with ratings below 175 watts. Operating position (horizontal or vertical) is critical with many of these lamps and should be checked before a lamp is installed.

High-Pressure Sodium (HPS) Lamps

This type of lamp is ideal for applications in warehouses, parking lots, and similar places where color

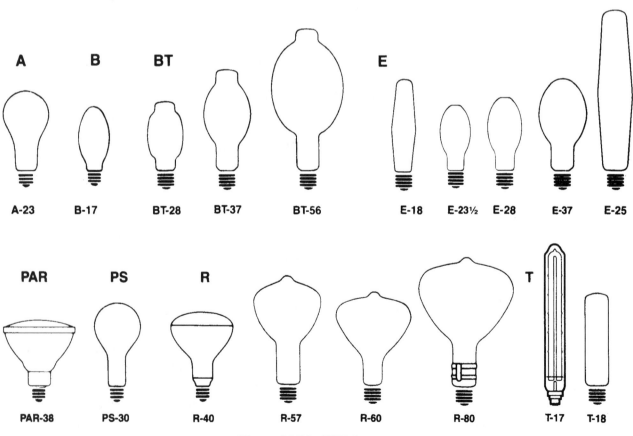

Figure 14-24   HID lamps.

recognition is necessary but high-quality color rendition is not required. The light output is orange in color. The lamp has a life rating equal to or better than that of any other HID lamp and has very stable light output over the life of the lamp. The efficacy is very good, ranging as high as 140 lumens per watt.

Low-Pressure Sodium (LPS) Lamps

This lamp has the highest efficacy of any of the lamps, ranging from 130 to greater than 180 lumens per watt. The light is monochromatic, containing energy in only a very narrow band of yellow. This lamp is usually used only in parking and storage areas where no color recognition is required. The lamp has a good life rating of 18,000 hours. It maintains a very constant light output throughout its life. The lamp is physically longer than HID lamps but generally is shorter than fluorescent lamps.

ENERGY SAVINGS

There is much concern about energy savings today. Manufacturers are required to design energy-saving products by certain dates. Consulting engineers are continually looking for these products as they design electrical systems. Electrical distributors stock these energy-saving products. It is up to you to search out these products. The following briefly discusses some of the more common energy-saving products.

Energy-Saving Ballasts

The market for magnetic (core-and-coil) ballasts is shrinking.

The National Appliance Energy Conservation Amendment of 1988, Public Law 100-357 prohibited manufacturers from producing ballasts having a power factor of less than 90 percent. Ballasts that meet or exceed the Federal standards for energy savings are marked with the letter E in a circle. Dimming ballasts and ballasts designed specifically for residential use were exempted.

Today's electronic ballasts are much more energy efficient than the older magnetic ballasts (core-and-coil). Energy-saving ballasts might cost more initially, but the payback is in the energy consumption saving over time.

The older fluorescent ballasts become very warm and might consume 14 to 16 watts, whereas an electronic ballast might consume 8 to 10 watts. Combined with energy-saving fluorescent lamps that use 32 or 34 watts instead of 40 watts, there is a considerable energy savings. You are buying light—not heat.

When installing fluorescent luminaires (fixtures), check the label on the ballast that shows the actual volt-amperes that the ballast and lamp will draw in combination. DO NOT attempt to use only lamp wattage when making load calculations, as this could lead to an overloaded branch-circuit. For example, a high-efficiency ballast might draw 42 volt-amperes total, whereas a older magnetic ballast might draw 102 volt-amperes.

The higher the power factor rating of a ballast, the more energy efficient the ballast. Look for a power factor rating in the mid- to high 90s.

Energy Saving Lamps

The National Energy Policy Act of 1992 enacted restrictions on lamps. In October 1995, the common 4-ft, 40-watt T12 linear medium bipin fluorescent lamp was eliminated. These discontinued lamps were directly replaced by energy-efficient 34-watt T12 lamps.

Some lamps may be designated F40T12/ES but draw 34 watts instead of 40 watts. The *ES* stands for "energy-saving." *ES* is a generic designation. Manufacturers may use other designations such as *SS* for SuperSaver, *EW* for Econ-o-Watt, *WM* for Watt-Miser, and others.

The older, high-wattage incandescent R30, R40, and PAR38 lamps were also discontinued and replaced with lower-wattage lamps.

T12 lamps are still found in 4-ft shop lights and square luminaires (fixtures) that use U-tube lamps. Most newer square luminaires (fixtures) have U-tube T8 lamps. In new commercial installations, the T8 lamp has taken over from the T12 lamp.

Energy-saving fluorescent lamps use up to 80 percent less energy than an incandescent lamp of similar brightness. Fluorescent lamps can last thirteen times or more longer than incandescent lamps.

It has been estimated that the total electric bill savings across the country will exceed $250 billion over the next fifteen years.

REVIEW QUESTIONS

Refer to the *National Electrical Code*® or the working drawings when necessary. Where applicable, responses should be written in complete sentences.

Describe the lamps specified by the following designations:

1. 100G40

   100 _____ , G _____ , 40 _____ .

2. 150A23

   150 _____ , A _____ , 23 _____ .

3. 120PAR/FL

   120 _____ , PAR _____ , FL _____ .

4. F40SPEC35/RS

   F40 _____ , SPEC _____ , 35 _____ , RS _____ .

5. MH70

   MH _____ , 70 _____ .

Give a brief definition for Questions 6 through 11.

6. Color rendering index _____

   _____

7. Efficacy _____

   _____

8. Footcandle _____

   _____

9. Illuminance _____

   _____

10. Lumen _____

    _____

11. Luminous intensity _____

    _____

For Questions 12 through 17, rank the lamp types according to their efficacy. Assume that the best efficacy is used in each case. Give a 1 to the lamp with the highest efficacy.

12. _____ Filament

13. _____ Fluorescent

14. _____ HPS

15. _____ LPS

16. _____ Mercury

17. _____ Metal halide

Define the light-producing element in each of the following three lamp types.

18. Fluorescent _____

_____

19. Incandescent _____

_____

20. Metal halide _____

_____

UNIT 15

Luminaires (Fixtures)

OBJECTIVES

After studying this unit, the student should be able to

- locate luminaires (fixtures) in a space.
- properly select and install luminaires (fixtures).
- discuss the attributes of several types of luminaires (fixtures).
- select and locate a luminaire (fixture) in a clothes closet.
- compute the lighting watts per square foot (square meter) for a space.

DEFINITIONS

In previous editions of the *National Electrical Code,®* the terms *luminaire* and *lighting fixture* have been used interchangeably. The Illuminating Engineering Society and the Institute of Electrical and Electronic Engineers recommend the use of *luminaire* because the term is an International Standard. Now that industry leaders are in agreement, *luminaire* is the preferred term.

The *National Electrical Code®* uses the term *luminaire* first, followed by the term *fixture* (lighting fixture) in parenthesis. In this text, both terms will be shown. There is a danger in using the term fixture alone, for in different contexts, it may have different meanings.

The *NEC®* defines a luminaire as *A complete lighting unit consisting of a lamp or lamps together with the parts designed to distribute the light, to position and protect the lamps and ballast (where applicable), and to connect the lamps to the power supply.**

The *NEC®* defines a lighting outlet as an outlet intended for the direct connection of a lampholder, a luminaire (lighting fixture), or a pendant cord terminating in a lampholder.

---

*Reprinted with permission from NFPA 70-2005.

INSTALLATION

The installation of luminaires (fixtures) is a frequent part of the work required for new building construction and for remodeling projects in which customers are upgrading the illumination of their facilities. To execute work of this sort, the electrician must know how to install luminaires (fixtures) and, in some cases, select the luminaires (fixtures).

The luminaires (fixtures) required for the commercial building are described in the specifications and indicated on the plans. The installation of luminaires (fixtures), lighting outlets, and supports is covered in *Article 410* of the *NEC.® Article 410* sets forth the basic requirements for the installation . . . commonly referred to as the "rough-in." *Article 410* also covers grounding, wiring, construction, lampholder installation and construction, lamps and ballasts, and special rules for recessed luminaires (fixtures).

The rough-in must be completed before the ceiling material can be installed. The exact location of the luminaires (fixtures) is rarely dimensioned on the electrical plans and, in some remodeling situations, there are no plans. In either case, the electrician should be able to rough in the outlet boxes and supports so that the luminaires (fixtures) will be correctly spaced in the area. If a single luminaire (fixture) is to be installed in an area, the center of the area can be found by drawing diagonals from each corner, Figure 15-1.

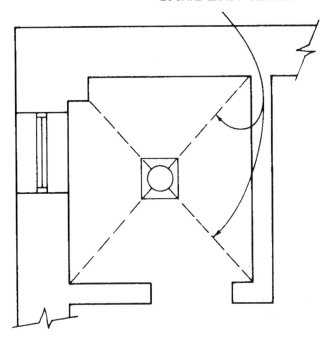

STRINGS STRETCHED
FROM CORNERS TO
LOCATE EXACT CENTER.

Figure 15-1   Luminaire (fixture) location.

When more than one luminaire (fixture) is required in an area, the procedure shown in Figure 15-2 should be followed. Uniform light distribution is achieved by spacing the luminaires (fixtures) so that the distances between the luminaires (fixtures) and between the luminaires (fixtures) and the walls follow these recommended spacing guides. The spacing ratios for specific luminaires (fixtures) are given in the data sheets published by each manufacturer. This number, usually between 0.5 and 1.5, when multiplied by the mounting height, gives the maximum distance that the luminaires (fixtures) may be separated and provides uniform illuminance on the work surface.

While doing luminaire (fixture) layouts, it may be necessary to apply good judgment in making the final decision. Consider the following situation where uniform illuminance is desired:

- The space is 20 ft (6 m) square.
- One-ft (300-mm) square ceiling tile will be installed.
- The floor-to-ceiling height is 9 ft (2.7 m).

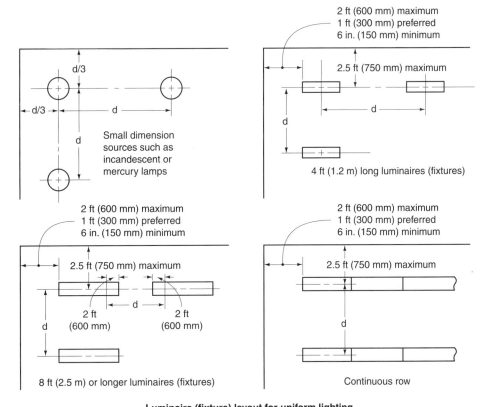

Luminaire (fixture) layout for uniform lighting
Note: d should not exceed spacing ratio times mounting height. Mounting height is from luminaire (fixture) to work plane for direct, semidirect, and general diffuse luminaires (fixtures) and from ceiling to work plane for semi-indirect and indirect luminaires (fixtures).

Figure 15-2   Luminaire (fixture) spacing.

- The work plane height is 2.5 ft (750 mm).
- The luminaire (fixture) is 1 ft (300 mm) square.
- The luminaire (fixture) spacing ratio is 1.1.

The work plane to luminaire (fixture) distance is:

$$9 \text{ ft} - 2.5 \text{ ft} = 6.5 \text{ ft}$$
$$(2.7 \text{ m} - 0.7 \text{ m} = 2 \text{ m})$$

The maximum center to center spacing distance is:

$$6.5 \text{ ft} \times 1.1 = 7.15 \text{ ft}$$
$$(2 \text{ m} \times 1.1 = 2.2 \text{ m})$$

If the ceiling is a 1-ft (300-mm) square grid, the maximum spacing becomes 7 ft (2.1 m).

The maximum distance from the center of the luminaire (fixture) to the wall should not be greater than:

$$\frac{7 \text{ ft}}{3} = 2 \text{ ft } 4 \text{ in.}$$

$$\frac{2.1 \text{ m}}{3} = 700 \text{ mm}$$

Working with the installer of the ceiling system, it is a given that the ceiling will be installed in a uniform layout. There will be either a full or a half grid at the wall. This limits the luminaire (fixture) placement at 1.5 ft (450 mm), 2 ft (600 mm), or 2.5 ft (750 mm) from center to wall.

Usually it will be decided, when there is an odd number of rows, that the center row of luminaires (fixtures) will be installed in the center of the space. This would place a 6-in. (150-mm) grid along each wall. For a uniform luminaire (fixture) layout, using nine luminaires (fixtures), they would be centered at 2 ft (600 mm), 10 ft (3 m), and 18 ft (5.5 m).

This layout exceeds the recommended spacing

distance. An alternative is to place the luminaires (fixtures) centered on 3 ft, 10 ft, and 17 ft (900 mm, 3 m, and 5.2 m), which would violate the recommended wall spacing distance. Another alternative is to exchange the luminaires (fixtures) for a style with a higher spacing ratio or to install additional luminaires (fixtures). Using sixteen luminaires (fixtures), they could be placed at 2 ft (600 mm), 7 ft (2.7 m), 13 ft (4 m), and 18 ft (5.5 m). This would not violate any of the recommendations. This type of compromise is a common occurrence in luminaire (fixture) layout where there is a grid ceiling system.

Supports

Both the lighting outlet and the luminaire (fixture) must be supported from a structural member of the building. To provide this support, a large variety of clamps and clips are available, Figure 15-3 and Figure 15-4. The selection of the type of support depends upon the way in which the building is constructed.

Surface-Mounted Luminaires (Fixtures)

For surface-mounted and pendant-hung luminaires (fixtures), the outlets and supports must be roughed in so that the luminaire (fixture) can be installed after the ceiling is finished. Support rods should be placed so that they extend about 1 in. (25 mm) below the finished ceiling. The support rod may be either a threaded rod or an unthreaded rod, Figure 15-5. If the luminaires (fixtures) are not available when the rough-in is necessary, luminaire (fixture) construction information should be requested from the manufacturer. The manufacturer can provide drawings that will indicate the exact dimensions of the mounting holes in the back of the luminaire (fixture), Figure 15-6.

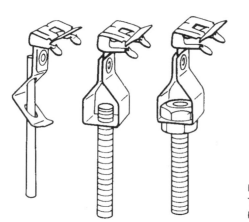

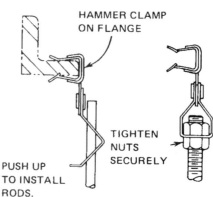

INSTALLATION INSTRUCTIONS

HAMMER CLAMP
ON FLANGE

TIGHTEN
NUTS
SECURELY

PUSH UP
TO INSTALL
RODS.

Figure 15-3   Rod hangers for connection to flange.

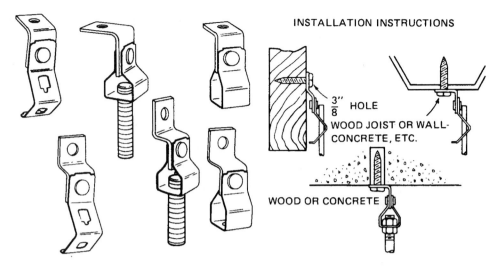

Figure 15-4   Rod hanger supports for flat surfaces.

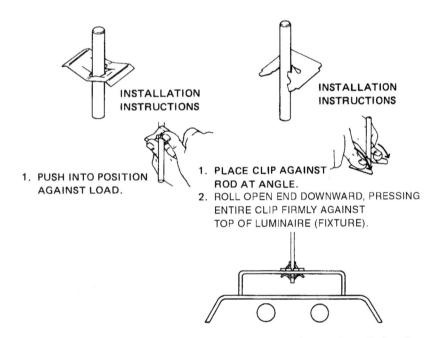

Figure 15-5   Luminaire (fixture) support using unthreaded rod.

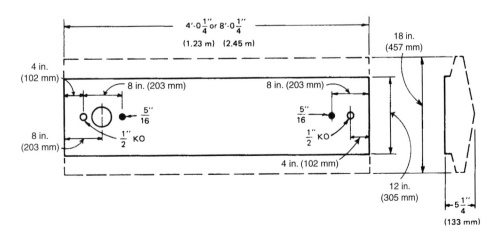

Figure 15-6   Luminaire (fixture) shop drawing indicating mounting holes.

Recessed Luminaires (Fixtures)

For recessed luminaires (fixtures), the outlet box will be located above the ceiling. This box must be accessible to the luminaire (fixture) opening by means of a metal raceway that is at least 18 in. (450 mm) long, but not more than 6 ft (1.8 m) long. Conductors suitable for the temperatures encountered are necessary. This information is marked on the luminaire (fixture). Branch-circuit conductors may be run directly into the junction box on listed prewired luminaires (fixtures) (*NEC® 410.67*), Figure 15-7.

Recessed luminaires (fixtures) are usually supported by adjustable rails (hangers) that are attached to the luminaires' (fixtures') rough-in box. These adjustable rails (hangers) are clearly shown in Figure 15-8A and Figure 15-8B.

LABELING

Always carefully read the label(s) on the luminaires (fixtures). The labels will provide, as appropriate, information similar to the following:

- wall mount only
- ceiling mount only
- maximum lamp wattage
- type of lamp
- access above ceiling required

- suitable for air-handling use
- for chain or hook suspension only
- suitable for operation in ambient not exceeding _____ °F (°C) temperature
- suitable for installation in poured concrete
- for installation in poured concrete only
- for line volt-amperes, multiply lamp wattage by _____
- suitable for use in suspended ceilings
- suitable for use in uninsulated ceilings
- suitable for use in insulated ceilings
- suitable for damp locations (such as bathrooms and under eaves)
- suitable for wet locations
- suitable for use as a raceway
- suitable for mounting on low-density cellulose fiberboard

The information on these labels, together with conformance with *NEC® Article 410* should result in a safe installation.

The Underwriters Laboratories *Electrical Construction Materials Directory* (known as the *Green Book*), the *General Information for Electrical Equipment Directory* (known as the *White Book*), and the luminaire (fixture) manufacturers' catalogs

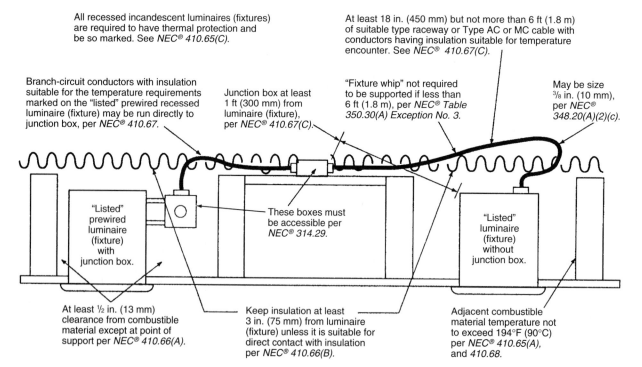

All recessed incandescent luminaires (fixtures) are required to have thermal protection and be so marked. See *NEC® 410.65(C)*.

At least 18 in. (450 mm) but not more than 6 ft (1.8 m) of suitable type raceway or Type AC or MC cable with conductors having insulation suitable for temperature encounter. See *NEC® 410.67(C)*.

Branch-circuit conductors with insulation suitable for the temperature requirements marked on the "listed" prewired recessed luminaire (fixture) may be run directly to junction box, per *NEC® 410.67*.

Junction box at least 1 ft (300 mm) from luminaire (fixture), per *NEC® 410.67(C)*.

"Fixture whip" not required to be supported if less than 6 ft (1.8 m), per *NEC® Table 350.30(A) Exception No. 3*.

May be size ⅜ in. (10 mm), per *NEC® 348.20(A)(2)(c)*.

"Listed" prewired luminaire (fixture) with junction box.

These boxes must be accessible per *NEC® 314.29*.

"Listed" luminaire (fixture) without junction box.

At least ½ in. (13 mm) clearance from combustible material except at point of support per *NEC® 410.66(A)*.

Keep insulation at least 3 in. (75 mm) from luminaire (fixture) unless it is suitable for direct contact with insulation per *NEC® 410.66(B)*.

Adjacent combustible material temperature not to exceed 194°F (90°C) per *NEC® 410.65(A)*, and *410.68*.

Figure 15-7   Requirements for installing recessed luminaires (fixtures).

and literature are also excellent sources of information on how to install luminaires (fixtures) properly. The Underwriters Laboratories lists, tests, and labels luminaires (fixtures) and publishes annual revisions to these directories.

Type IC Incandescent Recessed Luminaires (Fixtures)

Type IC luminaires (fixtures) are for installation in direct contact with insulation. They are recommended for homes. These luminaires (fixtures) operate under 194°F (90°C) when covered with insulation. Integral thermal protection is required, and they must be marked that thermal protection is provided. Watch the markings showing maximum lamp wattage and temperature rating for supply conductors. These luminaires (fixtures) may be installed in noninsulated locations. See Figure 15-8A.

Incandescent Recessed Luminaires (Fixtures) Not Marked IC

Recessed incandescent luminaires (fixtures) not marked Type IC and those marked for installing directly in poured concrete must not have insulation over the top of the luminaire (fixture). The insulation must be kept back at least 3 in. (75 mm) from the sides of the luminaire (fixture). Other combustible material must be kept at least 12 in. (300 mm) from the luminaire (fixture) except at the point of support. Watch for markings on the luminaire (fixture) for maximum clearances, maximum lamp wattage, and temperature ratings for the supply conductors. See Figure 15-8B.

Installing Recessed Luminaires (Fixtures)

The electrician must follow very carefully the requirements given in *NEC® Article 410 Parts XI and XII* for the installation and construction of recessed luminaires (fixtures). Of particular importance are the restrictions on conductor temperature ratings, luminaire (fixture) clearances from combustible materials, and maximum lamp wattage.

Recessed luminaires (fixtures) generate a considerable amount of heat within the enclosure and are a definite fire hazard if not wired and installed properly, Figure 15-7, Figure 15-8, and Figure 15-9. In addition, the excess heat will have an adverse effect on lamp and ballast life and performance.

If other than a Type IC luminaire (fixture) is being installed, the electrician should work closely with the installer of the insulation to be sure that the clearances, as required by the *NEC®* are followed.

To be absolutely sure that clearances are maintained, a box can be built around the recessed luminaire (fixture), Figure 15-10. The box will prevent the thermal insulation from coming into contact with the luminaire (fixture). Although this box ensures *NEC®* compliance, it does compromise the integrity of the ceiling thermal insulation. The heat loss around the luminaire (fixture) will be high, and may not be in conformance to EPA standards.

According to *NEC® 410.67(C)*, a tap conductor must be run from the luminaire (fixture) terminal connection to an outlet box. For this installation, the following conditions must be met:

- The conductor insulation must be suitable for the temperatures encountered.

Figure 15-8A   Incandescent recessed luminaire (fixture) marked IC.

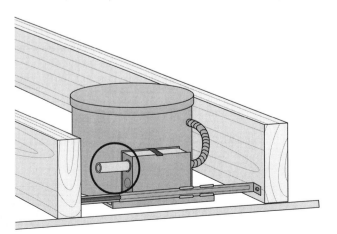

Figure 15-8B   Luminaire (fixture) not marked IC.

- The outlet box must be at least 1 ft (300 mm) from the luminaire (fixture).

- The tap conductor must be installed in a suitable raceway or cable such as Type AC or MC.

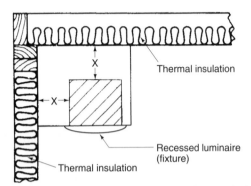

A luminaire (fixture) shall be installed to permit air circulation unless the luminaire (fixture) is identified for installation within thermal insulation. At least 3 in. (75 mm) of distance (see "**X**" in illustration) shall separate the luminaire (fixture) and the insulation.

Figure 15-9 Clearances for recessed luminaire (fixture).

Figure 15-10 A box built of fireproof material can be built around the luminaire (fixture) to prevent thermal insulation from coming into contact with the luminaire (fixture).

- The raceway shall be at least 18 in. (450 mm) but not more than 6 ft (1.8 m) long.

The branch-circuit conductors are run to the junction box. Here, they are connected to conductors from the luminaire (fixture). These wires have an insulation suitable for the temperature encountered at the lampholder. Locating the junction box at least 1 ft (300 mm) from the luminaire (fixture) ensures that the heat radiated from the luminaire (fixture) cannot overheat the wires in the junction box. The conductors must run through at least 18 in. (450 mm) of the metal raceway (but not to exceed 6 ft [1.8 m]) between the luminaire (fixture) and the junction box. Thus, any heat that is being conducted in the metal raceway will be dissipated considerably before reaching the junction box. Many recessed luminaires (fixtures) are factory equipped with a flexible metal raceway containing high-temperature wires that meet the requirements of *NEC® 410.67.*

Some recessed luminaires (fixtures) have a box mounted on the side of the luminaire (fixture) so that the branch-circuit conductors can be run directly into the box and then connected to the conductors entering the luminaire (fixture).

Additional wiring is unnecessary with these pre-wired luminaires (fixtures), Figure 15-11. It is important to note that *NEC® 410.11* states that branch-circuit wiring shall not be passed through an outlet box that is an integral part of an incandescent luminaire (fixture) unless the luminaire (fixture) is identified for through wiring.

A luminaire (fixture) may not be used as a raceway unless it is identified for that use, see *NEC® 410.31.*

If a recessed luminaire (fixture) is not prewired, the electrician must check the luminaire (fixture) for a label indicating what conductor insulation temperature rating is required.

Recessed luminaires (fixtures) are inserted into the rough-in opening and fastened in place by various devices. One type of support and fastening

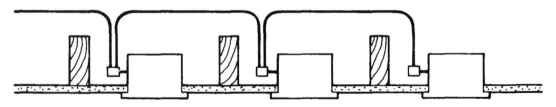

Figure 15-11 Installation permissible with prewired luminaire (fixture).

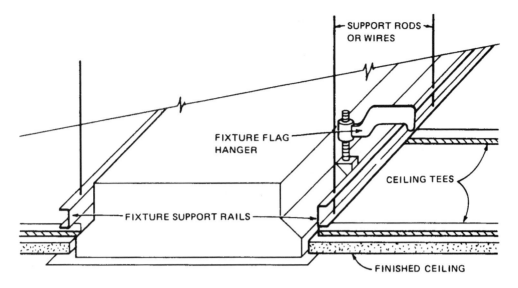

- SUPPORT RODS OR WIRES
- FIXTURE FLAG HANGER
- CEILING TEES
- FIXTURE SUPPORT RAILS
- FINISHED CEILING

Figure 15-12   Recessed luminaire (fixture) supported with flag hanger and support rails.

method for recessed luminaires (fixtures) is shown in Figure 15-12. The flag hanger remains against the luminaire (fixture) until the screw is turned. The flag then swings into position and hooks over the support rail as the screw is tightened.

Sloped Ceilings

Conventional recessed luminaires (fixtures) are suitable for installation in flat ceilings . . . not sloped ceilings. When selecting luminaires (fixtures) for installation in a sloped ceiling, be certain they are suitable for the particular slope such as a 2/12, 6/12, or 12/12 pitch. This information is available in the manufacturers' catalogs. Look for the marking "Sloped Ceilings."

Thermal Protection

All recessed incandescent luminaires (fixtures) must be equipped with factory-installed thermal protection, Figure 15-13. Marking on these luminaires (fixtures) must indicate this thermal protection. The ONLY exceptions to the *Code* rule are

1. if the luminaire (fixture) is identified for use and installed in poured concrete.

2. if the construction of the luminaire (fixture) is such that temperatures would be no greater than if the luminaires (fixtures) (lighting fixtures) had thermal protection. See *NEC® 410.65(C)* for the exact phrasing of the exceptions.

Thermal protection, Figure 15-13, will cycle on-off-on-off repeatedly until the heat problem is removed. These devices are factory-installed by the manufacturer of the luminaire (fixture).

Both incandescent and fluorescent recessed luminaires (fixtures) are marked with the temperature ratings required for the supply conductors if more than 140°F (60°C). In the case of fluorescent luminaires (fixtures), branch-circuit conductors within 3 in. of a ballast must have a temperature rating of at least 194°F (90°C). All fluorescent ballasts, including replacement ballasts, installed indoors must have integral thermal protection to be in compliance with *NEC® 410.73(E)(1)*. These thermally protected ballasts are designated as "Class P ballasts." This device will provide protection during normal operation, but it should not be expected to provide protection from the excessive heat that will be created by covering the luminaire (fixture) with insulation. For this reason, fluorescent luminaires (fixtures), just as incandescent luminaires (fixtures),

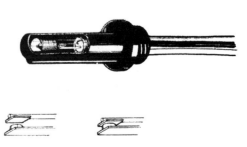

Figure 15-13   A thermal protector.

must have ½ in. (13 mm) clearance from combustible materials and 3 in. (75 mm) clearance from thermal insulation.

Because of the inherent risk of fire due to the heat problems associated with recessed luminaires (fixtures), always read the markings on the luminaire (fixture) and any instructions furnished with the luminaire (fixture) and consult *NEC® Article 410.*

Wiring

It was stated previously that it is very important to provide an exact rough-in for surface-mounted luminaires (fixtures). *NEC® 410.14(B)* emphasizes this fact by requiring that the outlet be accessible for inspection (without removing the surface-mounted luminaire). The installation of the outlet meets the requirements of *NEC® 410.14(B)* if the outlet is located so that the large opening in the back of the luminaire (fixture) can be placed over it, Figure 15-14.

To meet the requirements of *NEC® 410.31,* branch-circuit conductors with a rating of 194°F (90°C) may be used to connect luminaires (fixtures).

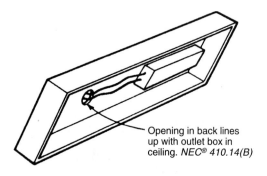

Opening in back lines up with outlet box in ceiling. *NEC® 410.14(B)*

Figure 15-14   Outlet installation for surface-mounted fluorescent luminaire (fixture).

However, these conductors must be of the single branch-circuit supplying the luminaires (fixtures). Therefore, all of the conductors of multiwire branch-circuits can be installed as long as these conductors are the grounded and ungrounded conductors of a single system. For example, when a building has a three-phase, four-wire supply, the neutral and three hot wires, one on each phase, may be installed in a luminaire (fixture) that has been approved as a raceway. This type of installation is suited to a long continuous row of luminaires (fixtures), Figure 15-15.

LOADING ALLOWANCE CALCULATIONS

The branch-circuits are usually determined when the luminaire (fixture) layout is completed. For incandescent luminaires (fixtures), the volt-ampere allowance for each luminaire (fixture) is based on the wattage rating of the luminaire (fixture). *NEC® 220.14(D)* stipulates that recessed incandescent luminaires (fixtures) be included at maximum rating. If an incandescent luminaire (fixture) is rated at 300 watts, it must be included at 300 watts even though a smaller lamp is to be installed. For fluorescent and HID luminaires (fixtures), the volt-ampere allowance is based on the rating of the ballast. In the past it was a general practice to estimate this value, usually on the high side. With recent advances in ballast manufacture and increased interest in reducing energy usage, the practice is to select a specific ballast type and to base the allowance on the operation of that ballast. The contractor is required to install an "as good as," or "better than," ballast. Several types of ballasts have been discussed in a previous unit. The design volt-ampere and the watts

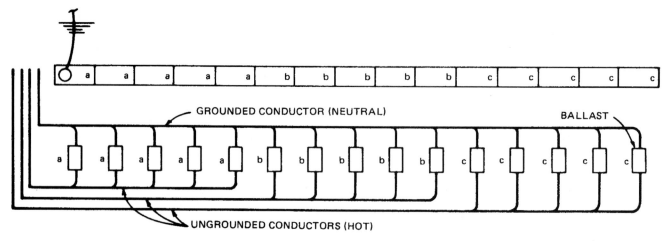

Figure 15-15   Multiwire circuit supplying continuous row of fluorescent luminaires (fixtures).

	Luminaire (Fixture)—Lamp Schedule
Style	Trade Size	Lamp	Ballast	Lens-Louver	Mounting	Description	VA	Watts
A	1½ by 4	2 F40/SPEC 30/RS	Energy saving	Wraparound	Surface		87	75
B	2 by 2	2 FB40/SPEC 30/6	Energy saving	Flat opal	Surface		87	75
C	4¼ by ¾	2 F40/SPEC 30/RS	Energy saving	Clear acrylic	Surface	Enclosed, gasketed	87	75
D	4 by ½	1 F48T12/CW	Standard	Luminous ceiling	Surface		74	64
E	¾	2 26W Quad T4	Compatible	Gold Alzak reflector	Recessed		144	74
F	4 by 2	4 F032/35K	Matching electronic	24 Cell lens-louver	Recessed		132	106
G	8 by 1	4 F032/35K	Matching electronic	24 Cell lens-louver	Recessed		132	106
H	8 by ¾	4 F40/SPEC 41/RS	Energy saving	Translucent acrylic	Surface		87	75
I	4 by ¾	2 F40/SPEC 41/RS	Energy saving	Translucent acrylic	Surface		87	75
J	1 by 1	1 150W, A21	NA	Fresnel lens	Recessed	IC rated	150	150
K	1⅓ by 1⅓	1 70W, HPS	Standard	Vandal resistant	Surface	With photo control	192	82
L	4 by 1	2 F40/CW	Energy saving	None	Surface or hung		87	75
M	1 by 1	12V 50W NFL	Transformer	Coilex baffle	Recessed	Adjustable spot	50	50
N	1¾ by ¼	1 60W, 120V	NA	None	Surface	Exposed lamp	60	60
O	4 by 1¾	3 F40/SPEC 41/RS	Two ballasts	Low brightness lens	Recessed		143	129
P	2 by 1¾	2 FB40/SPEC 30/6	One ballast	Low brightness lens	Recessed		87	75
Q	4 by 1¾	3 F40/SPEC 41/RS	Two ballasts	Small cell parabolic	Recessed		143	129

Table 15-1

for the luminaires (fixtures) selected for the commercial building are shown in Table 15-1. The following paragraphs discuss the styles of luminaires (fixtures) listed in Table 15-1 and scheduled to be used in the commercial building.

COMMERCIAL BUILDING LUMINAIRES (FIXTURES)

Style A

The Style A luminaire (fixture) is a popular fluorescent type that features a diffuser extending up the sides of the luminaire (fixture), Figure 15-16. These are often called wraparound lenses. This type of luminaire (fixture) provides good ceiling illumination, which is particularly important for low ceilings. The diffuser is usually available in either an acrylic or polystyrene material. Although a polystyrene diffuser is less expensive, it yellows as it ages and quickly becomes unattractive. Diffusers can be specified to be made with an acrylic material that does not yellow with age. The major disadvantage of the Style A luminaire (fixture) is the difficulty of locating replacements for yellow or broken diffusers. The ballast chosen for this luminaire (fixture) has an A sound rating and is of an energy-efficient style that, when operated with F40SPEC30/RS lamps, has a line current of 0.725 ampere at 120 volts (87 voltamperes) and a power rating of 75 watts. The Style A luminaire (fixture) is used in the beauty salon.

Style B

The Style B luminaire (fixture), another popular style, has solid metal sides, Figure 15-17. The bright sides of the Style A luminaire (fixture), when used on a low ceiling, may be objectionable to people

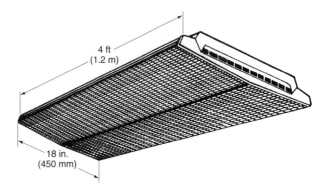

Figure 15-16   Style A fluorescent luminaire (fixture).

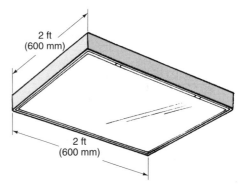

Figure 15-17   Style B fluorescent luminaire (fixture).

who must look at them for long periods of time. The Style B reduces this problem. The opal glass of the Style B luminaire (fixture) provides a soft diffusion of the light, but any flat diffuser may be used. Glass is easily cleaned and does not experience the same aging problems encountered by the plastic materials. The Style B was selected for the sales area of the bakery.

Style C

The Style C luminaire (fixture) is used where the possible contamination of the area is an important consideration, such as in the bakery, where food is prepared. Style C luminaires (fixtures) are suitable for use in bakeries, kitchens, slaughterhouses, meat markets, and food-packaging plants. The clear acrylic diffuser of this luminaire (fixture) protects the area in the event of a broken lamp. At the same time, the interior of the luminaire (fixture) is kept dry and free from dirt or dust, Figure 15-18. Raceways serving

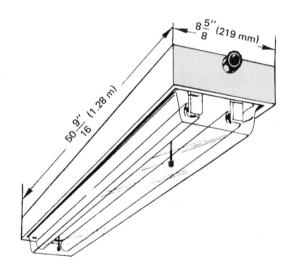

Figure 15-18   Style C fluorescent luminaire (fixture).

this luminaire (fixture) may enter from the top or from either end.

Style D

Luminous ceiling systems are used where a high level of diffuse light is required. The Style D system consists of fluorescent light strips (which may be ballasted for rapid-start, high-output, or very-high-output lamps) and a ceiling suspended 18 in. (450 mm) or more below the lamps, Figure 15-19. The ceiling may be of any translucent material, but usually consists of 2 ft by 2 ft (600 mm by 600 mm) or 2 ft by 4 ft (600 mm by 1200 mm) panels that are easily removed for cleaning and lamp replacement. An application of the Style D luminaire (fixture) is shown in the drugstore.

Style E

The Style E luminaire (fixture) is one of a type commonly referred to as downlights or recessed cans, Figure 15-20. It requires a ceiling opening of under 9 in. (225 mm) in diameter and has a height of 8 in. (200 mm). This fluorescent version uses two 26-watt quad-tube T4 lamps. A single unit will produce about 20 footcandles on a surface at a 9 ft (2.7 m) distance. Each lamp has a current rating of 0.6 ampere at 120 volts (72 volt-amperes) and uses 37 watts. The Style E luminaire (fixture) is used in several locations within the commercial building.

Style F

The Style F is a recessed fluorescent luminaire (fixture), Figure 15-21. It is commonly known as a troffer. This luminaire (fixture) is equipped with four F032/35K lamps and an electronic ballast. The lamps and the ballast are a matched set, or a system.

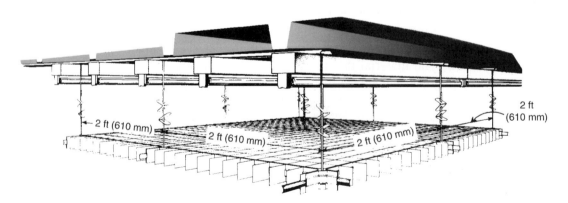

Figure 15-19   Style D fluorescent strip light.

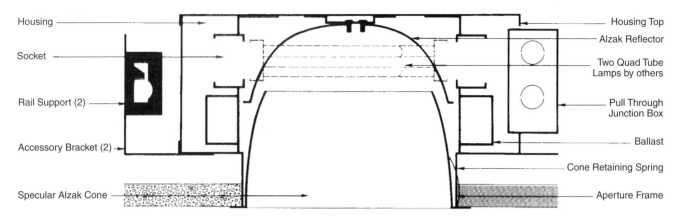

Figure 15-20   Style E fluorescent downlight.

Figure 15-21   Style F fluorescent luminaire (fixture).

This system has been developed to maximize the ratio of light output to watts input. Compared to standard F40T12/RS lamps and a magnetic ballast, this combination provides about 160 percent of light per watt. According to the manufacturer's data, each lamp in this luminaire (fixture) will initially produce 2900 lumen, with an efficacy of 110 lumens per watt.

This type of luminaire (fixture) is available with many styles of lenses and louvers. For this installation, a lens has been chosen that has the features of a lens but the appearance of a louver. A 2-in.-(50-mm)-deep

blade arrangement forms square light baffles on the surface of a lens. This gives a strong directionality to the light, concentrating the light downward to the work area. The Style F luminaire (fixture) and the similar Style G is used in the doctor's office. The lens-louver is also used in the insurance office.

Style G

The Style G luminaire (fixture) is identical to the Style F luminaire (fixture) except for the size. A single ballast is used with four F032/35K lamps. This is an identical luminaire (fixture) to the Style F except for the dimensions.

Style H

The Style H luminaire (fixture) shown in Figure 15-22 is designed to light corridors, the narrow areas between storage shelves, and other long, narrow spaces. Style H luminaires (fixtures) are available in slightly different forms from various manufacturers and are relatively inexpensive. This luminaire

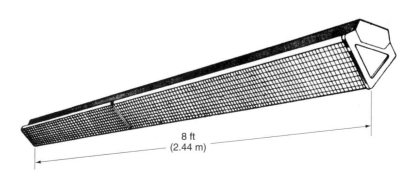

Figure 15-22   Styles H and I fluorescent luminaires (fixtures).

(fixture) has been chosen for use in the second-floor corridor. It has two F40 lamps placed in tandem (end to end). A two-lamp ballast serves the luminaire (fixture).

Style I

The Style I luminaire (fixture) is similar in appearance to Style H, but this luminaire (fixture) is equipped with four lamps and two ballasts. It is used in the drugstore where higher levels of illumination are desirable.

Style J

The Style J luminaire (fixture), Figure 15-23, is a recessed incandescent downlight luminaire (fixture) that can be installed in the opening left by removing or omitting a single 1 ft × 1 ft (300 mm) ceiling tile. This luminaire (fixture) is a Type IC, which has been approved to be covered with insula-

Figure 15-23   Style J incandescent luminaire (fixture).

tion. It has been specified to be equipped with a fresnel lens for wide distribution of the light and a 150-watt A21 lamp. The Style J downlight is used in exterior locations and in the doctor's office.

Style K

The Style K luminaire (fixture) is used on vertical exterior walls, where it provides a light pattern that covers a large area. The Style K luminaire (fixture) uses a high-intensity discharge (HID) source, such as metal halide or low-pressure sodium, Figure 15-24. This luminaire (fixture) provides reliable security illumination around a building. The Style K luminaire (fixture) is completely weatherproof and is equipped with a photoelectric cell to turn the light off during the day.

Style L

The Style L luminaire (fixture) is of open construction, as shown in Figure 15-25. This type is generally used in storage areas and other locations where it is not necessary to shield the lamps. It is often called an industrial luminaire (fixture). These luminaires (fixtures) are suspended from the ceiling on chains. In this type of installation, a rubber cord runs from the luminaire (fixture) to a receptacle outlet on the ceiling. The basement storage areas of both the bakery and the drugstore are good examples of the use of the Style L luminaire (fixture).

Style M

The Style M luminaire (fixture) is used to focus light on a specific object, Figure 15-26. This lumi-

Figure 15-24   Style K HID luminaire (fixture).

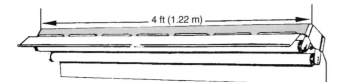

Figure 15-25   Style L fluorescent luminaire (fixture).

Figure 15-26   Style M MR–16 luminaire (fixture).

Figure 15-27   Style N linear incandescent luminaire (fixture).

naire (fixture) can be swiveled through 358 degrees laterally and 40 degrees from the zenith. This luminaire (fixture) uses an MR–6 lamp in sizes up to 50 watts and in a variety of light distribution patterns. The transformer is provided with the luminaire (fixture). At a distance of 4 ft (1.2 m), the 50-watt narrow flood lamp produces 26 footcandles. A dimmer switch, such as specified in the beauty salon, used to operate these luminaires (fixtures), must be approved for use with low-voltage illumination systems. The dimmer is connected in the supply to the transformers. An ordinary dimmer will, at best, provide sporadic operation. The beauty salon demonstrates an application of the Style M luminaire (fixture).

Style N

The Style N luminaire (fixture) is used in each of the rest rooms and is especially designed for installations where the lamp will be exposed to viewing, Figure 15-27. This luminaire (fixture) is installed on both sides of a mirror to provide excellent illumination of the face but to avoid the glare that can be a problem when conventional incandescent lamps are installed in a similar fashion. The lamp is a 20-in. (500-mm) linear incandescent with a 60-watt rating.

Style O

The Style O luminaire (fixture) is equipped with three F40T12/RS lamps and two ballasts. One of the ballasts supplies two lamps and the other one lamp. This arrangement provides for three levels of illuminance by using the single lamp in each luminaire (fixture), by using the pair of lamps in each luminaire (fixture), or by using all three lamps. In each arrangement, the illumination is uniformly distributed throughout the room. The luminaire (fixture) is fitted with a white louver and a lens. The louver, which was also used in the doctor's office, provides a high-quality light and has a pleasing effect on the appearance of the room. The Style O luminaire (fixture) is used in the staff and the reception areas of the insurance office.

Style P

The Style P luminaire (fixture) is similar to the Style O except that it is only 2 ft (600 mm) long and has a single ballast supplying two FB40/SPEC30/6 lamps. The FB40 is a lamp that was 4 ft (1.2 m) long but has been bent into a U-shape for use in luminaires (fixtures) that are only 2 ft (600 mm) long. This arrangement allows the use of the standard, rapid-start, two-lamp ballast. The Style P luminaires (fixtures) are used in the staff area of the insurance office to fill in areas of the room where the Style O was too long.

Style Q

The Style Q luminaire (fixture) is similar to the Style O except for the louvers. The lens specified for this room is a low-brightness lens especially designed for use in computer rooms. A high percentage of the light is directed downward to the horizontal surfaces. The Style Q was selected for the computer room as a minimum amount of light is produced on the computer monitor's vertical surface.

LUMINAIRES (FIXTURES) IN CLOTHES CLOSETS

Clothing, boxes, and other material normally stored in clothes closets are a potential fire hazard. These items may ignite on contact with the hot surface of an exposed light bulb. The bulb, in turn, may shatter and spray hot sparks and hot glass onto combustible materials. *NEC® 410.8* addresses the special requirements for installing in clothes closets.

It is significant to note that these rules cover ALL clothes closets . . . residential, commercial, and industrial. In Figure 15-28 and Figure 15-29 "A" represents the width of the storage space above the rod or 12 in. (300 mm), whichever is greater. "B"

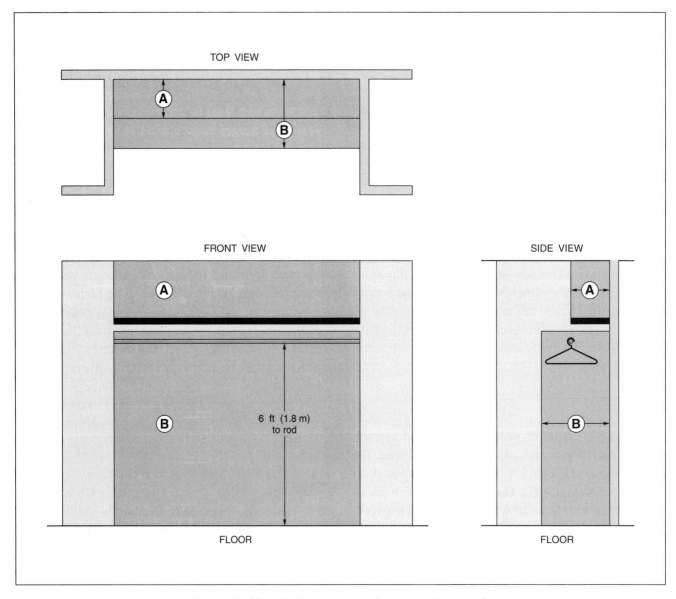

Figure 15-28   Clothes closet with one shelf and rod.

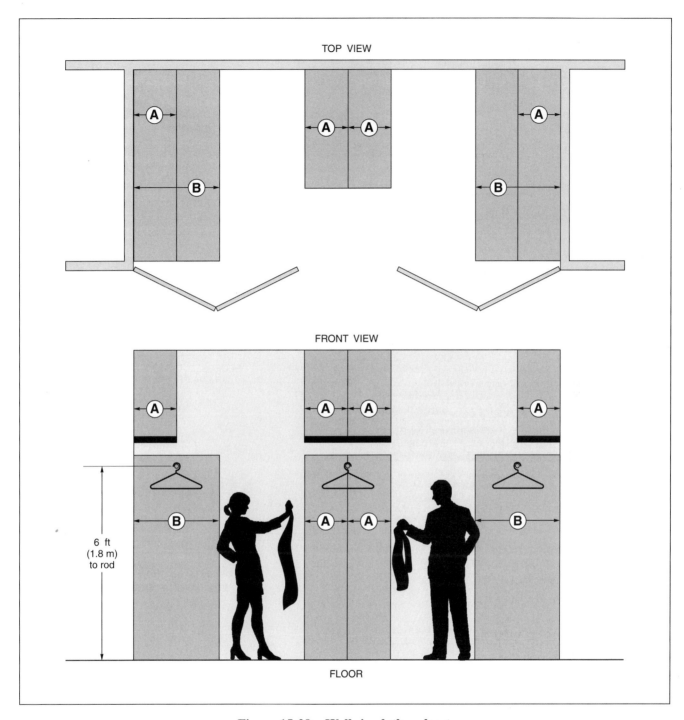

Figure 15-29   Walk-in clothes closet.

represents the depth of the storage space below the rod, which is 24 in. (600 mm) in depth and 6 ft (1.8 m), or to the highest rod, in height. See *NEC® 410.8(A)* and Figure 15-30, Figure 15-31, and Figure 15-32.

WATTS PER UNIT AREA CALCULATIONS

In some localities it is necessary that the illumination of a building comply with an energy code. These codes usually evaluate the illumination on the basis of the average watts per square foot (meter) of illumination load. Incandescent lamps are rated in watts, so it is just a matter of counting each type of luminaire (fixture), and then multiplying each by the appropriate lamp watts and adding the products. Fluorescent and HID luminaires (fixtures) are rated by the volt-amperes required for circuit design and the watts required to operate the luminaire (fixture).

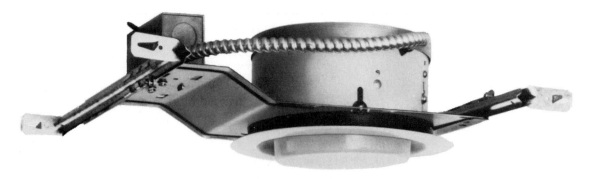

Figure 15-30   Recessed luminaire (fixture) with pull-chain.

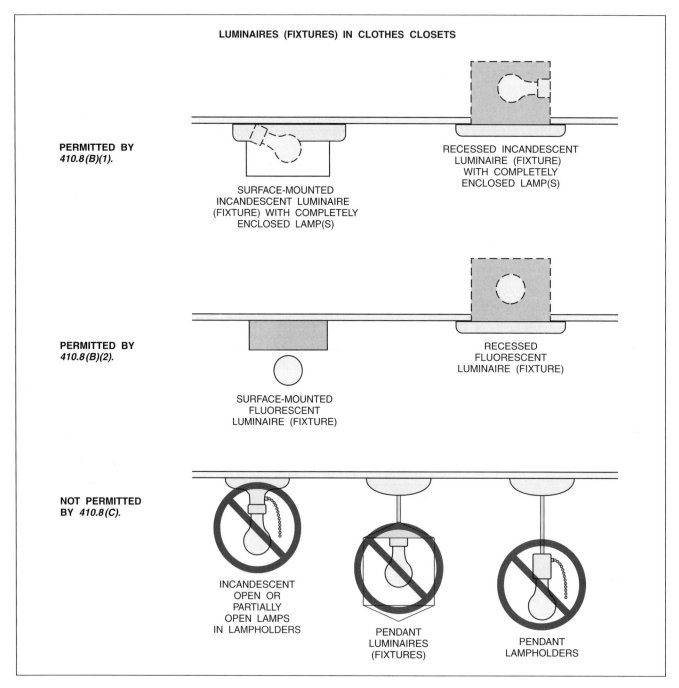

Figure 15-31   Luminaires (fixtures) in clothes closets.

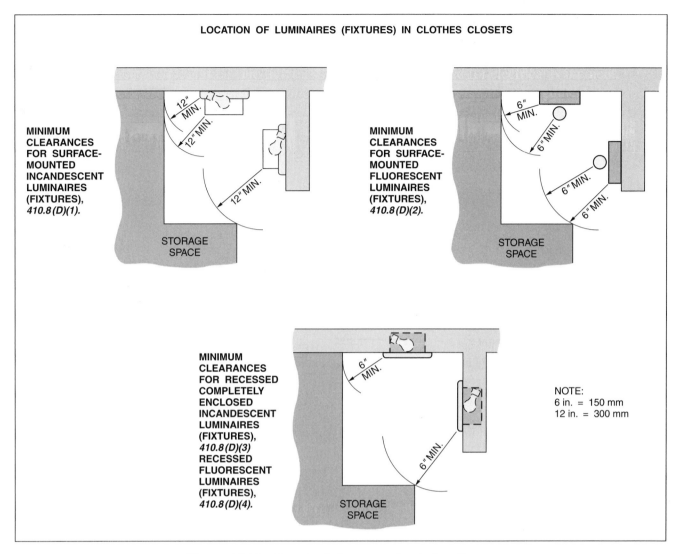

Figure 15-32   Required clearances in clothes closets.

The watt figure is never higher than the volt-ampere and is used to determine the watts per unit area for the building. The difference between watts and volt-amperes is determined by the quality of the ballast, the efficiency of the luminaire (fixture), and the efficacy of the lamp. Where the quality, the efficiency, and the efficacy are high, more light will be produced per watt, and thus fewer luminaires (fixtures) will be required to achieve the desired illumination at a given electrical load. After determining the total watts of load, this value is divided by the area to find the watts per unit area.

REVIEW QUESTIONS

Refer to the *National Electrical Code*® or the working drawings when necessary. Where applicable, responses should be written in complete sentences.

For Questions 1 and 2, six luminaires (fixtures), similar to Style J used in the commercial building, are to be installed in a room that is 12 ft (3.6 m) by 16 ft (4.8 m) with a 9-ft (2.8-m) floor-to-ceiling height. The spacing ratio for the luminaire (fixture) is 1:0.

1. The maximum distance that the luminaires (fixtures) can be separated and achieve uniform illuminance is _____ feet (_____ meters).

2. Three luminaires (fixtures) are to be installed in a row along the long side of the room. Draw a sketch indicating center-to-center and center-to-wall distances.

For Questions 3 through 6, two luminaires (fixtures), 8 ft (2.5 m) and 4 ft (1.2 m) with dimensions as shown in Figure 15-6, are to be installed in tandem (end to end). The end of the long luminaire (fixture) is to be 2 ft (600 mm) from the wall.

3. The center of the outlet box should be roughed in at _____ feet (_____ meters) from the wall.

4. The first support should be installed at _____ feet (_____ meters) from the wall.

5. The second support should be installed at _____ feet (_____ meters) from the wall.

6. The final support should be installed at _____ feet (_____ meters) from the wall.

Questions 7 and 8 concern the illumination in the beauty salon.

7. The loading allowance for the beauty salon illumination is _____ volt-amperes per square foot (_____ volt-amperes per square meter).

8. The volt-amperes per square foot (square meter) illumination loading for the beauty salon is _____ volt-amperes per square foot (_____ volt-amperes per square meter).

The following question pertains to illumination in general.

9. Of the luminaires (fixtures) selected for the commercial building, which style would be preferred for the illumination of shelving? _____

The remaining questions pertain to the installation of an illumination system in a clothes closet. Indicate compliance or violation and give the reason. In each case the luminaire (lighting fixture) is located on the ceiling with 10 in. (250 mm) clearance from the storage area.

10. A porcelain socket with a PL fluorescent lamp.   Compliance   Violation

11. A totally enclosed incandescent luminaire (fixture).   Compliance   Violation

12. A fluorescent strip luminaire (fixture) (bare lamp).   Compliance   Violation

UNIT 16

Emergency, Legally Required Standby, and Optional Standby Power Systems

OBJECTIVES

After studying this unit, the student should be able to

- select and size an emergency power system.
- install an emergency power system.
- differentiate the three types of special power systems addressed by the *NEC*.®

Many state and local codes require that equipment be installed in public buildings to ensure that electric power is provided automatically if the normal power source fails. The electrician should be aware of the special installation requirements of these systems.

There are three types of systems addressed by the *NEC*® that ensure that power is available in critical conditions:

1. *Article 700: Emergency Systems.* An emergency system consists of the circuits and equipment that supply, distribute, and control electricity for illumination, power, or both to required facilities when the normal electrical supply is interrupted. Emergency systems are installed where life safety is involved.

   Emergency systems will be legally required by a municipality, state, federal, other governmental agency, or other codes for specific types of buildings, such as health care facilities, hotels, theaters, and similar occupancies.

   The main purpose of an emergency system is to automatically come on to *supply, distribute, and control power and illumination essential for safety to human life* in the event that the normal

supply of power is interrupted. This would include such things as elevators, fire pumps, fire detection and alarm systems, communication systems, ventilation, lighting for the safe exiting of the building, and similar critical loads.

▶In order to minimize the possibility of power outages, overcurrent devices for the emergency system(s) shall be selectively coordinated with all supply-side overcurrent protective devices.◀ Achieving selectivity of overcurrent devices is covered in Unit 18.

2. *Article 701: Legally Required Standby Systems.* A legally required standby system automatically supplies power to selected loads such as illumination, power, or both. Loss of power to such loads is not life threatening. These systems will be required by a municipality, state, federal, or other governmental agency, or other codes.

   ▶In order to minimize the possibility of unwanted power outages, overcurrent devices for the legally required standby system(s) shall be selectively coordinated with all supply-side overcurrent protective devices.◀

3. *Article 702: Optional Standby Systems.* An optional standby system is just that—it is optional. Life safety is not an issue. No laws require it. Should loss of the normal supply of power occur, an optional standby system will re-establish power to selected loads until the normal power source comes back on. An optional standby system might be manual or automatic.

The preceding *NEC®* articles contain the how-to for on-the-job wiring requirements. The who, what, when, where, and why these systems are required are found in:

NFPA 99, *Standard for Health Care Facilities*

NFPA 101, *Life Safety Code*

NFPA 110, *Standard for Emergency and Standby Power Systems*

ANSI/IEEE Standard 446, *Recommended Practice for Emergency and Standby Power Systems for Industrial and Commercial Applications*

SOURCES OF POWER

When the need for emergency power is confined to a definite area, such as a stairway, and the power demand in this area is low, then self-contained battery-powered units are a convenient and efficient means of providing power. See *NEC® 700.12(A)*. In general, these units are wall-mounted and are connected to the normal source by permanent wiring methods. Under normal conditions, this regular source powers a regulated battery charger to keep the battery at full power. When the normal power fails, the circuit is completed automatically to one or more lamps that provide enough light to the area to permit its use, such as lighting a stairway sufficiently to allow people to leave the building. Battery-powered units are commonly used in stairwells, hallways, shopping centers, supermarkets, and other public structures.

If the power demand is high (excluding the operation of large motors), central battery powered systems are available. These systems usually operate at 32 volts and can service a large number of lamps.

SPECIAL SERVICE ARRANGEMENTS

NEC® 700.9(B) states that the wiring from an emergency source wiring shall be kept entirely independent of all other wiring and equipment and shall not enter the same raceway, cable, box, or cabinet with other wiring. Installing a separate service such as illustrated in Figure 16-1 meets the intent of this section.

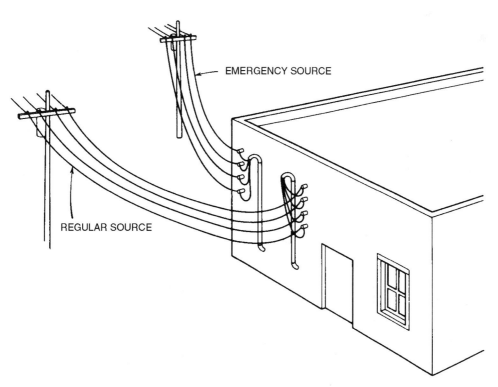

EMERGENCY SOURCE

REGULAR SOURCE

Figure 16-1 Separate services.

NEC® 701.10 allows the wiring of a legally required standby system to occupy the same raceways, cables, boxes, and cabinets with other general wiring. Furthermore, *NEC® 701.11(E)* permits the connection for this system to be made ahead of, but not within, the main service disconnect. Figure 16-2 illustrates this arrangement. The connections could be made in a junction box or to the service conductors outside of the building. Because the size of the tap conductors would be much smaller than the service conductors, cable limiters are installed at the point of the tap. These will limit the available fault-current to the withstand rating of the tap conductors, and the interrupting rating of the disconnect device.

EMERGENCY GENERATOR SOURCE

Generator sources may be used to supply emergency power. See *NEC® 700.12(B)*. The selection of such a source for a specific installation involves a consideration of the following factors:

- the engine type.
- the generator capacity.
- the load transfer controls.

A typical generator for emergency use is shown in Figure 16-3.

Engine Types and Fuels

The type of fuel to be used in the driving engine of a generator is an important consideration in the installation of the system. Fuels that may be used are LP gas, natural gas, gasoline, or diesel fuel. Factors affecting the selection of the fuel to be used include the availability of the fuel and local regulations governing the storage of the fuel. Natural gas and gasoline engines differ only in the method of supplying the fuel; therefore, in some installations, one of these fuels may be used as a standby alternative for the other fuel, Figure 16-4.

An emergency power source that uses gasoline, natural gas, or both usually has lower installation and operating costs than a diesel-powered source. However, the problems of fuel storage can be a deciding factor in the selection of the type of emergency generator. Gasoline is not only dangerous, but becomes stale after a relatively short period of time and thus cannot be stored for long periods. If natural gas is used, the British thermal unit (Btu) content must be greater than 1100 Btu per cubic foot. Diesel-powered generators require less maintenance and have a longer life. This type of diesel system is usually selected for installations having large power requirements because the initial costs closely approach the costs of systems using other fuel types.

Cooling

Smaller generator units are available with either air or liquid cooling systems. Units having a capacity greater than 15 kilowatts generally use liquid cooling. For air-cooled units, it is recommended that the heated air be exhausted to the outside.

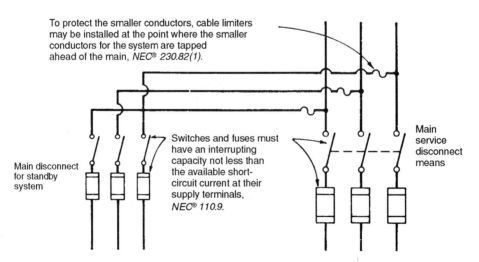

Figure 16-2   Connection ahead of service disconnect means for a legally required standby system.

Figure 16-3   Generator for emergency power supply.

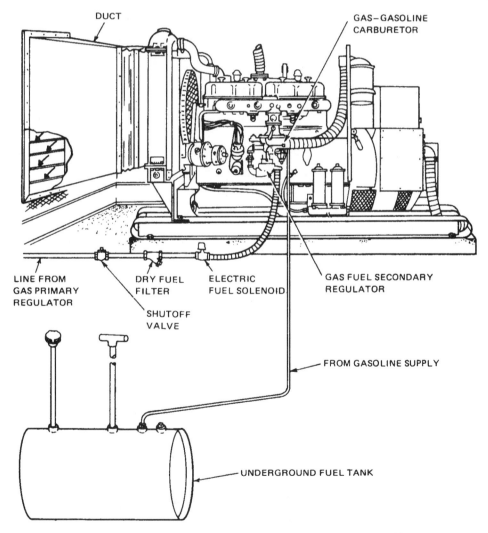

Figure 16-4   Generator powered by natural gas or gasoline.

In addition, a provision should be made to bring in fresh air so that the room where the generator is installed can be kept from becoming excessively hot. Typical installations of air-cooled emergency generator systems are shown in Figure 16-5 and Figure 16-6.

Generator Voltage Characteristics

Generators having any required voltage characteristic are available. A critical factor in the selection of a generator for a particular application is that it must have the same voltage output as the normal building supply system. For the commercial building covered in this text, the generator selected provides 208Y/120-volt, three-phase, 60-Hz power.

Capacity

It is an involved, but extremely important, task to determine the correct size of the engine-driven power system so that it has the minimum capacity necessary to supply the selected equipment. If the system is oversized, additional costs are involved in the installation, operation, and maintenance of the system. However, an undersized system may fail at the critical period when it is being relied upon to provide emergency power. To size the emergency system, it is necessary to determine initially all of the equipment to be supplied with emergency power. If motors are to be included in the emergency system, then it must be determined whether all of the motors can be started at the same time.

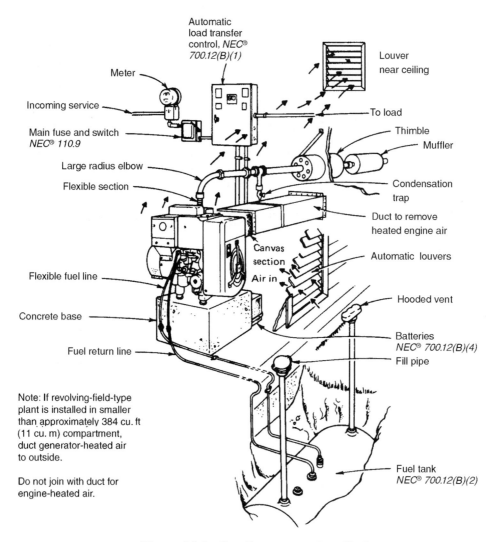

Figure 16-5   Small generator installation.

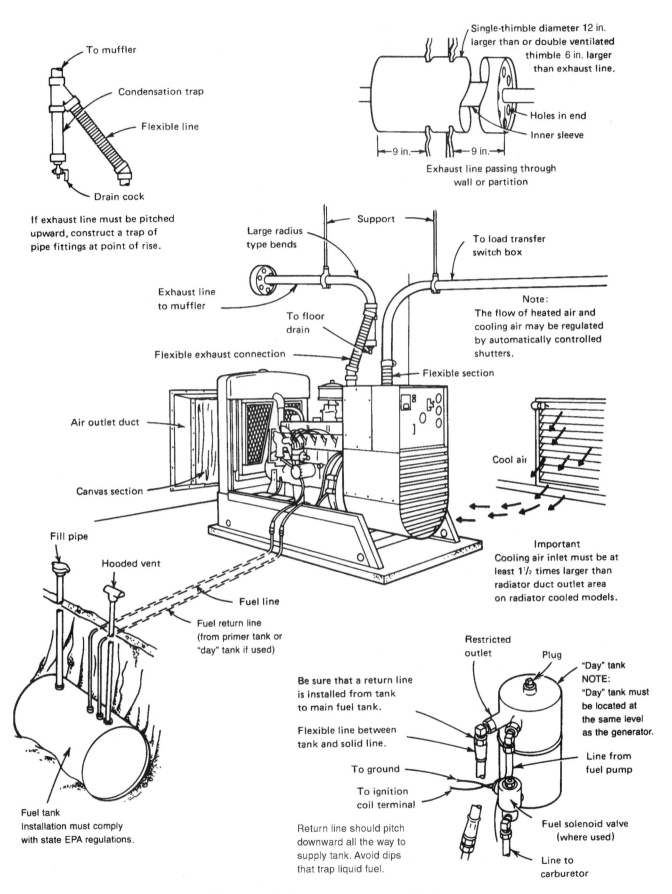

To muffler

Condensation trap

Flexible line

Drain cock

If exhaust line must be pitched
upward, construct a trap of
pipe fittings at point of rise.

Single-thimble diameter 12 in.
larger than or double ventilated
thimble 6 in. larger
than exhaust line.

Holes in end

Inner sleeve

9 in.   9 in.

Exhaust line passing through
wall or partition

Large radius
type bends

Support

To load transfer
switch box

Exhaust line
to muffler

To floor
drain

Note:
The flow of heated air and
cooling air may be regulated
by automatically controlled
shutters.

Flexible exhaust connection

Flexible section

Air outlet duct

Canvas section

Cool air

Important
Cooling air inlet must be at
least 1½ times larger than
radiator duct outlet area
on radiator cooled models.

Fill pipe

Hooded vent

Fuel line

Fuel return line
(from primer tank or
"day" tank if used)

Be sure that a return line
is installed from tank
to main fuel tank.

Flexible line between
tank and solid line.

To ground

To ignition
coil terminal

Return line should pitch
downward all the way to
supply tank. Avoid dips
that trap liquid fuel.

Fuel tank
Installation must comply
with state EPA regulations.

Restricted
outlet

Plug

"Day" tank
NOTE:
"Day" tank must
be located at
the same level
as the generator.

Line from
fuel pump

Fuel solenoid valve
(where used)

Line to
carburetor

Figure 16-6   Large generator installation.

This information is essential to ensure that the system has the capacity to provide the total starting inrush kilovolt-amperes required.

The starting kilovolt-amperes for a motor is equivalent to the locked rotor kilovolt-amperes of the motor. This value is determined by selecting the appropriate value from Table 16-1 and then multiplying this value by the horsepower. The locked-rotor kilovolt-amperes value is independent of the voltage characteristics of the motor. A 5-horsepower, code E motor requires a generator capacity of

$$5 \text{ hp} \times 4.99 \text{ kVA/hp} = 24.95 \text{ kVA}$$

If two motors are to be started at the same time, the emergency power system must have the capacity to provide the sum of the starting kilovolt-ampere values for the two motors.

For a single-phase motor rated at less than $\frac{1}{2}$ horsepower, Table 16-2 lists the approximate kilovolt-ampere values that may be used if exact information is not available. The power system for the commercial building must supply the following maximum kilovolt-amperes :

Five ⅙-hp, C-type motors		
5 × 1.85 kVA	=	9.25 kVA
One ½-hp, code L motor		
½ × 9.99 kVA	=	4.99 kVA
Lighting and receptacle load from emergency panel schedule	=	4.51 kVA
Total		18.75 kVA

Thus, the generator unit selected must be able to supply this maximum kilovolt-ampere load as well as the continuous kilovolt-ampere requirement.

Continuous kilovolt-ampere requirement:

Lighting and receptacle load	4.510 kVA
Motor load	3.763 kVA
Total	8.273 kVA

A check of manufacturers' data shows that a 12-kilovolt-ampere unit is available having a 20-kilovolt-ampere maximum rating for motor starting purposes and a 12-kilovolt-ampere continuous rating. A smaller generator may be installed if provisions, such as time delays on the contactors, can be made to prevent the motors from starting at the same

Table 430.7(B) Locked-Rotor Indicating Code Letters	
Code Letter	Kilovolt-Amperes per Horsepower with Locked Rotor
A	0–3.14
B	3.15–3.54
C	3.55–3.99
D	4.0–4.49
E	4.5–4.99
F	5.0–5.59
G	5.6–6.29
H	6.3–7.09
J	7.1–7.99
K	8.0–8.99
L	9.0–9.99
M	10.0–11.19
N	11.2–12.49
P	12.5–13.99
R	14.0–15.99
S	16.0–17.99
T	18.0–19.99
U	20.0–22.39
V	22.4 and up

Table 16-1

Reprinted with permission from NFPA 70-2005

Approximate kVA Values		
HP	Type	Locked-Rotor kVA
1/6	C	1.85
1/6	S	2.15
1/4	C	2.50
1/4	S	2.55
1/3	C	3.0
1/3	S	3.25
C = Capacitor-start motor		
S = Split-phase motor		

Table 16-2

time. See Table 16-2 for approximate locked-rotor kilovolt-amperes.

Derangement Signals

According to *NEC® 700.7*, a derangement signal, a signal device having both audible and visual alarms, shall be installed outside the generator room in a location where it can be readily and regularly observed. The purposes of a device such as the one shown in Figure 16-7 are to indicate any malfunction in the generator unit, any load on the system, or the correct operation of a battery charger.

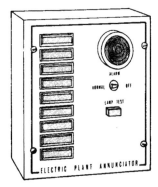

Figure 16-7 Audible and visual signal alarm annunciator.

Automatic Transfer Equipment

If the main power source fails, equipment must be provided to start the engine of the emergency generator and transfer the supply connection from the regular source to the emergency source, see *NEC® 700.6*. These operations can be accomplished by a control panel such as the one shown in Figure 16-8. A voltage-sensitive relay is connected to the main power source. This relay (transfer switch) activates the control cycle when the main source voltage

Figure 16-8   Automatic transfer control panel.

fails, Figure 16-9. The generator motor is started when the control cycle is activated. As soon as the motor reaches the correct speed, a set of contactors is energized to disconnect the load from its normal source and to connect it to the generator output.

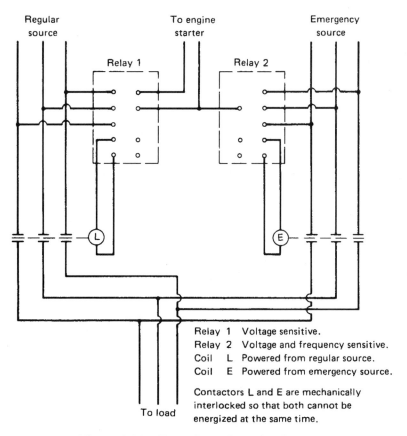

Relay 1 Voltage sensitive.
Relay 2 Voltage and frequency sensitive.
Coil   L Powered from regular source.
Coil   E Powered from emergency source.

Contactors L and E are mechanically interlocked so that both cannot be energized at the same time.

Figure 16-9   Transfer switch circuitry.

Wiring

The branch-circuit wiring of emergency systems must be separated from the standard system except for the special conditions noted. Figure 16-10 shows a typical branch-circuit installation for an emergency system. Key-operated switches, Figure 16-11, are installed to prevent unauthorized personnel from operating the lights.

Under certain conditions, emergency circuits are permitted to enter the same junction box as normal circuits, *NEC® 700.9(B)*. Figure 16-12 shows an exit light that contains two lamps: one lamp connected to

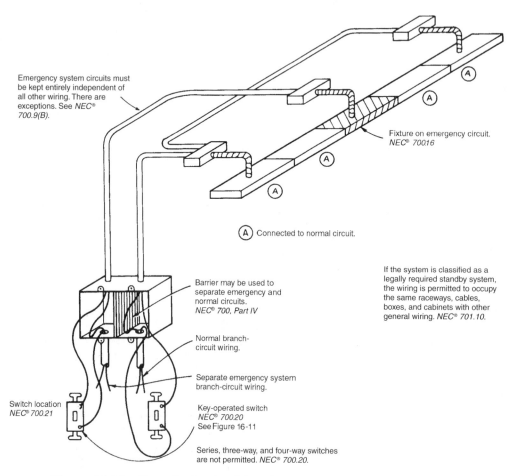

Emergency system circuits must be kept entirely independent of all other wiring. There are exceptions. See *NEC® 700.9(B)*.

Fixture on emergency circuit. *NEC® 70016*

Ⓐ Connected to normal circuit.

If the system is classified as a legally required standby system, the wiring is permitted to occupy the same raceways, cables, boxes, and cabinets with other general wiring. *NEC® 701.10*.

Barrier may be used to separate emergency and normal circuits. *NEC® 700, Part IV*

Normal branch-circuit wiring.

Separate emergency system branch-circuit wiring.

Switch location *NEC® 700.21*

Key-operated switch *NEC® 700.20* See Figure 16-11

Series, three-way, and four-way switches are not permitted. *NEC® 700.20*.

Figure 16-10   Branch-circuit wiring for emergency and other systems.

Figure 16-11   Key-operated switch for emergency lighting control.

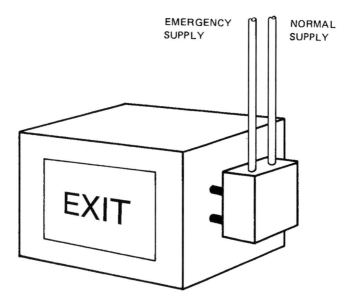

EMERGENCY
SUPPLY

NORMAL
SUPPLY

EXIT

Figure 16-12 Wiring for emergency and other systems.

the normal circuit and the other connected to the emergency circuit.

Sizing Generators When UPS Systems Are Involved

Uninterruptible power supply (UPS) systems are installed to provide "clean power" to critical loads such as computers. If the normal main power supply is lost, the UPS system kicks in with no interruption of power to the critical loads. However, loss of power also means that the standby generator will start. If the UPS system "sees" unstable power from a generator supply, it will reject that source of power, and the batteries will furnish the critical power until they are run down. The system will then shut down.

UPS systems loads are nonlinear loads. Non-linear loads are heavy in harmonic currents that cause overheating of the neutral conductor in a multiwire circuit where the neutral is common to more than one ungrounded phase conductor. Likewise, there will be overheating of the generator windings.

When nonlinear loads such as UPS systems are involved, leave the sizing to the experts! Generators must be oversized when supplying nonlinear loads to reduce heating and voltage waveform distortion. Contact the manufacturer of the generator for sizing recommendations. Different manufacturers have recommendations that vary from oversizing the gen-

erator 2 times, 3 times, and even as much as five times the UPS load. Call the manufacturer. Better to be safe than sorry!

Backfeed—A Real Hazard

Ask a power company's lineman about "back-feed." He knows all too well of the hazards of back-feed. There have been many injuries and deaths as a result of backfeed. It has been called "the Hidden Killer."

Think about it. The normal source of power might be coming through a step-down transformer; for example, a 2400/4160-volt primary stepped down to a 120/208-volt secondary. Now imagine that the normal source of power is lost. The generator kicks in. And guess what? A transformer works both ways: it can step down the voltage, or it can step up the voltage. Unless there is a properly connected transfer switch that disconnects the normal service conductors, the power from the generator will feed back into the service conductors . . . to the transformer . . . through the transformer onto the primary conductors. We now have a transformer that steps up the voltage from 120/208 volts to 2400/4160 volts. A lineman working on what he thinks is dead is now unknowingly working on a hot system. Other services connected to the secondary of that transformer will also be live. Always install listed transfer switches that perform their job of transferring power safely!

REVIEW QUESTIONS

Refer to the *National Electrical Code* or the working drawings when necessary. Where applicable, responses should be written in complete sentences.

There are three options for providing emergency electrical power to building systems. The options are listed here. Discuss the advantages and disadvantages of each.

1. A local source of power such as a battery or a motor-generator.

_____

_____

_____

_____

_____

2. A separate service.

_____

_____

_____

_____

_____

3. A connection ahead of the service main.

_____

_____

_____

_____

_____

For Questions 4 through 8, perform the following calculations and cite *NEC* references.

4. The minimum starting kilovolt-amperes for a 7½-horsepower, three-phase, 230-volt, code F motor is _____.

5. The maximum starting kilovolt-amperes for a 7½-horsepower, three-phase, 230-volt, code F motor is _____.

6. The minimum kilovolt-amperes of a generator that will be required to start, simultaneously, two 7½-horsepower, three-phase, 230-volt, code F motors is _____ .

7. The minimum kilovolt-amperes of a generator that will be required to start two selectively controlled, 7½-horsepower, three-phase, 230-volt, code F motors is

_____ .

8. List the special conditions for emergency circuit wiring that an electrician should know and those items that the occupant/owner should know.

UNIT 17

Overcurrent Protection: Fuses and Circuit Breakers

OBJECTIVES

After studying this unit, the student should be able to

- list and identify the types, classes, and ratings of fuses and circuit breakers.
- describe the operation of fuses and circuit breakers.
- develop an understanding of switch sizes, ratings, and requirements.
- define *interrupting rating*, short-circuit currents, I^2t, I_p, RMS, and current limitation.
- apply the *National Electrical Code®* to the selection and installation of overcurrent protective devices.
- use the time-current characteristics curves and peak let-through charts.

Overcurrent protection is one of the most important components of an electrical system. The overcurrent device opens an electrical circuit whenever an overload or short circuit occurs. Overcurrent devices in an electrical circuit may be compared to pressure relief valves on a boiler. If dangerously high pressures develop within a boiler, the pressure relief valve opens to relieve the high pressure. In a similar manner, the overcurrent device in an electrical system also acts as a "safety valve."

NEC® Article 240 sets forth the requirements for overcurrent protection. *NEC® 240.1 (FPN)* states that overcurrent protection for conductors and equipment is provided to open the circuit if the current reaches a value that will cause an excessive or dangerous temperature in conductors or conductor insulation. *NEC® 110.9* and *110.10* set forth requirements for interrupting rating and protection against fault current.

Two types of overcurrent protective devices are commonly used: fuses and circuit breakers. The Underwriters Laboratories, Inc. (UL) and the National Electrical Manufacturers Association

(NEMA) establish standards for the ratings, types, classifications, and testing procedures for fuses and circuit breakers.

As indicated in *NEC® 240.6*, the standard ampere ratings for fuses and nonadjustable circuit breakers are 15, 20, 25, 30, 35, 40, 45, 50, 60, 70, 80, 90, 100, 110, 125, 150, 175, 200, 225, 250, 300, 350, 400, 450, 500, 600, 700, 800, 1000, 1200, 1600, 2000, 2500, 3000, 4000, 5000, and 6000. Additional standard ratings for fuses are 1, 3, 6, 10, and 601 amperes.

Why does the *NEC®* list "standard ampere ratings"? There are many sections in the *Code* where permission is given to use *the next standard higher ampere rating overcurrent device*, or a requirement to use *a lower than standard ampere rating overcurrent device*. For example, in *NEC® 240.4(B)*, we find that if the ampacity of a conductor does not match a standard ampere rating fuse or breaker, then it is allowable to use the *next standard higher* for ratings of 800 ampere or less. Therefore, for a conductor that has an ampacity of 115 amperes, the *Code* permits the overcurrent protection to be sized at 125 amperes.

When protecting conductors where the over-current protection is greater than 800 amperes, the overcurrent protection must not be greater than the conductor's allowable ampacity. An example of this would be if three 500-kcmil copper conductors were installed in parallel. Each has an ampacity of 380 amperes. Therefore,

$$380 \times 3 = 1140 \text{ amperes.}$$

It would be a *Code* violation to protect these conductors with a 1200-ampere fuse. The correct thing to do is to round down to an ampere rating less than 1200 amperes. Fuse manufacturers provide 1100-ampere rated fuses.

DISCONNECT SWITCHES

Fused switches are available in ratings of 30, 60, 100, 200, 400, 600, 800, 1200, 1600, 2000, 2500, 3000, 5000, and 6000 amperes in both 250 and 600 volts. They are for use with copper conductors unless marked to indicate that the terminals are suitable for use with aluminum conductors. The switch's rating is based on 140°F (60°C) wire (14 AWG through 1 AWG) and 167°F (75°C) for wires 1/0 AWG and larger, unless otherwise marked. See *NEC® 110.14(C)*. Switches also may be equipped with ground-fault sensing devices. These will be clearly marked as such.

Switches may have labels that indicate their intended application, such as "Continuous Load Current Not To Exceed 80% Of The Rating Of The Fuses Employed In Other Than Motor Circuits."

- Switches intended for isolating use only are marked "For Isolation Use Only—Do Not Open Under Load."
- Switches that are suitable for use as service switches are marked "Suitable For Use As Service Equipment."
- When a switch is marked "Motor-Circuit Switch," it is for use only in motor circuits.
- Switches of higher quality may have markings such as: "Suitable For Use On A Circuit Capable Of Delivering Not More Than 100,000 [root mean squared] RMS Symmetrical Amperes, 600 Volts Maximum: Use Class T Fuses Having An Interrupting Rating Of No Less Than The Maximum Available Short Circuit Current Of The Circuit."

- Enclosed switches with horsepower ratings in addition to ampere ratings are suitable for use in motor circuits as well as for general use.
- Some switches have a dual-horsepower rating. The larger horsepower rating is applicable when using dual-element, time-delay fuses. Dual-element, time-delay fuses are discussed later in this unit (see Types of Fuses).
- Fusible bolted pressure contact switches are tested for use at 100 percent of their current rating, at 600 volts ac, and are marked for use on systems having available fault currents of 100,000, 150,000, and 200,000 RMS symmetrical amperes.

For more data regarding fused disconnect switches and fused power circuit devices (bolted pressure contact switches), see UL publications referred to in Unit 1 and the *National Electrical Code®* under "DISCONNECTING MEANS, SWITCHBOARDS, AND SWITCHES."

In the case of externally adjustable trip circuit breakers, the rating is considered to be the breaker's maximum trip setting, *NEC® 240.6*. The exceptions to this are:

1. if the breaker has a removable and sealable cover over the adjusting screws.
2. if it is located behind locked doors accessible only to qualified personnel.
3. if it is located behind bolted equipment enclosure doors.

In these cases, the adjustable setting is considered to be the breaker's ampere rating. This is an important consideration when selecting proper size phase and neutral conductors, equipment grounding conductors, overload relays in motor controllers, and other situations where sizing is based upon the rating or setting of the overcurrent protective device.

How High Should Disconnect Switches be Mounted?

▶*NEC® 240.24(A)* requires that overcurrent devices be readily accessible. This same *Code* reference requires that the center of the grip of a disconnect switch handle in its highest position be not more than 6 ft 7 in. (2.0 m) above the floor or working platform. The same rule applies to a circuit breaker.◀

FUSES AND CIRCUIT BREAKERS

For general applications, the voltage rating, the continuous current rating, the interrupting rating, and the speed of response are factors that must be considered when selecting the proper fuses and circuit breakers.

Voltage Rating

According to *NEC® 110.4* and *110.9*, the voltage rating of a fuse or circuit breaker shall be equal to or greater than the voltage of the circuit in which the fuse or circuit breaker is to be used. In the commercial building, the system voltage is 208Y/120-volt, wye connected. Therefore, fuses rated 250 volts or greater may be used. For a 480Y/277-volt, wye-connected system, fuses rated 600 volts would be used. Fuses and circuit breakers will generally work satisfactorily when used at any voltage less than the fuse's or circuit breaker's rating. For example, there would be no problem using a 600-volt fuse on a 208-volt system, although this practice may not be economically sound.

CAUTION: Fuses or circuit breakers that are marked ac should not be installed on dc circuits. Fuses and breakers that are suitable for use on dc circuits will be so marked.

Continuous Current Rating

The continuous current rating of a protective device is the amperes that the device can continuously carry without interrupting the circuit. The standard ratings are listed in *NEC® 240.6*. When applied to a circuit, the selection of the rating is usually based on the ampacity of the circuit conductors, although there are notable exceptions to this rule.

For example, referencing *NEC® Table 310.16*, we find that a size 8 AWG, Type THHN conductor has an allowable ampacity of 55 amperes. The proper overcurrent protection for this conductor would be 40-ampere if the terminals of the connected equipment are suitable for 140°F (60°C) conductors or 50-ampere if the terminals of the connected equipment are suitable for 167°F (75°C) conductors.

NEC® 240.4 lists several situations that permit the overcurrent protective device rating to exceed the ampacity of the conductor.

For example, the rating of a branch-circuit short-circuit and ground-fault protection fuse or circuit breaker on a motor circuit may be greater than the current rating of the conductors that supply the electric motor. This is permitted because properly sized motor overload protection (i.e., 125 percent of the motors' full-load current rating) will match the conductors' ampacity, which are also sized at 125 percent of the motors' full-load current rating.

Protection of Conductors

- When the overcurrent device is rated at 800 amperes or less:

  When the ampacity of a conductor does not match the ampere rating of a standard fuse or does not match the ampere rating of a standard circuit breaker that does not have an overload trip adjustment above its rating, the *Code* permits the use of the next higher standard ampere-rated fuse or circuit breaker. Adjustable trip circuit breakers were discussed previously. See *NEC® 240.4* and Figure 17-1.

- When the overcurrent device is rated above 800-ampere:

  When the ampere rating of a fuse or of a circuit breaker that does not have an overload trip adjustment exceeds 800 amperes, the conductor ampacity must be equal to or greater than the rating of the fuse or circuit breaker. The *Code* allows the use of fuses and circuit breakers that have ratings less than the standard sizes as listed in *NEC® 240.6*. Adjustable trip circuit breakers were discussed previously. See *NEC® 240.4* and Figure 17-2.

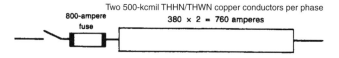

Figure 17-1 *NEC® 240.4(B)* **allows the use of an 800-ampere overcurrent device.**

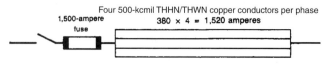

Figure 17-2 *NEC® 240.4(C)* **requires the use of a 1500-ampere overcurrent device.**

Interrupting Rating

Interrupting rating is defined in the *NEC®* as *the highest current at rated voltage that a device is intended to interrupt under standard test conditions.*

NEC® 110.9 states that *"Equipment intended to interrupt current at fault levels shall have an interrupting rating sufficient for the nominal circuit voltage and the current that is available at the line terminals of the equipment."**

Interrupting rating is a maximum rating, which is a measure of the fuse's or circuit breaker's ability to safely open an electrical circuit under fault conditions, such as an overload, short-circuit, or ground-fault.

Interrupting rating requirements are found in *NEC®* 240.60(C), 240.83(C), and 240.86 for series ratings.

Short-Circuit Current Rating

Electrical equipment may also be marked with a *short-circuit current rating*. A short-circuit current rating is determined by the manufacturer of the equipment in conformance to specific UL Standards. Electrical equipment shall not be connected to a system capable of delivering more fault current than the equipment's short-circuit current rating. Short-circuit current rating is referenced in *NEC®* 110.10.

The short-circuit current rating of all electrical equipment is based upon *how much current will flow* and *how long the current will flow.* Unit 17 and Unit 18 discuss the important issues of "Current Limitation" and "Peak Let-Through Current."

Overload currents have the following characteristics:

- They are greater than the normal current.

- They are contained within the normal conducting current path.

- If allowed to continue, they will cause overheating of the equipment, conductors, and the insulation of the conductors.

Short-circuit and ground-fault currents have the following characteristics:

- They flow outside the normal current path.

- They may be greater than the normal current.

- They may be less than the normal current.

---

*Reprinted with permission from NFPA 70-2005.

Short-circuit and ground-fault currents, which flow outside the normal current paths, can cause conductors to overheat. In addition, mechanical damage to equipment can occur as a result of the magnetic forces of the large current and arcing. Some short-circuit and ground-fault currents may be no larger than the normal load current, or they may be thousands of times larger than the normal load current.

The terms *interrupting rating* and *interrupting capacity* are used interchangeably in the electrical industry, but there is a subtle difference. Interrupting rating is the value of the test circuit capability. Interrupting capacity is the actual current that the contacts of a circuit breaker see when opening under fault conditions. You will also hear the term *AIC*, which means *amps interrupting capacity*. In Unit 18 you will learn how to calculate available short-circuit currents (fault currents) so that you will be able to properly apply breakers, fuses, and other electrical equipment for the available short-circuit current to which they might be subjected.

For example, the test circuit in Figure 17-3 is calibrated to deliver 14,000 amperes of fault current. The standard test circuit allows 4 ft (1.2 m) of conductor to connect between the test bus and the breaker's terminals. The standard test further allows 10 in. (25 mm) of conductor per pole to be used as the shorting wire. Therefore, when the impedance of the connecting conductors is taken into consideration,

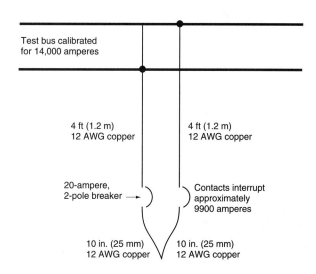

The label on this breaker is marked 14,000 amperes interrupting rating. Its actual interrupting capacity is approximately 9900 amperes. The connecting conductors have an ampacity that is the same as the breaker's ampere rating.

Figure 17-3   Typical laboratory test circuit for a molded-case circuit breaker.

the actual fault current that the contacts of the breaker see as they open is approximately 9900 amperes. The label on this circuit breaker is marked "14,000 amperes interrupting rating," yet its true interrupting capacity is 9900 amperes.

That is why it is important to adhere to the requirement of *NEC® 110.3(B)*, which states:

> *Installation and Use. Listed or labeled equipment shall be installed and used in accordance with any instructions included in the listing or labeling.**

Speed of Response

The time required for the fusible element of a fuse to open varies inversely with the magnitude of the current that flows through the fuse. In other words, as the current increases, the time required for the fuse to open decreases. The time-current characteristic of a fuse depends upon its rating and type. A circuit breaker also has a time-current characteristic. For the circuit breaker, however, there is a point at which the opening time cannot be reduced further due to the inertia of the moving parts within the breaker. The time-current characteristic of a fuse or circuit breaker should be selected to match the connected load of the circuit to be protected. Time-current characteristic curves are available from the manufacturers of fuses and circuit breakers.

TYPES OF FUSES

Dual-Element, Time-Delay Fuse

The dual-element, time-delay fuse, Figure 17-4, provides a time delay in the low-overload range to eliminate unnecessary opening of the circuit because of harmless overloads. This type of fuse is extremely responsive in opening on short circuits. A dual-element, time-delay fuse has one or more fusible elements (links) connected in parallel. Each of these links are then connected in series with an overload trigger assembly. The number of links depends upon the ampere rating of the fuse. The overload element(s) open when the current reaches a value of approximately 110 percent of the fuse's rating for a sustained period. The short-circuit element(s) opens when a short circuit or ground fault occurs. Dual-element, time-delay fuses carry

---

*Reprinted with permission from NFPA 70-2005.

500 percent of their rating for at least 10 seconds. For 250-volt fuses rated 30-ampere or less, the minimum time delay is 8 seconds. Figure 17-4 illustrates the fuse element's configuration and operation.

This fuse is a UL Class RK5 fuse. Classes of fuses are discussed later in this unit.

Dual-element fuses are ideal for use on motor circuits and other circuits having high-inrush characteristics. This type of fuse can be used as well for mains, feeders, subfeeders, and branch-circuits. Dual-element fuses may be used to provide backup protection for circuit breakers, bus duct, and other circuit components that lack an adequate interrupting rating, bracing, or withstand rating (see types of fuses covered later in this unit).

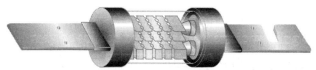

(A) The true dual-element fuse has distinct and separate overload and short-circuit elements.

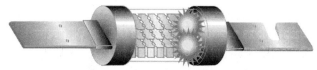

(B) Under sustained overload conditions, the trigger spring fractures the calibrated fusing alloy and releases the "connector."

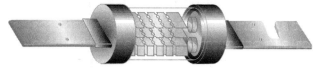

(C) The "open" dual-element fuse after opening under an overload condition.

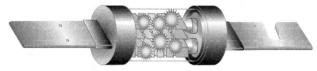

(D) Like the single-element fuse, a short-circuit current causes the restricted portions of the short-circuit elements to melt and arcing to burn back the resulting gaps until the arcs are suppressed by the arc-quenching material and increased arc resistance.

(E) The "open" dual-element fuse after opening under a short-circuit condition.

Figure 17-4   Cutaway views of a dual-element, time-delay fuse showing how the fuse operates on overload and short-circuit conditions. (*Courtesy* Cooper Bussmann, Inc.)

Using Fuses for Motor Overload Protection

Dual-element fuses can provide both motor overload protection and branch-circuit protection. Table 17-1 shows overload protection percentages per the *NEC*.

Sizing dual-element fuses slightly larger than the overload relay provides backup protection. This tight sizing of fuses means that if the overload relays fail to operate, the dual-element fuses will provide backup overload protection.

- Motor overload protection is based on the motor's nameplate rating in compliance with *NEC* 430.6(A)(2).

- Conductor size, disconnect switch size, branch-circuit, short-circuit, and ground-fault protection are based on the values in *NEC* Tables 430.247, 430.248, 430.249, and 430.250.

NEC 430.32 sets forth motor overload protection requirements. There are many questions to ask when searching for the proper level of motor overload protection. What is the motor's starting characteristics? Is the motor manually or automatically started? Is the motor continuous duty or intermittent duty? Does the motor have integral built-in protection? Is the motor impedance protected? Is the motor larger than one horsepower or is the motor one horsepower or less? Is the motor fed by a general-purpose branch-circuit? Is the motor cord- and plug-connected? Will the motor automatically restart if the overload trips? When these questions are answered, then look at *NEC* 430.32 for the level of overload protection needed.

Example:

What is the ampere rating of a dual-element, time-delay fuse that is to be installed to provide branch-circuit protection as well as overload protection for a motor with a 1.15 service factor and a nameplate current rating of 16 amperes?

$$16 \times 1.25 = 20 \text{ amperes}$$

Applying Fuses and Breakers on Motor Circuits

In recent years, energy-efficient motors have entered the scene. Energy-efficient motors have higher starting current characteristics than older style motors. These motors are referred to as Design B motors. Their high starting currents can cause nuisance opening of fuses and nuisance tripping of circuit breakers. Different manufacturers' same-size energy-efficient motors will have different starting current characteristics. Therefore, be careful when applying nontime-delay, fast-acting fuses on motor circuits. Because these fuses do not have the ability to handle the high inrush currents associated with energy-efficient motors, they cannot be sized at 125 percent of a motor's full-load ampere rating. They may have to be sized at 150 percent, 175 percent, 225 percent, or as high as 300 percent to 400 percent. Instant-trip circuit breakers will also experience nuisance tripping. Always check the time-current curves of fuses and breakers to make sure that they will handle the momentary motor starting inrush currents without nuisance opening or tripping. Additional information is found in Unit 7 of this text.

Using Fuses for Motor Branch-Circuit, Short-Circuit, and Ground-Fault Protection

NEC Table 430.52 shows that for a typical motor branch-circuit, short-circuit, and ground-fault protection, the maximum size permitted for dual-element fuses is based on a maximum of 175 percent of the full-load current of the motor. An exception to the 175 percent value is given in *NEC* 430.52, where permission is given to go as high as 225 percent of the motor full-load current if the lower value is not sufficient to allow the motor to start.

Example:

$$16 \times 1.75 = 28 \text{ amperes (maximum)}$$

The next higher standard rating is 30 amperes.

Motor Overload Protection		
	Maximum size in percentage of the motor's nameplate current rating *NEC* 430.32	If the size in left column does not result in a standard size, the absolute maximum is
Service factor not less than 1.15	125%	140%
Marked not more than 40°C temperature rise	125%	140%
All other motors	115%	130%

Table 17-1

If for some reason the 30-ampere fuse cannot handle the starting current, *NEC® 430.52(C)(1) Exception 2* allows the selection of a dual-element, time-delay fuse not to exceed:

$$16 \times 2.25 = 36 \text{ amperes (maximum)}$$

The next lower standard rating is 35 amperes.

Dual-Element, Time-Delay, Current-Limiting Fuses

Figure 17-5 shows dual-element, time-delay, current-limiting Class RK1 fuses that come under the UL Class RK1 category. They operate in the same manner as the Class RK5 dual-element, time-delay fuses previously discussed and illustrated in Figure 15-4. That is, they can handle five times their ampere rating for at least 10 seconds (8 seconds for 250-volt, 30-ampere or less ratings), which is desirable for motor and other high inrush loads. They open extremely fast under short-circuit or ground-fault situations.

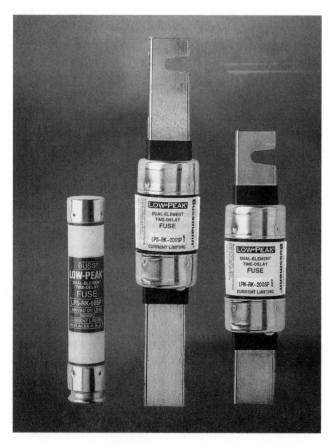

Figure 17-5　Dual-element, time-delay, current-limiting Class RK1 fuses. They provide time delay in the overload range and are fast-acting in the short-circuit range. (*Courtesy* Cooper Bussmann, Inc.)

Fast-Acting, Current-Limiting Fuses (Nontime-Delay)

The straight current-limiting fuse, shown in Figure 17-6, has an extremely fast response in both the low-overload and short-circuit ranges. When compared to other types of fuses, this type of fuse has the lowest energy let-through values. Current-limiting fuses are used to provide better protection to mains, feeders and subfeeders, circuit breakers, bus duct, switchboards, and other circuit components that lack an adequate interrupting rating, bracing, or withstand rating.

Current-limiting fuse elements can be made of silver or copper surrounded by a quartz sand arc-quenching filler.

A fast-acting, current-limiting fuse does not have the spring-loaded or loaded-link overload assembly found in dual-element fuses. To apply straight current-limiting fuses for motor circuits, refer to *NEC® Table 430.52* under the column "NON-TIME DELAY FUSE."

To be classified as "current limiting," *NEC® 240.2* states that when a fuse or circuit breaker is subjected to heavy (high-magnitude) fault currents, the fuse or breaker must reduce the fault current flowing into the circuit to a value less than the fault current that could have flowed into the circuit had there been no fuse or breaker in the circuit.

When used on motor circuits or other circuits having high current-inrush characteristics, the current-limiting nontime-delay fuses must be sized at a much higher rating than the actual load. That is, for a motor with a full-load current rating of 10 amperes, a 30- or 40-ampere fast-acting, current-limiting fuse may be required to start the motor. In this case, the fuse is considered to be the motor branch-circuit short-circuit protection as required in *NEC® Table 430.52* (Table 17-2).

Figure 17-6　Cutaway view of a current-limiting, fast-acting, single-element fuse. (*Courtesy* Cooper Bussmann, Inc.)

Table 430.52 Maximum Rating or Setting of Motor Branch-Circuit Short-Circuit and Ground-Fault Protective Devices

	Percentage of Full-Load Current			
Type of Motor	Nontime Delay Fuse[1]	Dual Element (Time-Delay) Fuse[1]	Instantaneous Trip Breaker	Inverse Time Breaker[2]
Single-phase motors	300	175	800	250
AC polyphase motors other than wound-rotor				
Squirrel cage — other than Design B energy-efficient	300	175	800	250
Design B energy-efficient	300	175	1100	250
Synchronous[3]	300	175	800	250
Wound rotor	150	150	800	150
Direct current (constant voltage)	150	150	250	150

Note: For certain exceptions to the values specified, see 430.54.

[1]The values in the Nontime Delay Fuse column apply to Time-Delay Class CC fuses.

[2]The values given in the last column also cover the ratings of nonadjustable inverse time types of circuit breakers that may be modified as in 430.52(C), Exception No. 1 and No. 2.

[3]Synchronous motors of the low-torque, low-speed type (usually 450 rpm or lower), such as are used to drive reciprocating compressors, pumps, and so forth, that start unloaded, do not require a fuse rating or circuit-breaker setting in excess of 200 percent of full-load current.

Table 17-2

Reprinted with permission from NFPA 70-2005

Types of Cartridge Fuses

The requirements governing cartridge fuses are contained in *NEC® Article 240, Part VI.* According to the *Code*, all cartridge fuses must be marked to show:

- ampere rating.
- voltage rating.
- interrupting rating when greater than 10,000 amperes.
- current-limiting type, if applicable.
- trade name or name of manufacturer.

The UL and CSA standards list Classes G, H, J, K1, K5, K9, L, RK1, RK5, T, or CC.

All fuses carrying the UL/CSA Class listings (Class G, H, J, K1, K5, K9, L, RK1, RK5, T, and CC) and plug fuses are tested on alternating-current circuits and are marked for ac use. When fuses are to be used on direct-current systems, the electrician should consult the fuse manufacturer because it may be necessary to reduce the fuse voltage rating and the interrupting rating to ensure safe operation.

The variables in the physical appearance of fuses include length, ferrule diameter, and blade length/width/thickness, as well as other distinctive

features. For these reasons, it is difficult to insert a fuse of a given ampere rating into a fuseholder rated for less amperage than the fuse. The differences in fuse construction also make it difficult to insert a fuse of a given voltage into a fuseholder with a higher voltage rating, *NEC® 240.60(B).* Figure 17-7 indicates several examples of the method of ensuring that fuses and fuseholders are not mismatched.

NEC® 240.60(B) also specifies that fuseholders for current-limiting fuses shall be designed so that they cannot accept fuses that are noncurrent limiting, Figure 17-8.

In general, low-voltage fuses may be used at system voltages that are less than the voltage rating of the fuse. For example, a 600-volt fuse combined

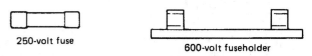

250-volt fuse   600-volt fuseholder

(A) A 250-volt fuse shall be constructed so that it cannot be inserted into a 600-volt fuseholder.

60-ampere fuse

30-ampere fuseholder

(B) A 60-ampere fuse is constructed so that it cannot be inserted into a 30-ampere fuseholder.

Figure 17-7   Examples of *NEC® 240.60(B)* requirements.

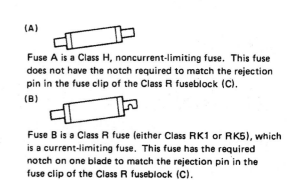

(A)

Fuse A is a Class H, noncurrent-limiting fuse. This fuse does not have the notch required to match the rejection pin in the fuse clip of the Class R fuseblock (C).

(B)

Fuse B is a Class R fuse (either Class RK1 or RK5), which is a current-limiting fuse. This fuse has the required notch on one blade to match the rejection pin in the fuse clip of the Class R fuseblock (C).

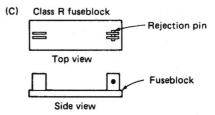

(C)   Class R fuseblock

Rejection pin

Top view

Fuseblock

Side view

Figure 17-8   Examples of *NEC® 240.60(B)* requirements.

with a 600-volt switch may be used on 575-volt, 480Y/277-volt, 208Y/120-volt, 240-volt, 50-volt, and 32-volt systems.

Class H.   Class H fuses, Figure 17-9 and Figure 17-10A, formerly were called *NEC®* or *Code* fuses.

Figure 17-9   Class H renewable link-type fuse. Note that the link is replaceable. ▶Renewable fuses are only permitted to be used as replacements on existing installations where there is no evidence of overfusing or tampering, *NEC® 240.60(D).*◀

Most low-cost, common, standard nonrenewable one-time fuses are Class H fuses. Renewable-type fuses also come under the Class H classification. Neither the interrupting rating nor the notation *Class H* appears on the label of a Class H fuse. This type of fuse is tested by the Underwriters Laboratories on circuits that deliver 10,000 amperes ac. Class H fuses are available with ratings ranging from 1 ampere to 600 amperes in both 250-volt ac and 600-volt ac types. Class H fuses are not current limiting.

A higher-quality, nonrenewable, one-time fuse is also available, called a Class K5 fuse, which has a 50,000-ampere interrupting rating.

Class K.   Class K fuses are grouped into three categories: K1, K5, and K9, Figure 17-10A through Figure 17-10D. These fuses may be UL listed or CSA certified with interrupting ratings in RMS symmetrical amperes in values of 50,000, 100,000, or 200,000 amperes. For each K rating, UL has assigned a maximum level of peak let-through current (I_p) and energy as given by I^2t. Class K fuses

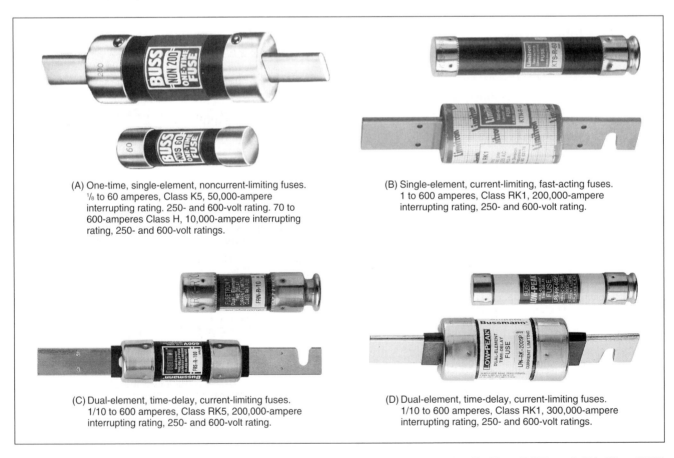

(A) One-time, single-element, noncurrent-limiting fuses. $\frac{1}{8}$ to 60 amperes, Class K5, 50,000-ampere interrupting rating. 250- and 600-volt rating. 70 to 600-amperes Class H, 10,000-ampere interrupting rating. 250- and 600-volt ratings.

(B) Single-element, current-limiting, fast-acting fuses. 1 to 600 amperes, Class RK1, 200,000-ampere interrupting rating, 250- and 600-volt rating.

(C) Dual-element, time-delay, current-limiting fuses. 1/10 to 600 amperes, Class RK5, 200,000-ampere interrupting rating, 250- and 600-volt rating.

(D) Dual-element, time-delay, current-limiting fuses. 1/10 to 600 amperes, Class RK1, 300,000-ampere interrupting rating, 250- and 600-volt ratings.

Figure 17-10   Various classes of fuses: (A) Classes H and K5, (B) Class RK1, (C) Class RK5, and (D) Class RK1 fuses. (*Courtesy* Cooper Bussmann, Inc.)

have varying degrees of current-limiting ability, depending upon the K rating. Class RK1 fuses have the greatest current-limiting ability and Class K9 fuses have the least current-limiting ability. A check of various fuse manufacturers' literature reveals that Class K9 fuses are no longer being manufactured. Class K fuses may be listed as time-delay fuses as well. In this case, UL requires that the fuses have a minimum time delay of 10 seconds at 500 percent of the rated current (8 seconds for 250 volts, 30 amperes). Class K fuses are available in ratings ranging from $\frac{1}{10}$ ampere to 600 amperes at 250- or 600-volts ac. Class K fuses have the same dimensions as Class H fuses.

Class J.   Class J fuses are current limiting and are so marked, Figure 17-11 A and B. They are listed by UL/CSA with a minimum interrupting rating of 200,000 RMS symmetrical amperes. Some have a special listing identified by the letters *SP*, and have an interrupting rating of 300,000 RMS symmetrical amperes. Certain Class J fuses are also considered to be dual-element, time-delay fuses, and are marked "time-delay." Class J fuses are physically smaller than Class H fuses. Therefore, when a fuseholder is

installed to accept a Class J fuse, it will be impossible to install a Class H fuse in the fuseholder, *NEC® 240.60(B)*.

The Underwriters Laboratories has assigned maximum values of I^2t and I_p that are slightly less than those for Class RK1 fuses. Both fast-acting, current-limiting Class J fuses and time-delay, current-limiting Class J fuses are available in ratings ranging from 1 ampere to 600 amperes at 600 volts ac.

Class L.   Class L fuses, Figure 17-12A and Figure 17-12B, are listed by UL/CSA in sizes ranging from 601 amperes to 6000 amperes at 600 volts ac. These fuses have specified maximum values of I^2t and I_p. They are current-limiting fuses and have a minimum interrupting rating of 200,000 RMS symmetrical amperes. These bolt-type fuses are used in bolted pressure contact switches. Class L fuses are available in both a fast-acting, current-limiting type and a time-delay, current-limiting type. Both types of Class L fuses meet UL requirements.

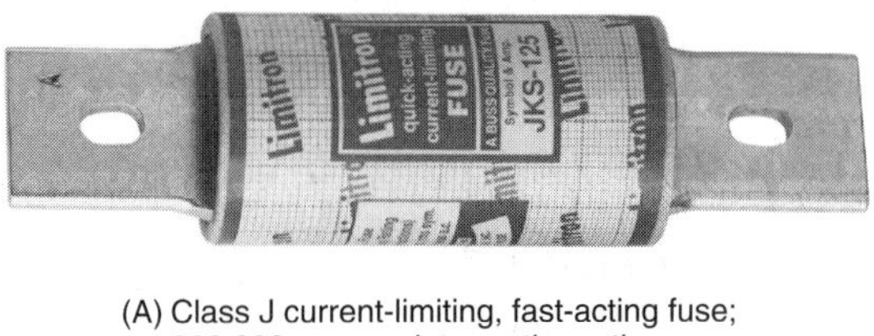

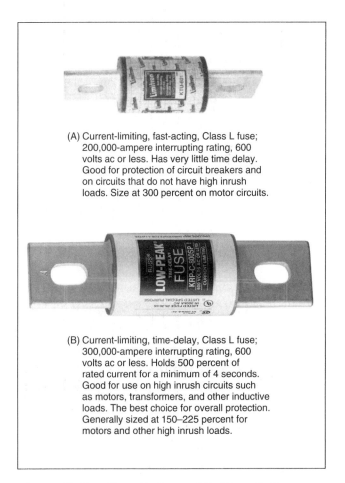

(A) Current-limiting, fast-acting, Class L fuse; 200,000-ampere interrupting rating, 600 volts ac or less. Has very little time delay. Good for protection of circuit breakers and on circuits that do not have high inrush loads. Size at 300 percent on motor circuits.

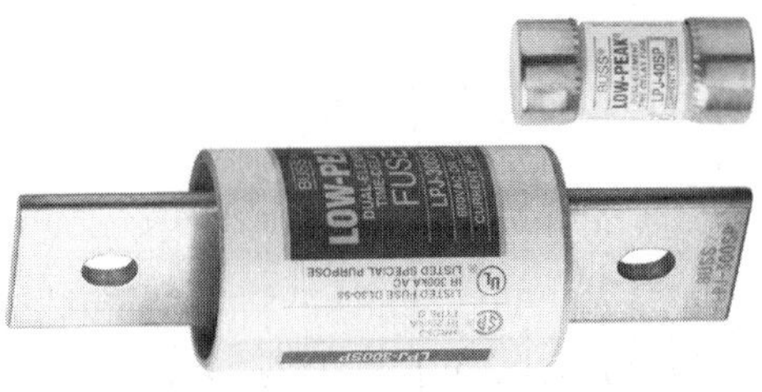

(B) Current-limiting, time-delay, Class L fuse; 300,000-ampere interrupting rating, 600 volts ac or less. Holds 500 percent of rated current for a minimum of 4 seconds. Good for use on high inrush circuits such as motors, transformers, and other inductive loads. The best choice for overall protection. Generally sized at 150–225 percent for motors and other high inrush loads.

(A) Class J current-limiting, fast-acting fuse; 200,000-ampere interrupting rating.

(B) Current-limiting, time-delay fuse. 1 to 600 amperes, Class J, 300,000-ampere interrupting rating, 600-volts rating.

Figure 17-11   Class J current-limiting fuses. (*Courtesy Cooper Bussmann, Inc.*)

Figure 17-12   Class L fuses. All Class L fuses are rated 600 volts. Listed in 601- to 6000-ampere rating.

Some Class L fuses have a special interrupting rating of 300,000 RMS symmetrical amperes. The fuse's label will indicate the part number followed by the letters "SP."

Class T. Class T fuses, Figure 17-13, are current-limiting fuses and are so marked. These fuses are UL listed with an interrupting capacity of 200,000 RMS symmetrical amperes. Class T fuses are physically smaller than Class H or Class J fuses. The configuration of this type of fuse limits its use to fuseholders and switches that will reject all other types of fuses.

Class T fuses rated 600 volts have electrical characteristics similar to those of fast-acting Class J fuses and are tested in a similar manner by Underwriters Laboratories. Class T fuses rated at 300 volts have lower peak let-through currents and I^2t values than comparable Class J fuses. Many series-rated panelboards are listed by Underwriters Laboratories with Class T mains. Because Class T fuses do not have a lot of time-delay (the ability to hold momentary inrush currents such as occurs when a motor is started), they are sized according to the nontime-delay fuse column in *NEC® Table 430.52*.

UL/CSA presently lists Class T fuses in sizes from 1 ampere to 1200 amperes. Common applications for Class T fuses are for mains, feeders, and branch-circuits.

Class T 300-volt fuses may be used on 120/240-volt, single-phase; 208Y/120-volt, three-phase, four-wire wye; and 240-volt, three-phase, three-wire, delta systems (ungrounded or corner grounded). The *NEC®* permits 300-volt Class T

fuses to be installed in single-phase, line-to-neutral circuits supplied from three-phase, four-wire solidly grounded neutral systems where the line-to-neutral voltage does not exceed 300 volts. The *NEC®* does not permit the use of 300-volt Class T fuses for line-to-line or line-to-line-to-line applications on 480Y/277-volt, three-phase, four-wire wye systems. Class T 600-volt fuses may be used on 480Y/277-volt, three-phase, four-wire wye; 480-volt, three-phase, three-wire; and any of the systems where Class T 300-volt fuses are permitted. See *NEC® 240.60(A)*.

Class G. Class G fuses, Figure 17-14, are cartridge fuses with small physical dimensions. Ratings ½ through 20 amperes are rated 600 volts ac. Ratings 25 through 60 are rated 480 volts ac. They are UL and CSA listed at an interrupting rating of 100,000 amperes. To prevent overfusing, Class G fuses are size-limiting within the four categories assigned to their ampere ratings. For example, a 15-ampere fuseholder will accept ½- through 15-ampere Class G fuses. A 20-ampere fuseholder will accept 20-ampere Class G fuses, a 30-ampere fuseholder will accept 25- and 30-ampere Class G fuses, and a 60-ampere fuseholder will accept 35-, 40-, 50-, and 60-ampere Class G fuses.

Class G fuses are current limiting. They may be used for the protection of ballasts, electric heat, and similar loads. They are UL listed for branch-circuit protection.

Ampere Rating	Dimensions
0–15	$^{13}/_{32}$ in. × $1^5/_{16}$ in.
16–20	$^{13}/_{32}$ in. × $1^{13}/_{32}$ in.
21–30	$^{13}/_{32}$ in. × $1^5/_8$ in.
31–60	$^{13}/_{32}$ in. × $2^1/_4$ in.

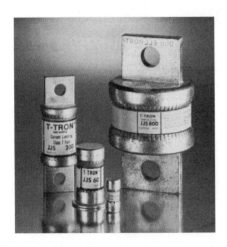

Figure 17-13   Class T current-limiting, fast-acting fuse; 200,000-ampere interrupting rating.

Figure 17-14   Class G fuses. (*Courtesy* Cooper Bussmann, Inc.)

Class R. The Class R fuse is a nonrenewable cartridge type and has a minimum interrupting rating of 200,000 RMS symmetrical amperes. The peak let-through current (I_p) and the total clearing energy (I^2t) values are specified for the individual case sizes. The values of I^2t and I_p are specified by UL based on short-circuit tests at 50,000, 100,000, and 200,000 amperes.

Class R fuses are divided into two subclasses: Class RK1 and Class RK5. The Class RK1 fuse has characteristics similar to those of the Class K1 fuse. The Class RK5 fuse has characteristics similar to those of the Class K5 fuse. These fuses must be marked either Class RK1 or RK5. In addition, they are marked to be current limiting.

Some Class RK1 fuses have a special interrupting rating of 300,000 RMS symmetrical amperes.

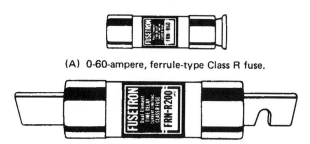

(A) 0-60-ampere, ferrule-type Class R fuse.

(B) 61-600-ampere, knife blade-type Class R fuse.

Figure 17-15   Class R cartridge fuses. (*Courtesy Cooper Bussmann, Inc.*)

The fuse's label will indicate the part number followed by the letters *SP*.

The ferrule-type Class R fuse has a rating range of 1/10 ampere to 60 amperes and can be distinguished by the annular ring on one end of the case, Figure 17-15A. The knife blade-type Class R fuse has a rating range of 61 amperes to 600 amperes and has a slot in the blade on one end, Figure 17-15B. When a fuseholder is designed to accept a Class R fuse, it will be impossible to install a standard Class H or Class K fuse (these fuses do not have the annular ring or slot of the Class R fuse). The requirements for noninterchangeable cartridge fuses and fuseholders are covered in *NEC® 240.60(B)*. However, the Class R fuse can be installed in older style fuse clips on existing installations. As a result, the Class R fuse may be called a one-way rejection fuse.

Electrical equipment manufacturers will provide the necessary rejection-type fuseholders in their equipment, which is then tested with a Class R fuse at short-circuit current values such as 50,000, 100,000, or 200,000 amperes. Each piece of equipment will be marked accordingly.

Cube Fuse. A new type of fuse, shown in Figure 17-16, is finding its way into original equipment manufacturers' (OEM) products. It is physically smaller than ordinary fuses and is dual-element, time-delay, having a 10-second opening time at 500 percent

Figure 17-16   A cube fuse. Note the blades for easy insertion into a fuseholder.

of its ampere rating. The cube fuse also has a 300,000-ampere interrupting capacity, has the characteristics of UL/CSA listed Class J fuses (fast acting under fault conditions), is rated 600 volts ac or less, has no exposed live parts, has open-fuse indication, mounts in a special fuseholder and rails designed for the purpose, provides Type 2 protection (no damage) for motor controllers, and is available in many ampere ratings from 1 ampere through 100 amperes.

Class CC. Class CC fuses are primarily used for control-circuit protection, for the protection of motor control circuits, ballasts, small transformers, and so on. They are UL listed as branch-circuit fuses. Class CC fuses are rated at 600 volts or less and have a 200,000-ampere interrupting rating in sizes from $\frac{1}{10}$ ampere through 30 amperes. These fuses measure $1\frac{1}{2}$ in. × $\frac{13}{32}$ in. (38.1 mm × 10.3 mm) and can be recognized by a button on one end of the fuse, Figure 17-17. This button is unique to Class CC fuses. When a fuseblock or fuseholder that has the matching Class CC rejection feature is installed, it is impossible to insert any other $1\frac{1}{2}$ in. × $\frac{13}{32}$ in. (38.1 mm × 10.3 mm) fuse. Only a Class CC fuse will fit into these special fuseblocks and fuseholders. A Class CC fuse can be installed in a standard fuseholder.

Plug Fuses

NEC® Article 240, Part V spells out the requirements for plug fuses.

Conventional plug fuses have a base referred to as an Edison base. This base is of the same shape as the base on a light bulb. Type S fuses have a special base.

Figure 17-17 Class CC fuses with rejection feature. (*Courtesy* **Cooper Bussmann, Inc.**)

The opening characteristics of plug fuses are available in three types, each serving a purpose:

1. The standard fuse link type does not have much time delay. It opens fast. It is used on lighting circuits and other nonmotor loads.

2. The loaded link type has a metal bead (heat sink) on the fuse element that gives it time delay to hold motor inrush starting currents.

3. The dual-element, time-delay has a spring-loaded short-circuit element plus an overload element connected in series with the short-circuit element. It is excellent for motor circuits.

Electricians will be most concerned with the following requirements for plug fuses, fuseholders, and adapters:

- They shall not be used in circuits exceeding 125 volts between conductors, except on systems having a grounded neutral with no conductor having more than 150 volts to ground. This situation is found in the 120/208-volt system in the commercial building covered in this text, or in the case of a 120/240-volt, single-phase system.

- They shall have ampere ratings of 0 to 30 amperes.

- They shall have a hexagonal configuration for ratings of 15 amperes and below.

- The screw shell must be connected to the load side of the circuit.

- Edison-base plug fuses may be used only as replacements in existing installations where there is no evidence of overfusing or tampering.

- All new installations shall use fuseholders requiring Type S plug fuses or fuseholders with a Type S adapter inserted to accept Type S fuses only.

- Type S plug fuses are classified 0 to 15 amperes, 16 to 20 amperes, and 21 to 30 amperes.

Prior to the installation of fusible equipment, the electrician must determine the ampere rating of the various circuits. An adapter of the proper size is then inserted into the Edison-base fuseholder; finally, the proper size of Type S fuse can be inserted into the fuseholder. The adapter makes the fuseholder nontamperable and noninterchangeable. For example, after a 15-ampere adapter is inserted

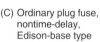

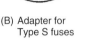

(A) Dual-element
Type S fuse

(B) Adapter for
Type S fuses

(C) Ordinary plug fuse,
nontime-delay,
Edison-base type

(D) Fusetron dual-element,
time-delay plug fuse,
Edison-base type

Figure 17-18   Plug fuse Types W, T, and S with adapter. (*Courtesy* Cooper Bussmann, Inc.)

into a fuseholder for a 15-ampere circuit, it is impossible to insert a Type S or an Edison-base fuse having a larger ampere rating into this fuseholder without removing the 15-ampere adapter. Adapters are designed so that they are extremely difficult to remove.

Type S fuses and suitable adapters, Figure 17-18, are available in a large selection of ampere ratings ranging from ¼ ampere to 30 amperes. When a Type S fuse has dual elements and a time-delay feature, it does not open unnecessarily under momentary overloads, such as the current surge caused by the startup of an electric motor. On heavy overloads or short circuits, this type of fuse opens very rapidly.

Summary of Fuse Applications

Table 17-3 provides rating and application information for the fuses supplied by one manufacturer. The table indicates the class, voltage rating, current range, interrupting rating, and typical applications for the various fuses. A table of this type may be used to select the proper fuse to meet the needs of a given situation.

TESTING FUSES

As mentioned at the beginning of this text, the Occupational Safety and Health Act (OSHA) clearly states that electrical equipment must not be worked on when it is energized. There have been too many injuries to those intentionally working on equipment hot, or thinking the power is off, only to find that it is still energized. If equipment is to be worked on hot, then proper training, fire-resistant clothing, and

protective gear (rubber blankets, insulated tools, goggles, rubber gloves, etc.) need to be used. A second person should be present when working on electrical equipment hot. OSHA has specific lockout and tag-out rules for working on energized electrical equipment.

When Power Is Turned On. On live circuits, extreme caution must be exercised when checking fuses. There are many different voltage readings that can be taken, such as line-to-line, line-to-ground, line-to-neutral, and so on.

Using a voltmeter, the first step is to be sure to set the scale to its highest voltage setting and then change to a lower scale after you are sure you are within the range of the voltmeter. For example, when testing what you believe to be a 120-volt circuit, it is wise to first use the 600-volt scale, then try the 300-volt scale, and then use the 150-volt scale — JUST TO BE SURE!

Taking a voltage reading across the bottom (load side) of fuses — either fuse-to-fuse, fuse-to-neutral, or fuse-to-ground — can show voltage readings because even though a fuse might have opened, there can be feedback through the load. You could come to a wrong conclusion. Taking a voltage reading from the line side of a fuse to the load side of a fuse will show open-circuit voltage if the fuse has blown and the load is still connected. This also can result in a wrong conclusion.

Reading from the line-to-load side of a good fuse should show zero voltage or else an extremely small voltage across the fuse.

Always read carefully the instructions furnished with electrical test equipment such as voltmeters, ohmmeters, and so on.

Bussmann Fuses: Their Ratings, Class, and Application**					
FUSE AND AMPERE RATING RANGE	Voltage Rating	Symbol	Industry Class	Interrupting Rating in Sym. RMS Amperes	Application All Types Recommended for Protection of General Purpose Circuits and Components
LOW-PEAK® time-delay FUSE 601–6000 A	600 V	KRP-C (SP)	L	300,000 A	Will hold 500% of its rating for at least 4 seconds. High Capacity Main, Feeder, Br. Ckts. and Large Motors Circuits. Has more time-delay than KTU. For motors, can be sized 150–225% of motor FLA.
LIMITRON® fast-acting FUSE 601–6000 A	600 V	KTU	L	200,000 A	High Capacity Main, Feeder, Br. Ckts. and Circuit Breaker Protection. For motors, size 300% of motor FLA.
LIMITRON® time-delay FUSE 601–4000 A	600 V	KLU	L	200,000 A	High Capacity Mains, Feeders, and Circuits for large motors. Has good time-delay characteristics. Will hold 500% of its rating for 10 seconds. Copper links, thus not as current-limiting as KTU or KRPC type.
FUSETRON® dual-element FUSE ¹⁄₁₀–600 A	250 V / 600 V	FRN-R / FRS-R	RK5 / RK5	200,000 A	Main, Feeder, Br. Ckts. and circuit breaker protection. Especially recommended for Motors, Welders, and Transformers and any loads having high inrush characteristics. Will hold 500% for 10 seconds. Size 115%–175% of motor FLA.
LOW-PEAK® dual-element FUSE ¹⁄₁₀–600 A	250 V / 600 V	LPN-RK (SP) / LPS-RK (SP)	RK1 / RK1	300,000 A	Main, Feeder, Br. Ckts. and circuit breaker protection. Especially recommended for Motors, Welders, and Transformers. (More current-limiting than Fusetron dual-element Fuse.) Will hold 500% for 10 seconds. Size at 115%–175% of motor FLA.
LIMITRON® fast-acting FUSE 1–600 A	250 V / 600 V	KTN-R / KTS-R	RK1 / RK1	200,000 A	Main, Feeder, Br. Ckts. and circuit breaker protection. Especially recommended for Circuit Breaker Protection. (High degree of current-limitation.) Because these fuses are fast-acting, size at 300% or more for motors.
TRON® fast-acting FUSE 1–1200 A (300 V) 1–800 A (600 V)	300 V / 600 V	JJN / JJS	T	200,000 A	Main, Feeder, Br. Ckts. and circuit breaker protection. Circuit Breaker Protection Small Physical Dimensions. Smaller than Class J. (High degree of current limitation.) Because these fuses are fast-acting, size at 300% or more for motors.
LOW-PEAK® time-delay FUSE 1–600 A	600 V	LPJ (SP)	J	300,000 A	Main, Feeder, Br. Ckts., and circuit breaker protection. Motor and Transformer Ckts. Size at 150%–225% for motors.
LIMITRON® fast-acting FUSE 1–600 A	600 V	JKS	J	200,000 A	Main, Feeder, and Br. Ckts. Circuit Breaker Protection. Because these fuses are fast acting, size at 300% or more for motors.
ONE-TIME FUSE ¹⁄₈–600 A	250 V / 600 V	NON / NOS	1–60 A, K5 65–600 A, H 1–60 A, K5 65–600 A, H	50,000 A 10,000 A 50,000 A 10,000 A	General Purpose — where fault currents are low. For motors, size at 300% or more.
SUPERLAG® renewable (replaceable) link fuse 1–600 A	250 V / 600 V	REN / RES	H	10,000 A	▶Renewable fuses are only permitted to be used as replacements on existing installations where there is no evidence of overfusing or tampering, NEC® 240.60(D).◀
SC® FUSE 1–60 A	600 V, 1–20 A 480 V, 25–60 A	SC	G	100,000 A	General Purpose Branch-Circuits. For motors, size at 300% or more.
Dual-element CUBE® fuse 1–100 A	600 V	TCF	J	300,000 A	Control panels on equipment. 10-second time-delay at 500% load.
FUSETRON® dual-element PLUG FUSE ³⁄₁₀–30 A, Edison Base	125 V	T	***	10,000 A	General Purpose Sizes ³⁄₁₀ A through 14 A excellent for motors. Size at 115%–125% for motors. Branch-circuit sizes 15-20-25-30 A.
FUSTAT® dual-element type S PLUG FUSE ³⁄₁₀–30 A	125 V	S	S****	10,000 A	General Purpose, Sizes ³⁄₁₀ A through 14 A excellent for motors. Size at 115%–125% for motors. Branch-circuit sizes 15-20-25-30 A.
ORDINARY PLUG FUSE Edison Base, ½–30 A	125 V	W	***	10,000 A	General Purpose. Less Time Delay than Types T and S. Not recommended for motor circuits unless sized at 300% or more of motor FLA.

NOTE: For motor circuits, refer to NEC® Article 430, especially Table 430.52.

** This table gives general application recommendations from one manufacturer's technical data. For critical applications, consult electrical equipment and specific fuse manufacturer's technical data.

*** Listed as Edison-Base Plug Fuse.

**** Difficult to change ampere ratings because of different thread on screw base for different ampere ratings.

Table 17-3

When Power Is Turned Off.   This is the safest way to test fuses. Remove the fuse from the switch, then take a resistance reading across the fuse using an ohmmeter. A good fuse will show zero or a very minimal resistance. An open (blown) fuse will generally show a high resistance reading.

Cable Limiters

Cable limiters are used quite often in commercial and industrial installations where parallel cables are used on service entrances and feeders.

Cable limiters differ in purpose from fuses, which are used for overload protection. Cable limiters are short-circuit devices that can *isolate* a faulted cable rather than having the fault open the entire phase. They are selected on the basis of conductor size; for example, a 500-kcmil cable limiter would be used on a 500-kcmil conductor.

Cable limiters are available for cable-to-cable or cable-to-bus installation for either aluminum or copper conductors. The type of cable limiter illustrated in Figure 17-19 is for use with a 500-kcmil copper conductor and is rated at 600 volts with an interrupting rating of 200,000 amperes.

Cable limiters are also used where taps are made ahead of the mains, such as they are in the commercial building where the emergency system is tapped ahead of the main fuses (as shown in Figure 16-2).

Figure 17-20 is an example of how cable limiters may be installed on the service of the commercial building. A cable limiter is installed at each end of each 500-kcmil conductor. Three conductors per phase terminate in the main switchboard where the service is then split into two 800-ampere bolted pressure switches.

Figure 17-21 shows how cable limiters are used where more than one customer is connected to one transformer.

Cable limiters are permitted, *NEC® 230.82(1)*.

Cable limiters are installed in the hot phase conductors only. They are not installed in a grounded conductor.

Figure 17-19   Cable limiter for 500-kcmil copper conductor. (*Courtesy* Cooper Bussmann, Inc.)

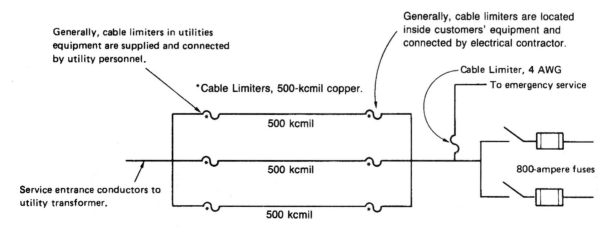

Figure 17-20   Use of cable limiters in a service entrance.

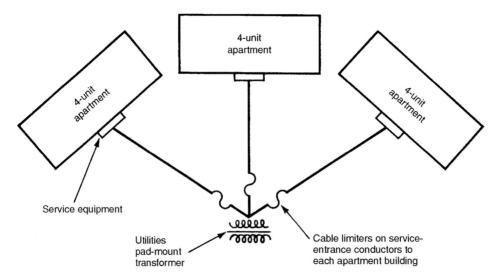

Figure 17-21 Some utilities install cable limiters where more than one customer is connected to one transformer. Thus, if one customer has problems in the main service equipment, the cable limiter on that service will open, isolating the problem from other customers on the same transformer. This is common for multiunit apartments, condos, and shopping centers.

DELTA, THREE-PHASE, CORNER GROUNDED "B" PHASE SYSTEM

See Unit 13 for a discussion of delta-connected, three-phase systems.

Fuses shall be installed in series with ungrounded conductors for overcurrent protection, *NEC® 240.20(A)*. This connection is sometimes called a corner ground. The *Code* prohibits overcurrent devices from being connected in series with any conductor that is intentionally grounded, *NEC® 240.22*.

However, there are certain instances where a three-pole, three-phase switch may be installed, where it is permitted to install fuses in two poles only. This would be in the case of a delta, three-phase, corner grounded "B" phase system, Figure 17-22. A device called a *solid neutral* can be installed in the "B" phase, as shown in the figure. Note that the switch has two fuses and one solid neutral installed.

Figure 17-22 shows a three-pole service-entrance switch. *NEC® 230.75* permits a busbar to be used as a disconnecting means for the grounded conductor. Thus, a two-pole service-entrance switch is permitted if the grounded conductor can be disconnected from the service-entrance conductors, Figure 17-23.

Solid Neutrals

A solid neutral, Figure 17-24, is made of copper bar that has exactly the same dimensions as a fuse for a given ampere rating and voltage rating. For example, a solid neutral rated at 100 amperes, 600 volts would be installed in a switch rated at 100 amperes, 600 volts.

Solid neutrals are generally used in retrofit situations.

Solid neutrals are installed in situations, for example, where a three-pole, three-fuse disconnect switch is used on a grounded "B" phase system, Figure 17-22. *NEC® 240.20* requires that "a fuse shall be connected in series with each ungrounded conductor." The solid neutral would be inserted into the grounded "B" phase leg instead of a fuse. Other examples of where a solid neutral might be used is where a three-pole, three-fuse disconnect switch is used on a three-wire, 240/120-volt, single-phase system, or where a fusible disconnect is used only as a disconnecting means, and where the user does not want any fuses to be installed in that particular switch.

TIME-CURRENT CHARACTERISTIC CURVES AND PEAK LET-THROUGH CHARTS

The electrician must have a basic amount of information concerning fuses and their application to be able to make the correct choices and decisions for everyday situations that arise on an installation.

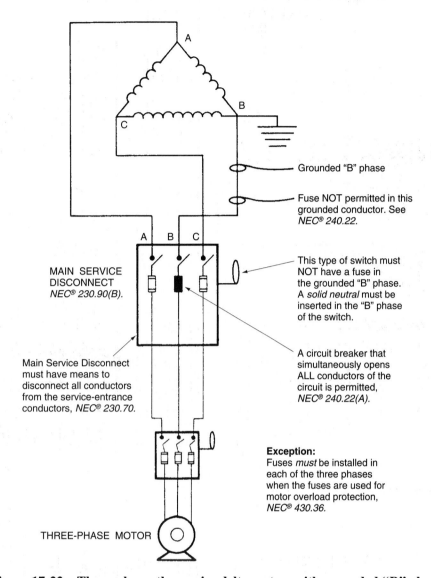

Grounded "B" phase

Fuse NOT permitted in this grounded conductor. See *NEC® 240.22.*

A
B
C

MAIN SERVICE DISCONNECT *NEC® 230.90(B).*

This type of switch must NOT have a fuse in the grounded "B" phase. A *solid neutral* must be inserted in the "B" phase of the switch.

Main Service Disconnect must have means to disconnect all conductors from the service-entrance conductors, *NEC® 230.70.*

A circuit breaker that simultaneously opens ALL conductors of the circuit is permitted, *NEC® 240.22(A).*

Exception:
Fuses *must* be installed in each of the three phases when the fuses are used for motor overload protection, *NEC® 430.36.*

THREE-PHASE MOTOR

Figure 17-22   Three-phase, three-wire delta system with grounded "B" phase.

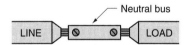

Neutral bus

LINE   LOAD

Figure 17-23 *NEC® 230.75* **allows this for disconnecting the grounded conductor at the service disconnect.**

Figure 17-24   Examples of a solid neutral.

The electrician must be able to use the following types of fuse data:

- time-current characteristic curves, including total clearing and minimum melting curves.

- peak let-through charts.

Fuse manufacturers furnish this information for each of the fuse types they produce.

The Use of Time-Current Characteristic Curves

The use of the time-current characteristic curves shown in Figure 17-25 and Figure 17-26 can be demonstrated by considering a typical problem. Assume that an electrician must select a fuse to protect a motor that is governed by *NEC® 430.32(A).*

AVERAGE MELTING TIME-CURRENT CHARACTERISTIC CURVES

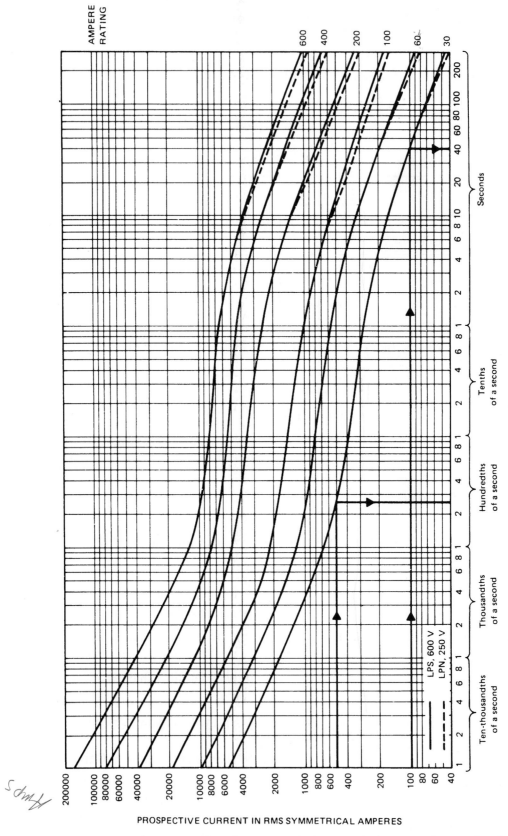

Figure 17-25   Average melting time-current characteristic curves for one family of current-limiting, dual-element, time-delay fuses. Time-current curves such as this are available from the manufacturers of fuses and circuit breakers.

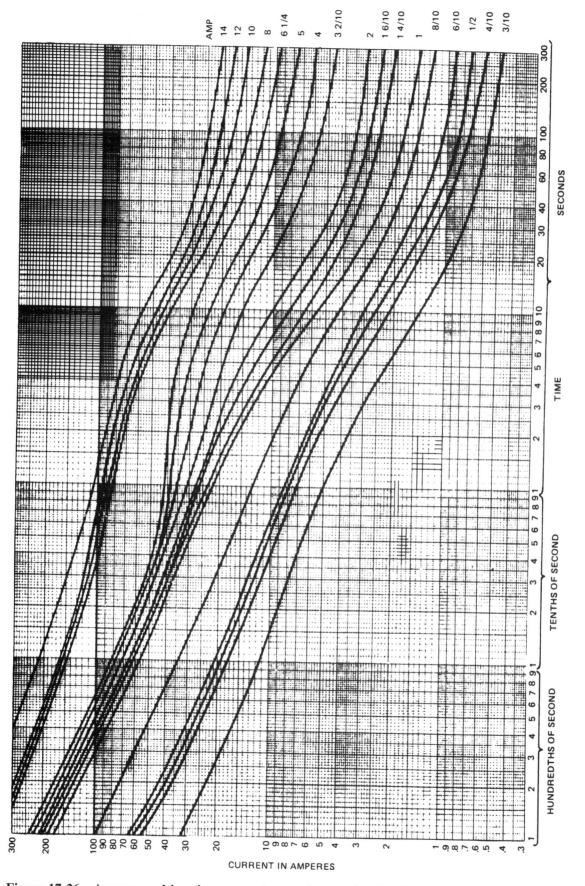

Figure 17-26   Average melting time-current curve for one family of time-delay fuses rated ³⁄₁₀ through 14 amperes.

This section indicates that the running overcurrent protection shall be based on not more than 125 percent of the full-load running current of the motor. The motor in this example is assumed to have a full-load current of 24 amperes. Therefore, the size of the required protective fuse is determined as follows:

$$24 \times 1.25 = 30 \text{ amperes}$$

Now the question must be asked: Is this 30-ampere fuse capable of holding the inrush current of the motor, which is approximately four to five times the running current, for a sufficient length of time for the motor to reach its normal speed? For this problem, the inrush current of the motor is assumed to be 100 amperes.

Refer to Figure 17-25. At the 100-ampere line, draw a horizontal line until it intersects the 30-ampere fuse line. Then draw a vertical line down to the base line of the chart. At this point, it can be seen that the 30-ampere fuse will hold 100 amperes for approximately 40 seconds. In this amount of time, the motor can be started. In addition, the fuse will provide running overload protection as required by *NEC® 430.32(A)*.

If the time-current curve for thermal overload relays in a motor controller is checked, it will show that the overload element will open for the same current in much less than 40 seconds. Therefore, in the event of an overload, the thermal overload elements will operate before the fuse opens. If the overload elements do not open for any reason, or if the contacts of the controller weld together, the properly sized dual-element fuse will open to reduce the possibility of motor burnout. The preceding method is a simple way to obtain backup or double-motor protection.

Referring again to Figure 17-25, assume that a short circuit of approximately 500 amperes occurs. Find the 500-ampere line at the left of the chart and then, as before, draw a horizontal line until it intersects the 30-ampere fuse line. From this intersection point, drop a vertical line to the base line of the chart. This value indicates that the fuse will open the fault in slightly over $^2/_{100}$ (0.02) second. On a 60-hertz system (60 cycles per second), one cycle equals 0.016 second. Therefore, a 30-ampere fuse will clear a 500-ampere fault in just over one cycle.

The Use of Peak Let-Through Charts

It is important to understand the use of peak let-through charts if one is to properly match the short-circuit current rating of electrical equipment with the let-through current values of the overcurrent protective devices (fuses or circuit breakers). Short-circuit current rating can be defined as "at a specified nominal voltage, the highest current at the equipment or component line terminals that the equipment or component can safely carry until an overcurrent device opens the circuit."

The short-circuit current ratings of electrical equipment (such as panelboards, bus duct, switchboard bracing, controllers, and conductors) and the interrupting ratings of circuit breakers are given in the published standards of the Underwriters Laboratories, NEMA, and the Insulated Cable Engineering Association (ICEA). The withstand ratings may be based either on the peak current (I_p) or on the RMS current.

Peak let-through charts give a good indication of the current-limiting effects of a current-limiting fuse or circuit breaker under "bolted fault" conditions. Current limitation is one of the principles used to obtain a series-rating listing for electrical equipment. Another principle used to obtain a series rating is referred to as "dynamic impedance," where the available fault current is further reduced because of the added impedance of the arc between the contacts of a circuit breaker as it is opening. Series rating of panels is discussed later in this unit.

To use charts such as Figure 17-27, find the 40,000-ampere point on the base line of the chart, draw a vertical line upward to Line A–B, and then move horizontally to the left margin where the value of 92,000 amperes is found. This is the peak current, I_p, at the first half cycle.

To find the peak let-through current of the 100-ampere fuse, start again at the 40,000-ampere point on the base line, go vertically upward to Line A–B, and then horizontally to the left margin, where the value of 10,400 is found. This is the peak let-through current, I_p, of this particular 100-ampere fuse.

To find the apparent RMS let-through current of this fuse, start again at the 40,000-ampere point on the base line, go vertically upward to Line A–B, and then horizontally to the left until Line A–B is reached, and then vertically downward to the base line. The value at this point on the base line is 4600 amperes. This is the apparent RMS amperes let-through current of this particular 100-ampere fuse.

Figure 17-28 shows how a current-limiting main overcurrent device can be applied to "Listed" series-rated panels, discussed in more detail a little later in this unit.

HOW TO USE THE LET-THROUGH CHARTS

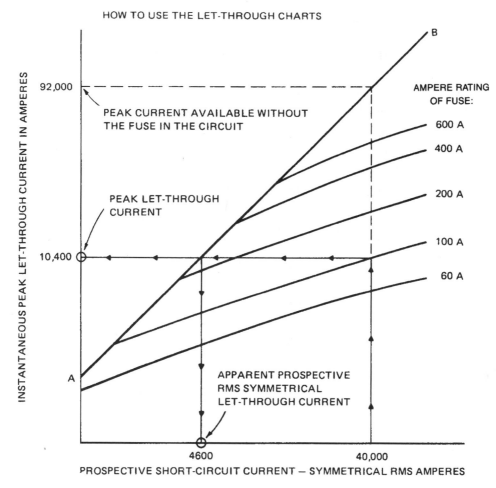

Figure 17-27   **Using the let-through charts to determine peak let-through current and apparent prospective RMS symmetrical let-through current.**

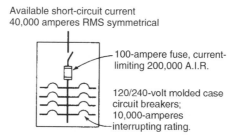

Available short-circuit current
40,000 amperes RMS symmetrical

100-ampere fuse, current-limiting 200,000 A.I.R.

120/240-volt molded case circuit breakers; 10,000-amperes interrupting rating.

Figure 17-28   **Example of a fuse used to protect circuit breakers. This installation meets the requirements of *NEC® 110.9* when the panelboard is listed as series-rated and the specified main fuses and branch-circuit breakers are used.**

Figure 17-29 shows a peak let-through chart for a family of dual-element fuses. Fuse and circuit breaker manufacturers provide charts of this type for various sizes and types of their fuses and circuit breakers.

What are these charts used for? Let's take an example of a 400-ampere busway that is marked as having 22,000-ampere bracing. The available fault current of the system has been determined to be approximately 50,000 amperes. In Figure 17-29, draw a vertical line upward from 50,000 amperes on the base line up to the 400-ampere fuse line. From this point, go horizontally to the left to Line A–B. From this point, draw a line vertically downward to the base line and read an apparent RMS current value of slightly more than 10,000 amperes. Therefore, the

Current Limitation Curves

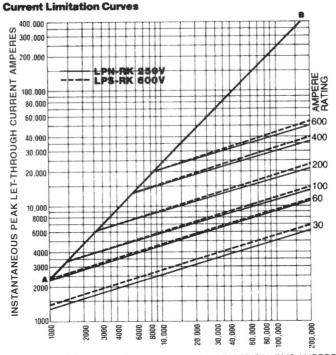

Figure 17-29   Current-limiting effect of a dual-element, time-delay Class RK1 fuse.

400-ampere busway with a bracing of 22,000 amperes is adequately protected by the 400-ampere, fast-acting, current-limiting fuse.

Table 17-4 shows how a manufacturer might indicate the maximum size and type of fuse that will protect given circuit breakers against fault currents that exceed the breakers' interrupting rating. This information is found in the UL *Recognized Components Directory* under "Circuit Breakers—Series Connected." Circuit breakers are tested according to the requirements of UL Standard 489, panelboards are tested to UL Standard 67, switchboards are tested to UL Standard 891, and motor controllers are tested to UL Standard 508.

TYPICAL TABLE SHOWING SIZES AND TYPES OF FUSES TO PROTECT CIRCUIT BREAKERS							
Branch Breaker			Main Fuse		Interrupting Rating		
Frame	Poles	Max. Amp.	Type	Max. Amp.	VAC	KA	Phase
ABC	2	15–100	J, T	200	120/240	200	1
DEF	2, 3	15–100	R	100	120/240	200	1
GHI	2, 3	70–225	J, T	400	600	100	1,3
JKL	1	15–100	J, T	200	277	200	1
MNO	2, 3	15–150	R	200	600	200	1,3
PQR	2, 3	70–400	L	1200	480	100	1,3
STU	2, 3	400	J, T, R	600	480	100	1,3
VWX	2, 3	400	J, T, R	400	480	200	1,3
SYX	1, 2	30	RK5	200	240	100	3

Typical table published by manufacturers showing the maximum sizes and types of fuses that may be used to protect circuit breakers against fault currents that exceed the breakers' maximum interrupting rating. This table was modeled from various manufacturers' "series-connected" technical information. Always refer to the specific manufacturer's data.

Table 17-4

CIRCUIT BREAKERS

In *NEC® Article 100* a circuit breaker is defined as *a device that is designed to open and close a circuit by nonautomatic means and to open the circuit automatically on a predetermined overcurrent without being damaged itself when properly applied within its rating.*

Molded-case circuit breakers are the most common type in use today, Figure 17-30. The tripping mechanism of this type of breaker is enclosed in a molded plastic case. The thermal-magnetic type of circuit breaker is covered in this unit.

Another type is the power circuit breaker, often referred to as an air-frame circuit breaker. They are larger and of heavier construction than molded-case and insulated-case breakers. They are available with many options such as draw-out mounting, electronic adjustable tripping and rating settings, and integral current limiters or fuses. Power circuit breakers are used in large industrial applications. Power circuit breakers can be maintained in the field. Molded-case circuit breakers are not intended to be maintained in the field, but to be replaced if there is a problem.

Still another type of circuit breaker is the insulated-case circuit breaker. It is a hybrid between a molded-case and a power (air-frame) circuit breaker, and is less costly than a power (air-frame) circuit breaker. Most of the features available on power breakers are also available on insulated-case circuit breakers.

Low-voltage power circuit breakers are listed under UL Standard 1066. Molded-case and insulated-case circuit breakers are listed under UL Standard

489. *NEC® 240.80* through *240.86* state the basic requirements for circuit breakers and are summarized as follows:

- Breakers shall be trip free so that if the handle is held in the On position, the internal mechanism trips the breaker to the Off position.

- The breaker shall clearly indicate whether it is in the On or Off position.

- The breaker shall be nontamperable; that is, it cannot be readjusted (to change its trip point or time required for operation) without dismantling the breaker or breaking the seal.

- The rating shall be durably marked on the breaker. For the smaller breakers with ratings of 100 amperes or less and 600 volts or less, this marking must be molded, stamped, or etched on the handle or other portion of the breaker that will be visible after the cover of the panel is installed.

- Every breaker having an interrupting rating other than 5000 amperes shall have its interrupting rating shown on the breaker.

- *NEC® 240.83(D)* and UL Standard 489 standard require that when a circuit breaker is to be used as a switch, the breaker must be so evaluated, listed, and marked. A typical use of a circuit breaker as a switch is in a panel to turn 120- or 277-volt lighting circuits on and off instead of installing separate switches. This is very common in commercial buildings. To switch fluorescent lighting circuits, breakers shall be marked SWD or HID. To switch high-intensity discharge (HID) lighting circuits, breakers shall be marked HID. Breakers not marked in this manner shall not be used as switches.

- A circuit breaker should not be loaded to more than 80 percent of its current rating for loads that are likely to be on for three hours or more, unless the breaker is marked otherwise, such as those breakers tested by the Underwriters Laboratories at 100 percent loading.

- If the voltage rating on a circuit breaker is marked with a single voltage (example: 240 volts), the breaker may be used in grounded or ungrounded systems where the voltage between any two conductors does not exceed the breaker's marked voltage rating. **CAUTION:** Do not use a two-pole, single-voltage-marked breaker

Figure 17-30   Molded-case circuit breakers.

to protect a three-phase, three-wire, corner-grounded circuit unless the breaker is marked with both one-phase and three-phase marking.

- If the voltage rating on a circuit breaker is marked with a slash voltage (example: 480/277 volts or 120/240 volts), the breaker may be used on a solidly grounded system where the voltage to ground does not exceed the breaker's lower voltage marking and where the voltage between any two conductors does not exceed the breaker's higher voltage marking.

- A cautionary FPN has been added, stating that: *Proper application of molded case circuit breakers on 3-phase systems, other than solidly grounded wye, particularly on corner-grounded delta systems, considers the circuit breaker's individual interrupting capacity.**

Thermal-Magnetic Circuit Breakers

A thermal-magnetic circuit breaker contains a bimetallic element. On a continuous overload, the bimetallic element moves until it unlatches the inner tripping mechanism of the breaker. Harmless momentary overloads do not cause the tripping of the bimetallic element. If the overload is heavy, or if a short circuit occurs, then the mechanism within the circuit breaker causes the breaker to interrupt the circuit instantly. The time required for the breaker to open the circuit completely depends upon the magnitude of the fault current and the mechanical condition of the circuit breaker. This time may range from approximately one-half cycle to several cycles.

Circuit breaker manufacturers calibrate and set the tripping characteristic for most molded-case breakers. Breakers are designed so that it is difficult to alter the set tripping point, *NEC® 240.82*. For certain types of breakers, however, the trip coil can be changed physically to a different rating. Adjustment provisions are made on some breakers to permit the magnetic trip range to be changed. For example, a breaker rated at 100 amperes may have an external adjustment screw with positions marked HI-MED-LO. The manufacturer's application data for this breaker indicates that the magnetic tripping occurs at 1500 amperes, 1000 amperes, or 500 amperes, respectively, for the indicated positions. These settings usually have a tolerance of ±10 percent.

The *ambient-compensated* type of circuit breaker is designed so that its tripping point is not affected by an increase in the surrounding temperature. An ambient-compensated breaker has two elements: the first element heats due to the current passing through it and because of the heat of the surrounding air; the second element is affected only by the ambient temperature. These elements act in opposition to each other. In other words, as the tripping element tends to lower its tripping point because of a high ambient temperature, the second element exerts an opposing force that stabilizes the tripping point. Therefore, current through the tripping element is the only factor that causes the element to open the circuit. Most breakers are calibrated for use in up to 104°F (40°C) ambient. This is marked on the breaker as 40°C.

Factors that can affect the proper operation of a circuit breaker include moisture, dust, vibration, corrosive fumes and vapors, and excessive tripping and switching. As a result, care must be taken when locating and installing circuit breakers and all other electrical equipment.

The interrupting rating of a circuit breaker is marked on the breaker label. The electrician should check the breaker carefully for the interrupting rating because the breaker may have several voltage ratings with a different interrupting rating for each. For example, assume that it is necessary to select a breaker having an interrupting rating at 240 volts of at least 50,000 amperes. A close inspection of the breaker may reveal the following data:

Voltage	Interrupting Rating
240 volts	65,000 amperes
480 volts	25,000 amperes
600 volts	18,000 amperes

Recall that for a fuse, the interrupting rating is marked on the fuse label. This rating is the same for any voltage up to and including the maximum voltage rating of the fuse.

The standard full-load ampere ratings of nonadjustable circuit breakers are the same as those for fuses according to *NEC® 240.6*. As previously mentioned, additional standard ratings of fuses are 1, 3, 6, 10, and 601 amperes.

The time-current characteristics curves for circuit breakers are similar to those for fuses. A typical circuit breaker time-current curve is shown in Figure 17-31.

---

*Reprinted with permission from NFPA 70-2005.

Current in percent of breaker trip unit rating

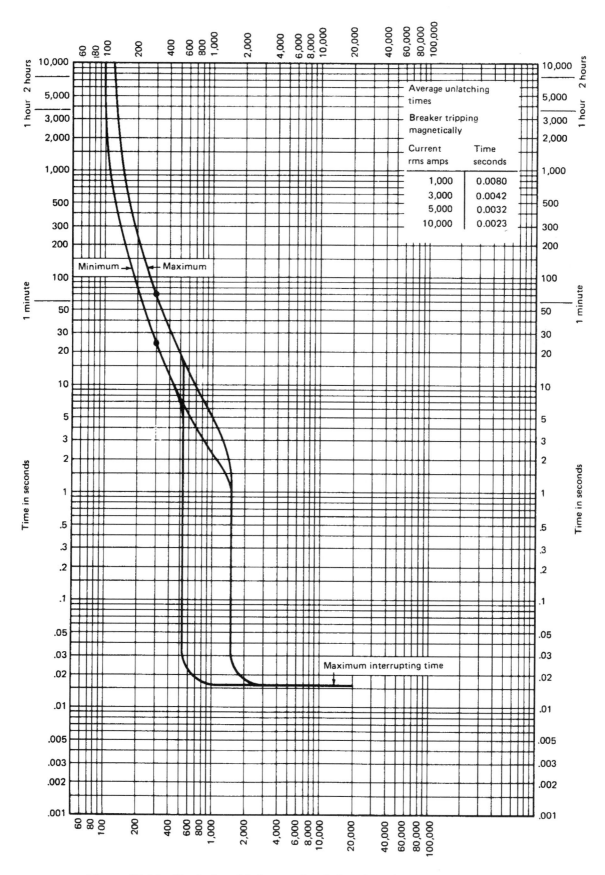

Figure 17-31   Typical molded-case circuit-breaker, time-current curve.

Note that the current is indicated in percentage values of the breaker trip unit rating. Therefore, according to this graph, the 100-ampere breaker being considered will:

1. carry its 100-ampere (100 percent) rating indefinitely.

2. carry 300 amperes (300 percent) for a minimum of 25 seconds and a maximum of 70 seconds.

3. unlatch its tripping mechanism in 0.0032 seconds (approximately one-quarter cycle) with a current of 5000 amperes.

4. interrupt the circuit in a maximum time of 0.016 seconds (one cycle) with a current of 5000 amperes (5000 percent).

The same time-current curve can be used to determine that a 200-ampere circuit breaker will:

1. carry its 200-ampere (100 percent) rating indefinitely.

2. carry 600 amperes (300 percent) for a minimum of 25 seconds and a maximum of 70 seconds.

3. unlatch its tripping mechanism in 0.0032 seconds (approximately one-quarter cycle) with a current of 5000 amperes.

4. interrupt the circuit in a maximum time of 0.016 seconds (one cycle) with a current of 5000 amperes (2500 percent).

This example shows that if a short circuit occurs in the magnitude of 5000 amperes, then both the 100-ampere breaker and the 200-ampere breaker installed in the same circuit will open together because they have the same unlatching times. In many instances, this action is the reason for otherwise unexplainable power outages (Unit 18). This is rather common when heavy (high value) fault currents occur on circuits protected with molded-case circuit breakers.

Common Misapplication

A common violation of *NEC®* 110.9 and 110.10 is the installation of a main circuit breaker (such as a 100-ampere breaker) that has a high interrupting rating (such as 50,000 amperes) while making the assumption that the branch-circuit breakers (with interrupting ratings of 10,000 amperes) are protected adequately against the 40,000-ampere short circuit, Figure 17-32.

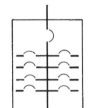

AVAILABLE SHORT-CIRCUIT CURRENT AT PANEL 40,000 AMPERES.

PANEL IS NOT "SERIES-RATED."

*100-AMPERE MAIN BREAKER WITH INTERRUPTING RATING OF 50,000 AMPERES.

MOLDED-CASE CIRCUIT BREAKERS; INTERRUPTING RATING OF 10,000 AMPERES.

*NOTE: UNLESS "SERIES-RATED," THE MAIN BREAKER CANNOT PROTECT THE BRANCH BREAKERS AGAINST A SHORT CIRCUIT OF 40,000 AMPERES; BRANCH BREAKERS ARE CAPABLE OF INTERRUPTING 10,000 AMPERES.

Figure 17-32   Violation of *NEC®* 110.9.

Standard circuit breakers with high interrupting ratings cannot protect a standard circuit breaker with a lower interrupting rating. This is confirmed in the UL *Electrical Construction Equipment Directory* in the circuit breaker section:

> *An interrupting rating on a circuit breaker included in a piece of equipment does not automatically qualify the equipment in which the circuit breaker is installed for use on circuits with higher available currents than the rating of the equipment itself.*

SERIES-RATED APPLICATIONS

Series-rated equipment is tested and listed by a nationally recognized testing laboratory (NRTL) as a total assembly. To establish a series-rating, the testing includes the panelboard and the fuses and/or the breakers in place. Many series-rating combinations are available. Read the label to find the series combination short-circuit rating.

A series-rated system basically is where the available fault current does not exceed the interrupting rating of the line-side overcurrent device but does exceed the interrupting rating of the load-side overcurrent device. The series-rating short-circuit current values will be found on the marking of the equipment.

▶To properly select a series rated system, you must use equipment that is tested and listed by a NRTL. For existing installations only, the components must be selected under engineering supervision. This means selected by a licensed professional engineer engaged primarily in the design or maintenance of electrical installations. See *NEC®* 240.86(A).◀

As discussed a little later, you must also consider motor contribution to the available fault current, since a motor acts much like a generator when a system becomes short-circuited. Instead of drawing current from the system, the motor acting as a generator "pumps" fault current back into the system. Roughly, the motor contribution is approximately the amount of current the motor would draw on start-up. Adding motor contribution to fault-current calculations is not unique to series-rated systems—it must be considered with all systems.

Where high available fault currents indicate the need for high interrupting breakers or fuses, a fully rated system is generally used. In a fully rated system, all breakers and fuses are rated for the fault current available at the point of application. Another less costly way to safely match a main circuit breaker or main fuses ahead of branch-circuit breakers is to use listed *series-rated* equipment. Series-rated equipment is accepted by most electrical inspectors across the country. Series-rated equipment is also referred to as *series-connected*.

Underwriters Laboratories lists series-rated panelboards with breaker main/breaker branches and fused main/breaker branches. Figure 17-33 shows a series-rated panel that may be used where the available fault current does not exceed 22,000 amperes. The panel has a 22,000 AIC main breaker and branch-circuit breakers rated 10,000 AIC. This series arrangement of breakers results in a cushioning effect when the main breaker contacts and the branch breaker contacts simultaneously open under fault conditions that exceed the interrupting rating of the branch-circuit breaker. The impedance of the two arcs in series reduces the fault current that the branch breakers actually see. The impedance of the arc is called "dynamic impedance."

A disadvantage of series-rated systems is that should a heavy fault current occur on any of the branch-circuits, the main breaker will also trip off,

causing a complete loss of power to the panel. Likewise, a series-rated main fuse/breaker branches panel would also experience a power outage, but the possibility exists that only one fuse would open, leaving one-half (for a single-phase panel) or two-thirds (for a three-phase, four-wire panel) of the power on.

For normal overload conditions on a branch-circuit, the branch breaker only will trip off, and should not cause the main breaker to trip or main fuses to open. Refer to Unit 18 for more data on the subject of selectivity and nonselectivity.

CAUTION: Be extremely careful when installing series-rated breakers. Do not mix different manufacturer's breakers. Do not use any breakers that have not been UL recognized as being suitable for use in combination with one another.

The UL *Recognized Component Directory* (*Yellow Book*) has this to say about series-connected breakers:

> *The devices covered under this category are incomplete in certain constructional features or restricted in performance capabilities and are intended for use as components of complete equipment submitted for investigation rather than for direct separate installation in the field. The final acceptance of the component is dependent upon its installation and use in complete equipment submitted to Underwriters Laboratories Inc.*

NEC® 240.86(B) requires that the manufacturer of a series-rated panel must legibly mark the equipment. The label will show the many ampere ratings and catalog numbers of those breakers that are suitable for use with that particular piece of equipment.

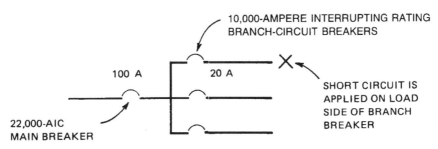

Figure 17-33   Series-rated circuit breakers. In this example, both the 20-ampere breaker and the 100-ampere main breaker trip off under high-level fault conditions.

NEC® 110.22 extends *240.86* by requiring that there be "field marking" of series-rated equipment. Most electrical inspectors expect the electrical contractor to attach this labeling to the equipment. For example, the main distribution equipment of the commercial building is located remote from the individual tenants' panelboards. The electrical contractor will affix legible and durable labels to each disconnect switch in the main distribution equipment, identifying the size and type of overcurrent protection that has been installed for the proper protection of the downstream panelboards, Figure 17-34.

This CAUTION—SERIES RATED SYSTEM labeling will alert future users of the equipment that they should not indiscriminately replace existing overcurrent devices or install additional overcurrent devices with sizes and types of fuses or circuit breakers not compatible with the series-rated system as originally designed and installed. Otherwise, the integrity of the series-rated system might be sacrificed, creating a hazard to life and property.

SERIES-RATED SYSTEMS WHERE ELECTRIC MOTORS ARE CONNECTED

Series-rated systems have an available fault current higher than the interrupting rating of the load-side breakers, and have an acceptable overcurrent device of higher interrupting rating ahead of the load side breakers. The series combinations' interrupting rating is marked on the end-use equipment, *NEC® 240.86(B)*.

Because electric motors contribute short-circuit current under fault conditions on an electrical system, *NEC® 240.86(C)* sets forth two requirements when series-rated systems are applied where electric motors are involved.

1. Do not connect electric motors between the load side of the higher rated overcurrent device and the line side of the lower rated overcurrent device.

```
CAUTION — SERIES COMBINATION
SYSTEM RATED _____ AMPERES
IDENTIFIED REPLACEMENT
COMPONENTS REQUIRED
```

Figure 17-34   Typical label for a series-rated system.

2. The sum of the connected motor full-load currents shall not exceed 1 percent of the interrupting rating of the lower rated circuit breaker. For example, if the interrupting rating of the lower rated breakers in a series-rated panel is 10,000 amperes, then the 1 percent motor load limitation is 100 amperes. Therefore, the full-load currents of all of the motors connected to this series-rated panel must not exceed 100 amperes.

CURRENT-LIMITING BREAKERS

A current-limiting circuit breaker will limit the let-through energy (I^2t) to something less than the I^2t of a one-half cycle symmetrical wave.

The label on this type of breaker will show the words "CURRENT-LIMITING." The label will also indicate the breaker's let-through characteristics or will indicate where to obtain the let-through characteristic data. This let-through data is necessary to ensure adequate protection of downstream components (wire, breakers, controllers, busbar bracing, and so on). Therefore, it is important when installing circuit breakers to ensure that not only the circuit breakers have the proper interrupting rating (capacity), but also that all of the components connected to the circuit downstream from the breakers are capable of withstanding the let-through current of the breaker. To use the previous problem as an example, this means that the branch-circuit conductors must be capable of withstanding 40,000 amperes for approximately one-half cycle (the opening time of the breaker).

COST CONSIDERATIONS

As previously discussed, there are a number of different types of circuit breakers to choose from. The selection of the type to use depends upon a number of factors, including the interrupting rating, selectivity, space, and cost. When an installation requires something other than the standard molded-case-type circuit breaker, the use of high interrupting types or current-limiting types may be necessary. The electrician and/or design engineer must complete a "short-circuit study" to ensure that the overcurrent protective devices provide proper protection against short-circuit conditions per *National Electrical Code®* requirements.

List Price Cost Comparisons		
Type	Interrupting Rating	List Price
Standard, plug-in	10,000	$165.00
Standard, bolt-on	22,000	322.00
Standard, bolt-on	100,000	390.00
Current-limiting	200,000	1163.00
Electronic solid-state having adjustable trips and ground-fault protection, 225 frame size	65,000	3323.00

Table 17-5

The list price cost comparisons in Table 17-5 are for various 100-ampere, 240-volt, panelboard-type circuit breakers.

Molded-case circuit breakers with an integral current-limiting fuse are available. The thermal element in this type of breaker is used for low overloads; the magnetic element is used for low-level short circuits; and the integral fuse is used for short circuits of high magnitude. The interrupting capacity of this breaker/fuse combination (sometimes called a limiter) is higher than that of the same breaker without the fuse.

The selection of the ampacity of circuit breakers for branch-circuits is governed by the requirements of *NEC® Article 240*.

MOTOR CIRCUITS

NEC® Table 430.52 shows that for motor branch-circuits, the maximum setting of a conventional inverse-time circuit breaker must not exceed 250 percent of the full-load current of the motor.

For an instantaneous-trip circuit breaker, permitted only in listed combination controllers, the maximum setting is 800 percent of the motors' full-load current, except for Design B motors. For Design B motors, the maximum setting is 1100 percent. The design data is found on the nameplate of the motor.

The exceptions in *NEC® 430.52(C)* recognize that the 800 percent and 1100 percent settings might not allow the motor to start, particularly in the case of energy-efficient Design B motors. Design B motors have a high starting inrush current that can cause an instantaneous-trip circuit breaker to nuisance trip or a fuse to open. If an "engineering evaluation" can demonstrate the need to exceed the percentages shown in *NEC® Table 430.52*, then:

- the 800 percent setting may be increased to a maximum of 1300 percent.
- the 1100 percent setting may be increased to a maximum of 1700 percent.

Additional discussion regarding motor circuit design is found in Unit 7.

HEATING, AIR-CONDITIONING, AND REFRIGERATION OVERCURRENT PROTECTION

Check the nameplate carefully—and do what it says! The nameplate on HVAC (heating air-conditioning and refrigeration) equipment might indicate "maximum size fuse," "maximum size fuse or circuit breaker," or "maximum size fuse or HACR circuit breaker."

In the past, circuit breakers were subjected to specific tests unique to HVAC equipment and were marked with the letters *HACR*. Today, no special additional tests are made. All currently listed circuit breakers are now suitable for HVAC application. These circuit breakers and HVAC equipment may or may not show the letters *HACR*. Because of existing inventories, it will take many years for the marking HACR to disappear from the scene. In the meantime, read and follow the information found on the nameplate of the equipment. See Figure 17-35, Figure 17-36, and Figure 17-37.

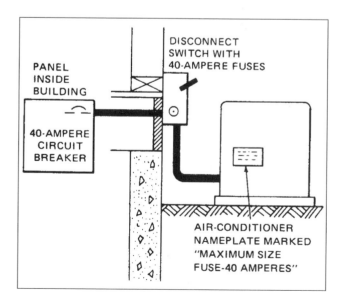

Figure 17-35 This installation complies with *NEC® 440.14*. The disconnect switch is within sight of the unit and contains the 40-ampere fuses called for on the air-conditioner nameplate as the branch-circuit protection.

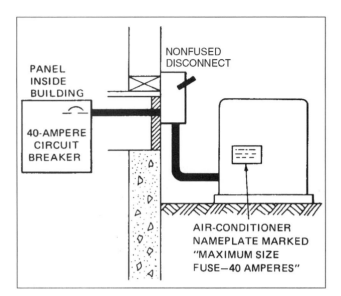

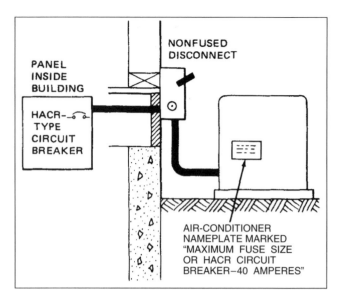

Figure 17-36 This installation violates *NEC® 110.3(B)*. Although the disconnect switch is within sight of the air conditioner, it does not contain fuses. Note that the branch-circuit protection is provided by the 40-ampere circuit breaker inside the building. Note also that the nameplate requires that the branch-circuit protection be via 40-ampere fuses maximum. If fused branch-circuit protection were provided at the panel inside the building, the installation would meet the *Code* requirements.

Figure 17-37 ▶In the past, air-conditioning equipment and circuit breakers were marked with the letters *HACR*. As time passes and current inventory is used up, this marking will disappear from the nameplates of equipment and on the circuit breaker label. In the meantime, follow the manufacturer's installation instructions and the data found on the nameplate. Refer to *NEC® 110.3(B)*.◀

REVIEW QUESTIONS

Refer to the *National Electrical Code®* or the working drawings when necessary. Where applicable, responses should be written in complete sentences.

1. Why must overcurrent protection be provided?

   _____

   _____

   _____

   _____

Several factors should be understood when selecting overcurrent protective devices. In your own words, define each of the following as they apply to overcurrent devices.

2. Voltage rating:

   _____

   _____

   _____

   _____

3. Continuous current rating:

_____

_____

_____

_____

4. Interrupting rating:

_____

_____

_____

_____

5. Speed of response:

_____

_____

_____

_____

6. Current limitation:

_____

_____

_____

_____

7. Peak let-through:

_____

_____

_____

_____

NEC® *Table 430.52* gives four options for selecting the *maximum* size motor branch-circuit, short-circuit, and ground-fault protective devices. These options are listed here. For each option, determine the correct rating or setting for a 25-horsepower, three-phase, 230-volt motor. The motor is marked "Design B."

8. Nontime-delay fuses:

_____

_____

9. Dual-element, time-delay fuses:

_____

_____

10. Instantaneous-trip breakers:

_____

_____

11. Inverse time breakers:

_____

_____

12. Define the following terms:
    a. I^2t _____
    b. I_p _____
    c. RMS _____

13. Class J fuses (will fit) (will not fit) into standard fuse clips. (Circle the correct answer.)

Using the chart from Figure 17-25, find the opening time for a 60-ampere fuse under the following conditions:

14. A 300-ampere load will cause the fuse to open in approximately _____ seconds.

15. A 400-ampere load will cause the fuse to open in approximately _____ seconds.

16. An 800-ampere short circuit will cause the fuse to open in approximately _____ seconds.

17. A 5000-ampere short circuit will cause the fuse to open in approximately _____ seconds.

Using the chart from Figure 17-29, determine the approximate current values for a 60-ampere, 250-volt fuse when the prospective short-circuit current is 40,000 amperes.

18. The instantaneous peak let-through current is approximately _____ amperes.

19. The apparent RMS current is approximately _____ amperes.

20. A section of plug-in busway nameplate indicates that the busway is braced for 14,000 amperes. The available fault current is 30,000 amperes. Using Figure 17-29, determine if a 200-ampere, 250-volt, dual-element fuse will limit the current sufficiently to protect the busway against a possible fault of 30,000 amperes. Explain.

_____

_____

_____

21. In a thermal-magnetic circuit breaker, overloads are sensed by the _____ element, and short circuits are sensed by the _____ element.

22. A cable limiter is
    a. a short-circuit device only.   T   F
    b. not to be used for overload protection of conductors.   T   F
    c. generally connected to both ends of large paralleled conductors so
       that if a fault occurs on one of the conductors, that faulted cable is
       isolated from the system.   T   F

23. Listed series-rated equipment allows the branch-circuit breakers to have a lower interrupting rating than the main. What two sections of the *National Electrical Code*® cover the marking of series-rated systems with the words CAUTION — SERIES RATED SYSTEM? _____

_____

Using the chart in Figure 17-31, provide the unlatching time and interrupting time (also referred to as *opening time*) for a 50-ampere circuit breaker under the following stated conditions. Unlatching time and interrupting time are critical when designing a selectively coordinated electrical system.

24. For a load of 300 amperes, the minimum interrupting time is _____ seconds.

25. For a load of 300 amperes, the maximum interrupting time is _____ seconds.

26. For a 1000-ampere fault, the average unlatching time is _____ seconds.

27. For a 5000-ampere fault, the average unlatching time is _____ seconds.

28. The ampere range of Class L fuses is from _____ to _____ amperes.

29. The voltage rating of a plug fuse is _____ volts.

30. The interrupting rating of a Class J fuse is _____ amperes.

31. When the nameplate of HVAC equipment is marked "Maximum Size Fuse 50-Amperes," is it permitted to connect the equipment to a 50-ampere circuit breaker? (Explain.) _____

_____

UNIT 18

Short-Circuit Calculations and Coordination of Overcurrent Protective Devices

OBJECTIVES

After studying this unit, the student should be able to

- perform short-circuit calculations using the point-to-point method.
- calculate short-circuit currents using the appropriate tables and charts.
- define the terms *coordination*, *selective systems*, and *nonselective systems*.
- use time-current curves.

The student must understand the intent of *NEC®* *110.9* and *110.10*. That is, to ensure that the fuses and/or circuit breakers selected for an installation are capable of interrupting the current at the rated voltage under any condition (overload, short circuit, or ground fault) with complete safety to personnel and without damage to the panel, load center, switch, or electrical equipment in which the protective devices are installed.

An overloaded condition resulting from a miscalculation of load currents will cause a fuse to open or a circuit breaker to trip in a normal manner. However, a miscalculation, a guess, or ignorance of the magnitude of the available short-circuit currents may result in the installation of breakers or fuses having inadequate interrupting ratings. Such a situation can occur even though the load currents in the circuit are checked carefully. Breakers or fuses having inadequate interrupting ratings need only be subjected to a short circuit to cause them to explode, resulting in injury to personnel and serious damage to the electrical equipment. The interrupting rating

of an overcurrent device is its maximum rating and must not be exceeded.

In any electrical installation, individual branch-circuits are calculated as has been discussed previously in this text. After the quantity, size, and type of branch-circuits are determined, these branch-circuit loads are then combined to determine the size of the feeder conductors to the respective panelboards. Most consulting engineers will specify that a certain number of spare branch-circuit breakers be installed in the panelboard, plus a quantity of spaces that can be used in the future.

For example, a certain calculation may require a minimum of sixteen 20-ampere branch-circuits. The specification might call for a twenty-four circuit panelboard with sixteen active circuits, four spares, and four spaces.

The next step is to determine the interrupting rating requirements of the fuses or circuit breakers to be installed in the panel. *NEC®* *110.9* is an all-encompassing section that covers the interrupting rating requirements for services, mains, feeders, subfeeders, and branch-circuit overcurrent of devices. For various types of equipment, normal currents can be determined by checking the equipment name-plate current, voltage, and wattage ratings. In addition, an ammeter can be used to check for normal and overloaded circuit conditions.

A standard ammeter must not be used to read short-circuit current, as this practice will result in damage to the ammeter and possible injury to personnel.

SHORT-CIRCUIT CALCULATIONS

The following sections will cover several of the basic methods of determining available short-circuit

currents. As the short-circuit values given in the various tables are compared with the actual calculations, it will be noted that there are slight variances in the results. These differences are due largely to (1) rounding off the numbers in the calculations and (2) variations in the resistance and reactance data used to prepare the tables and charts. For example, the value of the square root of three (1.732) is used frequently in three-phase calculations. Depending on the accuracy required, values of 1.7, 1.73, or 1.732 can be used.

In actual practice, the available short-circuit current at the load side of a transformer is less than the values shown in Problem 1. However, this simplified

Determining the Short-Circuit Current at the Terminals of a Transformer Using the Impedance Formula

PROBLEM 1:

Assume that the three-phase transformer installed by the utility company for the commercial building has a rating of 300 kilovolt-amperes at 208Y/120 volts with an impedance of 2 percent (from the transformer nameplate). The available short-circuit current at the secondary terminals of the transformer must be determined. To simplify the calculations, it is also assumed that the utility can deliver unlimited short-circuit current to the primary of the transformer. In this case, the transformer primary is known as an *infinite bus* or an *infinite primary*.

The first step is to determine the normal full-load current rating of the transformer:

$$I \text{ (at the secondary)} = \frac{kVA \times 1000}{E \times 1.73} = \frac{300 \times 1000}{208 \times 1.73}$$
$$= 834 \text{ amperes normal full load}$$

Using the impedance value given on the nameplate of the transformer, the next step is to find a multiplier that can be used to determine the short-circuit current available at the secondary terminals of the transformer.

The factor of 0.9 shown in the following calculations reflects the fact that the transformer's actual impedance might be 10 percent less than marked on the nameplate and would be a worst-case condition. In electrical circuits, the lower the impedance, the higher the current.

If the transformer is marked 2 percent impedance:

Then, multiplier $= \dfrac{100}{2 \times 0.9} = 55.556$

and, $834 \times 55.556 = 46,334$ amperes of short-circuit current.

If the transformer is marked 1 percent impedance:

Then, multiplier $= \dfrac{100}{1 \times 0.9} = 111.111$

and, $834 \times 111.111 = 92,667$ amperes of short-circuit current.

If the transformer is marked 4 percent impedance:

Then, multiplier $= \dfrac{100}{4 \times 0.9} = 27.778$

and, $834 \times 27.778 = 23,167$ amperes of short-circuit current.

method of finding the available short-circuit currents will result in values that are conservative.

The actual impedance value on a UL-listed 25-kilovolt-ampere or larger transformer can vary plus or minus 10 percent from the transformer's marked impedance. The actual impedance for a transformer with a marked impedence of 2 percent could be as low as 1.8 percent and as high as 2.2 percent. This will affect the available fault-current calculations.

For example, in Problem 1, the marked impedance is reduced by 10 percent to reflect the transformer's possible actual impedance. The calculations show this worst-case scenario. All short-circuit examples in this text have the marked transformer impedance values reduced by 10 percent.

Another factor that affects fault-current calculations is voltage. Utility companies are allowed to vary voltage to their customers within a certain range. This might be plus or minus 10 percent for power services and 5 percent for lighting services. The higher voltage will result in a larger magnitude of fault current.

Another source of short-circuit current comes from electric motors that are running at the time the fault occurs. This is covered later in this unit.

Thus, it can be seen that no matter how much data we plug into our fault-current calculations, there are many variables that are out of our control. What we hope for is to arrive at a result that is reasonably accurate so that our electrical equipment is reasonably safe insofar as interrupting ratings and withstand ratings are concerned.

In addition to the methods of determining available short-circuit currents that are provided in the following discussion, there are computer software programs that do the calculations. These programs are fast, particularly when there are many points in a system to be calculated.

Determining the Short-Circuit Current at the Terminals of a Transformer Using Tables

Table 18-1 is a table of the short-circuit currents for a typical transformer. NEMA and transformer manufacturers publish short-circuit tables for many sizes of transformers having various impedance values. Table 18-1 provides data for a 300-kilovolt-ampere, three-phase transformer with an impedance of 2 percent. According to the table, the symmetrical short-circuit current is 42,090 amperes at the secondary terminals of a 208Y/120-volt transformer (refer to the zero-foot row of the table). This value is on the low side because the manufacturer that developed the table did not allow for the ± impedance variation allowed by the UL standard. Table 18-2 indicates that the available fault current at the secondary of a 120/208-volt, three-phase, 300-kilovolt-ampere transformer with a 1.11 percent impedance is 83,383 amperes.

Determining the Short-Circuit Current at Various Distances from a Transformer Using the Table

The amount of available short-circuit current decreases as the distance from the transformer increases, as indicated in Table 18-1. See Problem 2.

Determining Short-Circuit Currents at Various Distances from Transformers, Switchboards, Panelboards, and Load Centers Using the Point-to-Point Method

A simple method of determining the available short-circuit currents (also referred to as fault current) at various distances from a given location is the point-to-point method. Reasonable accuracy is obtained when this method is used with three-phase and single-phase systems.

PROBLEM 2:

For a 300-kilovolt-ampere transformer with a secondary voltage of 208 volts, find the available short-circuit current at a main switch that is located 25 ft (7.5 m) from the transformer. The main switch is supplied by four 750-kcmil copper conductors per phase in steel conduit.

Refer to Table 18-1 and read the value of 39,090 amperes in the column on the right-hand side of the table for a distance of 25 ft (7.5 m).

Symmetrical Short-Circuit Currents at Various Distances from a 300-kVA, Three-Phase, 2 Percent Impedance Transformer

WIRE-SIZE (COPPER)

VOLTS	DIST (FT)	(MM)	#14	#12	#10	#8	#6	#4	#1	0	00	000	2-000	0000	250	2-250	3-300	3-350	350	2-350	3-350	3-400	500	2-500	750	4-750
208	0	0	42090	42090	42090	42090	42090	42090	42090	42090	42090	42090	42090	42090	42090	42090	42090	42090	42090	42090	42090	42090	42090	42090	42090	42090
	5	1.5	6910	10290	14730	19970	25240	29840	34690	35770	36640	37340	39610	37930	38270	40100	40870	40960	38840	40410	40960	41030	39300	40650	39650	41460
	10	3	3640	5610	8460	12350	17090	22230	29030	30760	32210	33410	37340	34420	35030	38270	39710	39870	36040	38840	39870	40010	36850	39300	37480	40840
	25	7.5	1500	2360	3670	5650	8430	12150	18930	21170	23240	25090	31710	26750	27780	33590	36560	36930	29550	34780	36930	37230	31020	35730	32190	39090
	50	15	760	1200	1890	2950	4530	6810	11740	13670	15610	17510	25090	19320	20520	27780	32250	32850	22660	29550	32850	33340	24520	31020	26050	36480
	100	30	380	600	960	1510	2350	3610	6610	7920	9320	10810	17510	12320	13380	20520	26010	26850	15400	22660	26850	27530	17250	24520	18860	32190
	200	60	190	300	480	760	1190	1860	3510	4280	5140	6090	10810	7110	7860	13380	18860	19590	9360	15400	19590	20370	10820	17250	12150	26050
	500	150	80	120	190	310	480	760	1460	1800	2180	2630	4990	3130	3500	6510	10030	10770	4290	7820	10770	11400	5100	9120	5870	16570
	1000	300	40	60	100	150	240	380	740	910	1110	1350	2630	1620	1820	3500	5650	6140	2250	4290	6140	6560	2710	5100	3160	10310
	5000	1500	10	10	20	30	50	80	150	180	230	280	550	330	380	740	1260	1380	470	930	1380	1490	570	1130	670	2560
240	0	0	37820	37820	37820	37820	37820	37820	37820	37820	37820	37820	37820	37820	37820	37820	37820	37820	37820	37820	37820	37820	37820	37820	37820	37820
	5	1.5	7750	11330	15810	20720	25260	28940	32560	33340	33960	34460	36080	34870	35120	36420	36960	37020	35520	36640	37020	37070	35840	36800	36090	37370
	10	3	4140	6320	9400	13430	18040	22670	28230	29560	30660	31550	34460	32290	32730	35120	36140	36260	33470	35520	36260	36350	34060	35840	34510	36930
	25	7.5	1720	2700	4180	6360	9360	13190	19640	21620	23380	24920	30240	26260	27090	31650	33860	34130	28480	32530	34130	34340	29610	33230	30510	35680
	50	15	870	1380	2160	3360	5130	7620	12730	14630	16480	18230	24920	19850	20900	27090	30610	31060	22740	28480	31060	31430	24300	29610	25570	33780
	100	30	440	700	1100	1730	2680	4100	7380	8770	10220	11720	18230	13220	14240	20900	25600	26280	16150	22740	26280	26830	17860	24300	19310	30510
	200	60	220	350	550	880	1370	2130	3990	4830	5770	6790	11720	7880	8650	14240	19200	20030	10190	16150	20030	20710	11650	17860	12960	25570
	500	150	90	140	220	350	560	870	1670	2050	2490	2990	5610	3540	3960	7230	10890	11620	4820	8590	11620	12250	5700	9920	6520	17200
	1000	300	40	70	110	180	280	440	850	1050	1280	1550	2990	1850	2080	3960	6300	6820	2560	4820	6820	7270	3070	5700	3570	11130
	5000	1500	10	10	20	40	60	90	170	210	260	320	630	380	430	860	1440	1580	540	1070	1580	1710	660	1290	770	2910
480	0	0	18910	18910	18910	18910	18910	18910	18910	18910	18910	18910	18910	18910	18910	18910	18910	18910	18910	18910	18910	18910	18910	18910	18910	18910
	5	1.5	10450	12820	14750	16150	17080	17690	18200	18310	18400	18470	18690	18520	18550	18730	18800	18810	18610	18760	18810	18810	18650	18780	18690	18850
	10	3	6750	9170	11630	13580	15400	16530	17540	17740	17910	18040	18470	18150	18210	18550	18690	18710	18320	18610	18710	18720	18400	18650	18470	18800
	25	7.5	3180	4740	6770	9150	11520	13570	15690	16160	16540	16840	17830	17100	17250	18040	18380	18410	17490	18180	18410	18440	17690	18280	17840	18630
	50	15	1680	2590	3900	5680	7840	10170	13190	13960	14600	15120	16840	15560	15820	17250	17870	17940	16260	17490	17940	18000	16610	17690	16890	18360
	100	30	860	1350	2090	3180	4680	6600	9820	10810	11690	12460	15120	13130	13540	15820	16930	17060	14240	16260	17060	17170	14810	16610	15260	17840
	200	60	440	690	1080	1680	2560	3810	6370	7320	8240	9110	12460	9930	10450	13540	15300	15530	11370	14240	15530	15710	12150	14810	12780	16890
	500	150	180	280	440	700	1080	1670	3040	3640	4290	4960	8010	5560	6140	9360	11820	12190	7050	10320	12190	12500	7880	11150	8600	14550
	1000	300	90	140	220	350	550	860	1620	1970	2370	2800	4960	3270	3610	6140	8520	8940	4300	7050	8940	9290	4960	7880	5560	11830
	5000	1500	20	30	40	70	110	180	340	420	510	620	1210	750	840	1610	2600	2820	1040	1980	2820	3020	1250	2350	1450	4730
600	0	0	15130	15130	15130	15130	15130	15130	15130	15130	15130	15130	15130	15130	15130	15130	15130	15130	15130	15130	15130	15130	15130	15130	15130	15130
	5	1.5	10210	11790	12920	13690	14180	14500	14770	14820	14870	14900	15010	14930	14940	15040	15070	15080	14970	15050	15080	15080	15000	15060	15010	15100
	10	3	7270	9270	11010	12350	13280	13890	14410	14520	14610	14680	14900	14730	14770	14940	15020	15020	14820	14970	15020	15030	14870	15000	14900	15070
	25	7.5	3740	5370	7280	9230	10920	12200	13410	13670	13870	14040	14570	14170	14250	14680	14850	14870	14380	14750	14870	14890	14490	14800	14570	14980
	50	15	2040	3080	4500	6270	8170	9950	11940	12400	12770	13060	14040	13310	13460	14250	14590	14620	13700	14380	14620	14650	13900	14490	14050	14840
	100	30	1060	1650	2510	3730	5290	7080	9650	10350	10930	11420	13060	11840	12090	13460	14080	14150	12510	13700	14150	14210	12850	13900	13120	14570
	200	60	540	850	1330	2040	3040	4390	6840	7640	8390	9050	11420	9640	10010	12090	13160	13290	10640	12510	13290	13390	11160	12850	11580	14050
	500	150	220	350	550	860	1320	2010	3550	4180	4830	5480	8180	6110	6530	9210	10960	11210	7310	9900	11210	11400	7990	10470	8560	14050
	1000	300	110	170	280	440	680	1050	1950	2360	2800	3270	5480	3760	4110	6530	8540	8860	4780	7310	8860	9120	5410	7990	5970	10940
	5000	1500	20	40	60	90	140	220	420	520	640	770	1470	910	1030	1930	3030	3270	1260	2340	3270	3480	1510	2750	1740	5180

Table 18-1

Short-Circuit Currents Available from Various Size Transformers				
Voltage and Phase	kVA	Full Load Amps	% Impedance†† (Nameplate)	Short-Circuit Amps†
120/240 1 ph.*	25	104	1.58	11,574
	37½	156	1.56	17,351
	50	209	1.54	23,122
	75	313	1.6	32,637
	100	417	1.6	42,478
	167	695	1.8	60,255
120/208 3 ph.**	25	69	1.6	4,791
	50	139	1.6	9,652
	75	208	1.11	20,821
	100	278	1.11	27,828
	150	416	1.07	43,198
	225	625	1.12	62,004
	300	833	1.11	83,383
	500	1388	1.24	124,373
	750	2082	3.5	66,095
	1000	2776	3.5	88,167
	1500	4164	3.5	132,190
	2000	5552	5.0	123,377
	2500	6950	5.0	154,444
277/480 3 ph.	112½	135	1.0	15,000
	150	181	1.2	16,759
	225	271	1.2	25,082
	300	361	1.2	33,426
	500	601	1.3	51,362
	750	902	3.5	28,410
	1000	1203	3.5	38,180
	1500	1804	3.5	57,261
	2000	2406	5.0	53,461
	2500	3007	5.0	66,822

* Single-phase values are L–N values at transformer terminals. These figures are based on change in turns ratio between primary and secondary, 100,000 kVA primary, zero feet from terminals of transformer, 1.2 (%x) and 1.5 (%R) multipliers for L–N vs. L–L reactance and resistance values and transformer X/R ratio = 3.

** Three-phase, short-circuit currents based on "infinite" primary.

†† UL listed transformers 25 kVA or greater have a ±10% impedance tolerance. Short-circuit amps reflect a worst-case condition.

† Fluctuations in system voltage will affect the available short-circuit current. For example, a 10 percent increase in system voltage will result in a 10 percent increase in the available short-circuit currents shown in the table.

Table 18-2

The following procedure demonstrates the use of the point-to-point method:

Step 1. Determine the full-load rating of the transformer in amperes from the transformer nameplate, tables, or the following formulas:

a. For three-phase transformers:

$$I_{FLA} = \frac{kVA \times 1000}{E_{L\text{-}L} \times 1.73}$$

where $E_{L\text{-}L}$ = Line-to-line voltage

b. For single-phase transformers:

$$I_{FLA} = \frac{kVA \times 1000}{E_{L\text{-}L}}$$

Step 2. Find the percent impedance (Z) on nameplate of transformer.

Step 3. Find the transformer multiplier "M_1":

$$M_1 = \frac{100}{\text{transformer \% impedance (Z)} \times 0.9}$$

Note: Because the marked transformer impedance can vary ±10% per the UL Standard 1561, the 0.9 factor takes this into consideration to show worst-case conditions.

Step 4. Determine the transformer let-through short-circuit current at the secondary terminals of the transformer. Use tables or the following formula:

a. For three-phase transformers (L-L-L):

$$I_{SCA} = \text{transformer}_{FLA} \times \text{multiplier "M}_1\text{"}$$

b. For single-phase transformers (L-L):

$$I_{SCA} = \text{transformer}_{FLA} \times \text{multiplier "M}_1\text{"}$$

c. For single-phase transformers (L-N):

$$I_{SCA} = \text{transformer}_{FLA} \times \text{multiplier "M}_1\text{"} \times 1.5$$

Note: The 1.5 factor is explained in Step 5 in the paragraph marked "*X*."

Step 5. Determine the "f" factor:

a. For three-phase faults:

$$f = \frac{1.73 \times L \times I_{L\text{-}L\text{-}L}}{N \times C \times E_{L\text{-}L}}$$

b. For single-phase, line-to-line (L-L) faults on single-phase, center-tapped transformers:

$$f = \frac{2 \times L \times I_{L\text{-}L}}{N \times C \times E_{L\text{-}L}}$$

c. For single-phase, line-to-neutral (L-N) faults on single-phase, center-tapped transformers:

$$f = \frac{2 \times L \times I_{L\text{-}N}}{N \times C \times E_{L\text{-}N}}$$

where . . .

L = the length of the circuit to the fault, in feet.

I = the available fault current in amperes at the beginning of the circuit.

C = the constant derived from Table 18-3 for the specific type of conductors and wiring method.

E = the voltage, line-to-line or line-to-neutral. See Step 4a, b, and c to decide which voltage to use.

N = the number of conductors in parallel.

X = the fault current in amperes at the transformer terminals. At the secondary terminals of a single-phase, center-tapped transformer, the L-N fault current is higher than the L-L fault current. At some distance from the terminals, depending upon the wire size and type, the L-N fault current is lower than the L-L fault current.

This can vary from 1.33 to 1.67 times. These figures are based on the change in the turns ratio between the primary and secondary, infinite source impedance, a distance of zero feet from the terminals of the transformer, and 1.2 percent reactance (X) and 1.5 percent resistance (R) for the L-N versus L-L resistance and reactance values. For simplicity, in Step 4c we used an approximate multiplier of 1.5. First, do the L-N calculation at the transformer secondary terminals, Step 4c, and then proceed with the point-to-point method. See Figure 18-2 for an example.

Step 6. After finding the "f" factor, refer to Table 18-4 and locate in Chart M the appropriate value of the multiplier "M$_2$" for the specific "f" value. Or, calculate as follows:

$$M_2 = \frac{1}{1+f}$$

Step 7. Multiply the available fault current at the beginning of the circuit by the multiplier "M$_2$" to determine the available symmetrical fault current at the fault.

$$I_{SCA} \text{ at fault } = I_{SCA} \text{ at beginning of circuit} \times \text{"M}_2\text{"}$$

Motor Contribution. All motors running at the instant a short circuit occurs contribute to the short-circuit current. The amount of current from the motors is equal approximately to the starting (locked rotor) current for each motor. This current value depends upon the type of motor, its characteristics, and its code letter. Refer to *NEC® Table 430.7(B)*. It is common practice to multiply the full-load ampere rating of the motor by 4 or 5 to obtain a close approximation of the locked-rotor current and provide a margin of safety. For energy-efficient motors, multiply the motor's full-load current rating by 6 to 8 times for a reasonable approximation of fault-current contribution. The current contributed by running motors at the instant a short circuit occurs is added to the value of the short-circuit current at the main switchboard prior to the start of the point-to-point calculations for the rest of the system. To simplify the following problems, motor contributions have not been added to the short-circuit currents.

AWG or kcmil	Copper Conductors Three Single Conductors						Copper Conductors Three Conductor Cable					
	Steel Conduit			Nonmagnetic Conduit			Steel Conduit			Nonmagnetic Conduit		
	600V	5KV	15KV	600V	5KV	15KV	600V	5KV	15KV	600V	5KV	15KV
14	389	—	—	389	—	—	389	—	—	389	—	—
12	617	—	—	617	—	—	617	—	—	617	—	—
10	981	—	—	981	—	—	981	—	—	981	—	—
8	1557	1551	1557	1556	1555	1558	1559	1557	1559	1559	1558	1559
6	2425	2406	2389	2430	2417	2406	2431	2424	2414	2433	2428	2420
4	4779	3750	3695	3825	3789	3752	3830	3811	3778	3837	3823	3798
3	4760	4760	4760	4802	4802	4802	4760	4790	4760	4802	4802	4802
2	5906	5736	5574	6044	5926	5809	5989	5929	5827	6087	6022	5957
1	7292	7029	6758	7493	7306	7108	7454	7364	7188	7579	7507	7364
1/0	8924	8543	7973	9317	9033	8590	9209	9086	8707	9472	9372	9052
2/0	10755	10061	9389	11423	10877	10318	11244	11045	10500	11703	11528	11052
3/0	12843	11804	11021	13923	13048	12360	13656	13333	12613	14410	14118	13461
4/0	15082	13605	12542	16673	15351	14347	16391	15890	14813	17482	17019	16012
250	16483	14924	13643	18593	17120	15865	18310	17850	16465	19779	19352	18001
300	18176	16292	14768	20867	18975	17408	20617	20051	18318	22524	21938	20163
350	19529	17385	15678	22736	20526	18672	22646	21914	19821	24904	24126	21982
400	20565	18235	16365	24296	21786	19731	24253	23371	21042	26915	26044	23517
500	22185	19172	17492	26706	23277	21329	26980	25449	23125	30028	28712	25916
600	22965	20567	17962	28033	25203	22097	28752	27974	24896	32236	31258	27766
750	24136	21386	18888	28303	25430	22690	31050	30024	26932	32404	31338	28303
1000	25278	22539	19923	31490	28083	24887	33864	32688	29320	37197	35748	31959

AWG or kcmil	Aluminum Conductors Three Single Conductors						Aluminum Conductors Three Conductor Cable					
	Steel Conduit			Nonmagnetic Conduit			Steel Conduit			Nonmagnetic Conduit		
	600V	5KV	15KV	600V	5KV	15KV	600V	5KV	15KV	600V	5KV	15KV
14	236	—	—	236	—	—	236	—	—	236	—	—
12	375	—	—	375	—	—	375	—	—	375	—	—
10	598	—	—	598	—	—	598	—	—	598	—	—
8	951	950	951	951	950	951	951	951	951	951	951	951
6	1480	1476	1472	1481	1478	1476	1481	1480	1478	1482	1481	1479
4	2345	2332	2319	2350	2341	2333	2351	2347	2339	2353	2349	2344
3	2948	2948	2948	2958	2958	2958	2948	2956	2948	2958	2958	2958
2	3713	3669	3626	3729	3701	3672	3733	3719	3693	3739	3724	3709
1	4645	4574	4497	4678	4631	4580	4686	4663	4617	4699	4681	4646
1/0	5777	5669	5493	5838	5766	5645	5852	5820	5717	5875	5851	5771
2/0	7186	6968	6733	7301	7152	6986	7327	7271	7109	7372	7328	7201
3/0	8826	8466	8163	9110	8851	8627	9077	8980	8750	9242	9164	8977
4/0	10740	10167	9700	11174	10749	10386	11184	11021	10642	11408	11277	10968
250	12122	11460	10848	12862	12343	11847	12796	12636	12115	13236	13105	12661
300	13909	13009	12192	14922	14182	13491	14916	14698	13973	15494	15299	14658
350	15484	14280	13288	16812	15857	14954	15413	15490	15540	16812	17351	16500
400	16670	15355	14188	18505	17321	16233	18461	18063	16921	19587	19243	18154
500	18755	16827	15657	21390	19503	18314	21394	20606	19314	22987	22381	20978
600	20093	18427	16484	23451	21718	19635	23633	23195	21348	25750	25243	23294
750	21766	19685	17686	23491	21769	19976	26431	25789	23750	25682	25141	23491
1000	23477	21235	19005	28778	26109	23482	29864	29049	26608	32938	31919	29135

Ampacity	Plug-In Busway		Feeder Busway		High Imped. Busway
	Copper	Aluminum	Copper	Aluminum	Copper
225	28700	23000	18700	12000	—
400	38900	34700	23900	21300	—
600	41000	38300	36500	31300	—
800	46100	57500	49300	44100	—
1000	69400	89300	62900	56200	15600
1200	94300	97100	76900	69900	16100
1350	119000	104200	90100	84000	17500
1600	129900	120500	101000	90900	19200
2000	142900	135100	134200	125000	20400
2500	143800	156300	180500	166700	21700
3000	144900	175400	204100	188700	23800
4000	—	—	277800	256400	—

Table 18-3   C Values

Chart to Convert "f" Values to "M₂" Values When Using the Point-to-Point Method			
f	M_2	f	M_2
0.01	0.99	1.20	0.45
0.02	0.98	1.50	0.40
0.03	0.97	2.00	0.33
0.04	0.96	3.00	0.25
0.05	0.95	4.00	0.20
0.06	0.94	5.00	0.17
0.07	0.93	6.00	0.14
0.08	0.93	7.00	0.13
0.09	0.92	8.00	0.11
0.10	0.91	9.00	0.10
0.15	0.87	10.00	0.09
0.20	0.83	15.00	0.06
0.30	0.77	20.00	0.05
0.40	0.71	30.00	0.03
0.50	0.67	40.00	0.02
0.60	0.63	50.00	0.02
0.70	0.59	60.00	0.02
0.80	0.55	70.00	0.01
0.90	0.53	80.00	0.01
1.00	0.50	90.00	0.01
		100.00	0.01

$$M_2 = \frac{1}{1+f}$$

Table 18-4

SHORT-CIRCUIT CURRENT VARIABLES

Phase-to-Phase-to-Phase Fault (L-L-L)

The three-phase fault-current value determined in Step 7 is the *approximate* current that will flow if the three hot phase conductors of a three-phase system are shorted together in what is commonly referred to as a bolted fault. This is the worst-case condition.

Phase-to-Phase Fault (L-L)

To obtain the *approximate* short-circuit current values when two hot conductors of a three-phase system are shorted together, use 87 percent of the three-phase current value. In other words, if the three-phase current value is 20,000 amperes when the three hot lines are shorted together (L-L-L value), then the short-circuit current to two hot lines shorted together (L-L) is approximately:

$$20,000 \times 0.87 = 17,400 \text{ amperes}$$

Phase-to-Neutral (Ground) (L-N) or (L-G)

For solidly grounded three-phase systems, such as the 208Y/120-volt system that supplies the commercial building, the phase-to-neutral (ground)

bolted short-circuit value can vary from 25 percent to 125 percent of the L-L-L bolted short-circuit current value.

It is rare that the L-N (or L-G) fault current would exceed the L-L-L fault current. Therefore, it is common practice to consider the L-N (or L-G) short-circuit current value to be the same as the L-L-L short-circuit current value.

In summary:

L-L-L bolted short-circuit current = 100 percent
L-L   bolted short-circuit current = 87 percent
L-N   bolted short-circuit current = 100 percent

Example:

If the three-phase (L-L-L) fault current has been calculated to be 20,000 amperes, then the L-N fault current is approximately:

$$20,000 \times 1.00 = 20,000 \text{ amperes}$$

The main concern is to provide the proper interrupting rating for the overcurrent protective devices and adequate withstand rating for the equipment. Therefore, for most three-phase electrical systems, the line-line-line bolted fault-current value will provide the desired level of safety.

Arcing Fault Multipliers (Approximate)

There are times when one wishes to know the values of arcing faults. These values vary considerably. The following are acceptable approximations.

Type of Fault	480 Volts	208 Volts
L-L-L	0.89	0.12
L-L	0.74	0.02
L-G (L-N)	0.38*	0

*Some reference books indicate this value to be 0.19.

Example:

What is the approximate line-to-ground arcing fault value on a 480/277-volt system where the line-line-line fault current has been calculated to be 30,000 amperes?

Solution: $30,000 \times 0.38 = 11,400$ amperes

Fault Current at Main Switchboard

As shown in Problem 3, the fuses or circuit breakers located in the main switchboard of the commercial building must have an interrupting capacity of at least 40,311 RMS symmetrical amperes. It is good practice to install protective

PROBLEM 3:

It is desired to find the available short-circuit current at the main switchboard of the commercial building. Once this value is known, the electrician can provide overcurrent devices with adequate inter-rupting ratings and the proper bus barbracing within the switchboard (see *NEC® 110.9* and *110.10*). Figure 18-1 shows the actual electrical system for the commercial building. As each of the following steps in the point-to-point method is examined, refer to the tables given in Table 18-1, Table 18-2, Table 18-3, and Table 18-4 to determine the necessary values of *C, f,* and M.

Step 1. $I_{FLA} = \dfrac{kVA \times 1000}{E_{L\text{-}L} \times 1.73} = \dfrac{300 \times 1000}{208 \times 1.73} = 834$ amperes

Step 2. $M_1 = \dfrac{100}{Z\% \times 0.9} = \dfrac{100}{1.8} = 55.556$

Step 3. $I_{SCA} = I_{FLA} \times M = 834 \times 55.556 = 46,334$

Step 4. $\dfrac{1.73 \times L \times I_{L\text{-}L\text{-}L}}{3 \times C \times E_{L\text{-}L}} = \dfrac{1.73 \times 25 \times 46,334}{3 \times 22,185 \times 208} = 0.145$

Step 5. $M_2 = \dfrac{1}{1+f} = \dfrac{1}{1 + 0.145} = 0.87$

Step 6. The short-circuit current available at the line-side lugs:

$I = I_{SCA} \times M_2 = 46,334 \times 0.87 = 40,311$ RMS symmetrical amperes.

(You could also refer to Table 18-4, then approximate the multiplier "M_2" by noting that "f" value of 0.145 is close to 0.15, where "M" equals 0.87.)

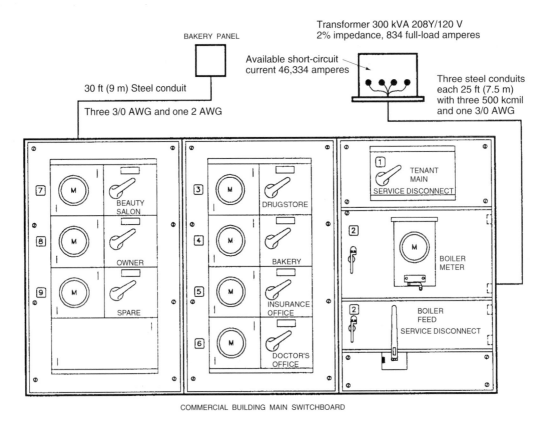

COMMERCIAL BUILDING MAIN SWITCHBOARD

Figure 18-1   Electrical system for the commercial building.

devices having an interrupting rating at least 25 percent greater than the actual calculated available fault current. This practice generally provides a margin of safety to permit the rounding off of numbers, as well as compensating for a reasonable amount of short-circuit contribution from any electrical motors that may be running at the instant the fault occurs.

The fuses specified for the commercial building have an interrupting rating of 200,000 amperes (see the Specifications). In addition, the switchboard bracing is specified to be 50,000 amperes.

In the commercial building, series-rated equipment may be installed. Current-limiting fuses not to exceed the maximum size and type specified by the manufacturer of the equipment are installed in the main switchboard for the protection of the feeder conductors and panelboards. Circuit breakers having an interrupting rating of 10,000 amperes are installed in the panelboards. The requirements for this arrangement are set in the Specifications for the commercial building.

If series-rated equipment is not installed, the panelboards must be fully rated for the available fault current at each location. Series-rated equipment may be selected under engineering supervision or by using tested combinations.

If noncurrent-limiting overcurrent devices (standard molded-case circuit breakers) are to be installed in the main switchboard, breakers having adequate interrupting ratings must be installed in the panelboards. A short-circuit study must be made for each panelboard location to determine the value of the available short-circuit current.

The cost of circuit breakers increases as the interrupting rating of the breaker increases. The most economical protection system generally results when current-limiting fuses are installed in the main switchboard to protect the breakers in the panelboards.

See Problem 4 as an example of short-circuit calculations for a single-phase transformer.

Review of Short-Circuit Requirements

1. To meet the requirements of *NEC® 110.9* and *110.10*, it is absolutely necessary to determine the available fault currents at various points on the electrical system. If a short-circuit study is not done, the selection of overcurrent devices may be in error, resulting in a hazard to life and property.

2. To install a fully rated system, use fuses and circuit breakers that have an interrupting rating not less than the available fault current at the point of application.

3. To install a series-rated system, use listed series-rated electrical equipment where the available fault current exceeds the short-circuit rating of the downstream lower-rated circuit breakers to be installed in the system. Series-rated electrical equipment is available with fuse/breaker and breaker/breaker combinations. The lower interrupting rated circuit breakers are protected by a higher interrupting rated main device. Tested in combination, this arrangement is called series-rated and is addressed in *NEC® 240.86*. Series-rated electrical equipment is marked with its maximum short-circuit rating and the type of overcurrent devices permitted to be used in the specific series-rated combination. Series-rated systems may also be designed and selected under engineering supervision. See Problem 4.

▶To properly select a series-rated system, you must use equipment that is tested and listed by an NRTL, or for existing installations only, select the components under engineering supervision. This means they are selected by a licensed professional engineer engaged primarily in the design or maintenance of electrical installations. ◀ Refer to *NEC® 240.86*.

For an easy-to-use, computer version of doing short-circuit calculations, visit the Bussmann Web

---

PROBLEM 4. Single-Phase Transformer:

This problem is illustrated in Figure 18-2. The point-to-point is used to determine the currents for both line-to-line and line-to-neutral faults for a 167-kilovolt-ampere, 2 percent impedance transformer on a 120/240-volt, single-phase system. The marked transformer impedance can vary by ±10 percent per UL Standard 1561. To show the worst-case condition, this is accounted for at the beginning of the calculations by reducing the impedance 10 percent using this formula: $\dfrac{100}{2 \times 0.9} = M_1$

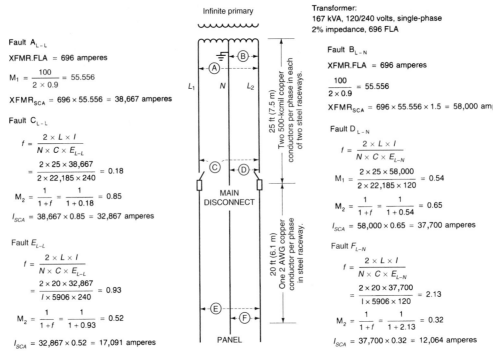

Figure 18-2   Point-to-point calculations for a single-phase, center-tapped transformer. Calculations show L-L and L-N values.

site at http://www.bussmann.com. Point to "Application Info" in the left column, and in the drop-down window click on "Software." In the following drop-down window click on "Short Circuit Calculations."

Problem 4 presents an excellent exercise in doing a short-circuit calculation, whether manually or by using a computer program.

COORDINATION OF OVERCURRENT PROTECTIVE DEVICES

Although this text cannot cover the topic of electrical system coordination (selectivity) in detail, the following material will provide the student with a working knowledge of this important topic. See Problem 4.

What Is Coordination?

Electrical system overcurrent protective devices can be coordinated "selective" or "non selective."

▶The *NEC*® defines selective coordination as: *Localization of an overcurrent condition to restrict outages to the circuit or equipment affected, accomplished by the choice of overcurrent protective devices and their ratings or settings.**◀

A situation known as nonselective coordination occurs when a fault on a branch-circuit opens not

only the branch-circuit overcurrent device but also the feeder overcurrent device, Figure 18-3. Nonselective systems are installed unknowingly and cause needless power outages in portions of an electrical system that should not be affected by a fault.

A selectively coordinated system, Figure 18-4, is one in which only the overcurrent device immediately upstream from the fault opens. Obviously, the installation of a selective system is much more desirable than a nonselective system.

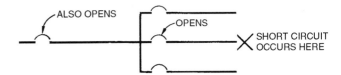

THE FAULT ON THE BRANCH-CIRCUIT TRIPS BOTH THE BRANCH-CIRCUIT BREAKER AND THE FEEDER CIRCUIT BREAKER. AS A RESULT, POWER TO THE PANEL IS CUT OFF, AND CIRCUITS THAT SHOULD NOT BE AFFECTED ARE NOW OFF.

Figure 18-3   Schematic of a nonselective system.

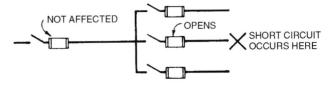

ONLY THE BRANCH-CIRCUIT FUSE OPENS. ALL OTHER CIRCUITS AND THE MAIN FEEDER REMAIN ON.

Figure 18-4   Schematic of a selective system.

*Reprinted with permission from NFPA 70-2005.

The importance of selectivity in an electrical system is covered extensively throughout *NEC® Article 517*. This article pertains to health care facilities, where maintaining electrical power is extremely important. The unexpected loss of power in certain areas of hospitals, nursing care centers, and similar health care facilities can be catastrophic.

The importance of selectivity is also emphasized in *NEC® 240.12* (Electrical Systems Coordination), *NEC® 230.95* (Ground-Fault Protection for Equipment), *NEC® 620.62* (Selective coordination for elevators, escalators, and moving walks), *NEC® 700.27* (coordination for emergency circuits), and *NEC® 701.18* (Optional Standby Systems). These sections refer to installations where additional hazards would be introduced should a nonorderly shutdown occur. The *Code* defines coordination, in part, as the proper localizing of a fault condition to restrict outages to the equipment affected.

Some local electrical codes require that all circuits, feeders, and mains in buildings such as schools, shopping centers, assembly halls, nursing homes, retirement homes, churches, restaurants, and any other places of public occupancy be selectively coordinated so as to minimize the dangers associated with total power outages.

Nonselectivity of an electrical system is not considered good design practice and is generally accepted only as a trade-off for a low-cost installation.

It is advisable to check with the authority enforcing the *Code* before proceeding too far in the selection of overcurrent protective devices for a specific installation.

By knowing how to determine the available short-circuit current and ground-fault current, the electrician can make effective use of the time-current curves and peak let-through charts (Unit 17) to find the length of time required for a fuse to open or a circuit breaker to trip.

What Causes Nonselectivity?

In Figure 18-5, a short circuit in the range of 3000 amperes occurs on the load side of a 20-ampere breaker. The magnetic trip of the breaker is adjusted permanently by the manufacturer to unlatch at a current value equal to 10 times its rating or 200 amperes. The feeder breaker is rated at 100 amperes; the magnetic trip of this breaker is set by the manufacturer to unlatch at a current equal to 10 times its rating or 1000 amperes. This type of breaker generally cannot be adjusted in the field. Therefore, a current of 200 amperes or more will cause the 20-ampere breaker to trip instantly. In addition, any current of 1000 amperes or more will cause the 100-ampere breaker to trip instantly.

For the breakers shown in Figure 18-5, a momentary fault of 3000 amperes will trip (unlatch) both breakers. Because the flow of current in a series circuit is the same in all parts of the circuit, the 3000-ampere fault will trigger both magnetic trip mechanisms. The time-current curve shown in Figure 17-31 indicates that for a 3000-ampere fault, the unlatching time for both breakers is 0.0042 second and the interrupting time for both breakers is 0.016 second.

The term *interrupting time* refers to the time it takes for the circuit breakers' contacts to open, thereby stopping the flow of current in the circuit. Refer to Figure 18-5 and Figure 18-6.

This example of a nonselective system should make apparent to the student the need for a thorough study and complete understanding of time-current curves, fuse selectivity ratios, and unlatching time data for circuit breakers. Otherwise, a blackout may occur, such as the loss of exit and emergency lighting. The student must be able to determine available short-circuit currents (1) to ensure the proper selection of protective devices with adequate interrupting ratings and (2) to provide the proper coordination as well.

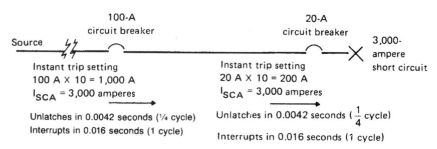

Figure 18-5 Nonselective system verification.

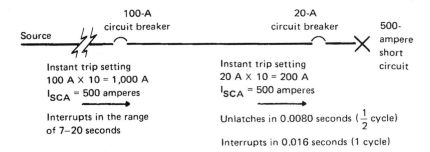

Figure 18-6   Selective system verification.

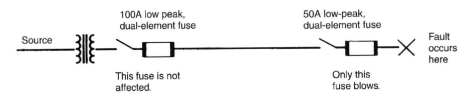

Figure 18-7   Selective system using fuses.

Figure 18-6 shows an example of a selective circuit. In this circuit, a fault current of 500 amperes trips the 20-ampere breaker instantly (the unlatching time of the breaker is approximately 0.0080 second, and its interrupting time is 0.016 second). The graph in Figure 17-31 indicates that the 100-ampere breaker interrupts the 500-ampere current in a range from 7 to 20 seconds. This relatively lengthy trip time range is due to the fact that the 500-ampere fault acts upon the current thermal trip element only and does not affect the magnetic trip element, which operates on a current of 1000 amperes or more.

Selective System Using Fuses

The proper choice of the various classes and types of fuses is necessary if selectivity is to be achieved, Figure 18-7. Indiscriminate mixing of fuses of different classes, time-current characteristics, and even manufacturers may cause a system to become nonselective.

To ensure selective operation under low-overload conditions, it is necessary only to check and compare the time-current characteristic curves of fuses. Selectivity occurs in the overload range when the curves do not cross one another. See Problem 5.

Fuse manufacturers publish Selectivity Guides, similar to the one shown in Table 18-5, to be used for short-circuit conditions. When using these guides, selectivity is achieved by maintaining a specific amperage ratio between the various classes and types of fuses. A selectivity chart is based on any fault current up to the maximum interrupting ratings of the fuses listed in the chart.

PROBLEM 5:
    It is desired to install 100-ampere, dual-element fuses in a main switch and 50-ampere dual-element fuses in the feeder switch. Is this combination of fuses "selective"?
    Refer to the Selectivity Guide in Table 18-5 using the *line* and *column* marked Ⓐ. Because 100:50 is a 2:1 ratio, the installation is "selective." In addition, any fuse of the same type having a rating of less than 50 amperes will also be "selective" with the 100-ampere main fuses. Thus, for any short circuit or ground-fault on the load side of the 50-ampere fuses, only the 50-ampere fuses will open.

	Ratios for Selectivity									
LINE-SIDE FUSE	**LOAD-SIDE FUSE**								Ⓐ	
	KRP-CSP LOW-PEAK time-delay Fuse 601–6000A Class L	KTU LIMITRON fast-acting Fuse 601–6000A Class L	KLU LIMITRON time-delay Fuse 601–4000A Class L	KTN-R, KTS-R LIMITRON fast-acting Fuse 0–600A Class RK1	JJS, JJN TRON fast-acting Fuse 0–1200A Class T	JKS LIMITRON quick-acting Fuse 0–600A Class J	FRN-R, FRS-R FUSETRON dual-element Fuse 0–600A Class RK5	LPN-RK-SP, LPS-RK-SP LOW-PEAK dual-element Fuse 0–600A Class RK1	LPJ-SP LOW-PEAK time-delay Fuse 0–600A Class J	SC Type Fuse 0–60A Class G
KRP-CSP LOW-PEAK time-delay Fuse 601–6000A Class L	2:1	2:1	2.5:1	2:1	2:1	2:1	4:1	2:1	2:1	N/A
KTU LIMITRON fast-acting Fuse 601–6000A Class L	2:1	2:1	2.5:1	2:1	2:1	2:1	6:1	2:1	2:1	N/A
KLU LIMITRON time-delay Fuse 601–4000A Class L	2:1	2:1	2:1	2:1	2:1	2:1	4:1	2:1	2:1	N/A
KTN-R, KTS-R LIMITRON fast-acting Fuse 0–600A Class RK1	N/A	N/A	N/A	3:1	3:1	3:1	8:1	3:1	3:1	4:1
JJN, JJS TRON fast-acting Fuse 0–1200A Class T	N/A	N/A	N/A	3:1	3:1	3:1	8:1	3:1	3:1	4:1
JKS LIMITRON quick-acting Fuse 0–600A Class J	N/A	N/A	N/A	3:1	3:1	3:1	8:1	2:1	2:1	4:1
FRN-R, FRS-R FUSETRON dual-element Fuse 0–600A Class RK5	N/A	N/A	N/A	1.5:1	1.5:1	1.5:1	2:1	1.5:1	1.5:1	1.5:1
LPN-RK-SP, LPS-RK-SP LOW-PEAK dual-element Fuse 0–600A Class RK1 Ⓐ	N/A	N/A	N/A	3:1	3:1	3:1	8:1	2:1	2:1	4:1
LPJ-SP LOW-PEAK time-delay Fuse 0–600A Class J	N/A	N/A	N/A	3:1	3:1	3:1	8:1	2:1	2:1	4:1
SC Type Fuse 0–60A Class G	N/A	N/A	N/A	2:1	2:1	2:1	4:1	3:1	3:1	2:1

N/A = NOT APPLICABLE

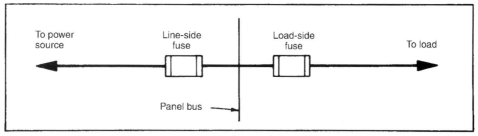

Selectivity Guide. This chart is one manufacturer's selectivity guide. All manufacturers of fuses have similar charts, but the charts may differ in that the ratios for given combinations of fuses may not be the same as in this chart. It is important to use the technical data for the particular type and manufacturer of the fuses being used. Intermixing fuses from various manufacturers may result in a nonselective system.

Table 18-5

Selective System Using Circuit Breakers

Circuit-breaker manufacturers publish time-current characteristic curves and unlatching information, Figure 17-31.

For normal overload situations, a circuit breaker having an ampere rating lower than the ampere rating of an upstream circuit breaker will trip. The upstream breaker will not trip. The system is "selective."

For low-level faults less than the instantaneous-trip setting of an upstream circuit breaker, a circuit breaker having an ampere rating lower than the ampere rating of the upstream breaker will trip, and the upstream breaker will not trip. The system is "selective," Figure 18-6.

For fault-current levels above the instantaneous-trip setting of the upstream circuit breaker, both the branch-circuit breaker and the upstream circuit breaker will trip off. The system is "nonselective," Figure 18-5.

There are no ratio selectivity charts for breakers as there are for fuses, Table 18-5.

SINGLE PHASING

The *National Electrical Code®* requires that all three-phase motors be provided with running overcurrent protection in each phase in the motor circuit, as stated in *NEC® Article 430, Part III.* A line-to-ground fault generally will open one fuse. There will be a resulting increase of 173–200 percent in the line current in the remaining two connected phases. This increased current will be sensed by the motor fuses and overload relays when such fuses and relays are sized at 125 percent or less of the full-load current rating of the motor. Thus, the fuses and/or the overload relays will open before the motor windings are damaged. When properly matched, the overload relays will open before the fuses. The fuses offer backup protection if the overload relays fail to open for any reason.

A line-to-line fault in a three-phase motor will open two fuses. In general, the operating coil of the motor controller will drop out, thus providing protection to the motor winding.

To reduce single-phasing problems, each three-phase motor must be provided with individual overload protection through the proper sizing of the overload relays and fuses. Phase-failure relays are also available. However, can other equipment be affected by a single-phasing condition?

In general, loads that are connected line-to-neutral or line-to-line, such as lighting, receptacles, and electric heating units will not burn out under a single-phasing condition. In other words, if one main fuse opens, then two-thirds of the lighting, receptacles, and electric heat will remain on. If two main fuses open, then one-third of the lighting remains on, and a portion of the electric heat connected line-to-neutral will stay on.

Nothing can prevent the occurrence of single-phasing. What must be detected is the increase in current that occurs under single-phase conditions.

This is the purpose of motor overload protection as required in *NEC® Article 430, Part III.*

It is essential to maintain some degree of lighting in occupancies such as stores, schools, offices, and health care facilities (such as nursing homes). A total blackout in these public structures has the potential for causing extensive personal injury due to panic. A loss of one or two phases of the system supplying a building should not cause a complete power outage in the building.

REVIEW QUESTIONS

Refer to the *National Electrical Code®* or the working drawings when necessary. Where applicable, responses should be written in complete sentences.

Refer to Table 18-1, which shows the symmetrical short-circuit currents for a 300-kilovolt-ampere transformer with 2 percent impedance.

1. The current on a 208-volt system at a distance of 50 ft using 1 AWG conductors will be _____ amperes.

2. The current on a 480-volt system at a distance of 100 ft using 500-kcmil conductors will be _____ amperes.

3. In your own words, define and give a hypothetical example of a selectively coordinated system.

_____

_____

_____

_____

_____

_____

4. Identify some common causes of nonselectivity.

_____

_____

_____

_____

Indicate whether the following systems would be selective or nonselective. Support your answer.

5. A KRP-C 2000-ampere fuse is installed on a feeder that serves an LPS-R 600-ampere fuse.

_____

_____

_____

_____

6. A KTS-R 400-ampere fast-acting fuse is installed on a feeder that serves an FRS-R 200-ampere dual-element fuse.

_____

_____

_____

_____

7. A panelboard is protected by a main 225-ampere circuit breaker and contains several 20-ampere breakers for branch-circuit protection. The breakers are all factory set to unlatch at 10 times their rating. Circle your response.

   a. For a 500-ampere short circuit or ground fault on the load side of a 20-ampere breaker, only the 20-ampere breaker will trip off.   T   F

   b. For a 3000-ampere short circuit or ground fault on the load side of the 20-ampere breaker, the 20-ampere breaker and the 225-ampere breaker will trip off.   T   F

Refer to the following drawing to make the requested calculations. Use the point-to-point method and show all calculations.

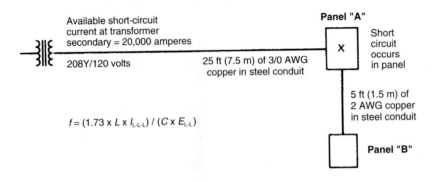

8. Calculate the short-circuit current at panelboard A.

9. Calculate the short-circuit current at panelboard B.

Calculate the following fault-current values for a three-phase L-L-L bolted short circuit that has been calculated to be 40,000 amperes.

10. The approximate value of a line-to-line fault will be _____ amperes.

11. The approximate value of a line-to-ground fault will be _____ amperes.

UNIT 19

Equipment and Conductor Short-Circuit Protection

OBJECTIVES

After studying this unit, the student should be able to

- understand that all electrical equipment has a withstand rating.
- discuss the withstand rating of conductors.
- understand important *National Electrical Code®* sections that pertain to interrupting rating, available short-circuit current, current-limitation, effective grounding, bonding, and temperature limitation of conductors.
- understand what the term *ampere squared seconds* means.
- perform calculations to determine how much current a copper conductor can safely carry for a specified period of time before being damaged or destroyed.
- refer to charts to determine conductor withstand ratings.
- understand that the two forces present when short circuits or overloads occur are *thermal* and *magnetic*.
- discuss the 10-ft (3-mm) tap conductor maximum overcurrent protection.

All electrical equipment, including switches, motor controllers, conductors, bus duct, panelboards, load centers, switchboards, and so on, has the capability to withstand a certain amount of electrical energy for a given amount of time before damage to the equipment occurs. This gives rise to the term *component short-circuit current rating.* This term is used in the *NEC®* and in numerous UL Standards. The term is synonymous with the term *withstand rating.*

Underwriters Laboratories standards specify certain test criteria for the preceding equipment. For example, switchboards must be capable of withstanding a given amount of fault current for at least three cycles. Both the amount of current and the length of time for the test are specified.

Simply stated, component short-circuit current rating is the ability of the equipment to hold together for the time it takes the overcurrent protective device to respond to the fault condition.

Acceptable damage levels are well defined in the various Underwriters Standards for electrical equipment. Where the equipment is intended to actually break the current, such as with a fuse or circuit breaker, the equipment is marked with its interrupting rating. Electrical equipment manufacturers conduct exhaustive tests to determine the withstand and/or interrupting ratings of their products.

Equipment tested and listed is often marked with the size and type of overcurrent protection required. For example, the label on a motor controller might indicate a maximum size fuse for different sizes of thermal overloads used in that controller. This marking indicates that fuses must be used for the overcurrent protection. A circuit breaker would not be permitted, see *NEC® 110.3(B)*.

The capability of a current-limiting overcurrent device to limit the let-through energy to a value less than the amount of energy that the electrical system is capable of delivering means that the equipment can be protected against fault-current values of high magnitude. Equipment manufacturers specify current-limiting fuses to minimize the potential damage that might occur in the event of a high-level short circuit or ground fault.

The electrical engineer and/or electrical contractor must perform short-circuit studies and then determine the proper size and type of current-limiting overcurrent protective device, which can be used ahead of the electrical equipment that does not have an adequate withstand or interrupting rating for the available fault current to which it will be subjected. The calculation of short-circuit currents is covered in Unit 18.

Study the normal circuit, overloaded circuit, short-circuit, ground fault, and open-circuit diagrams (Figure 19-1A through Figure 19-1E), and observe how Ohm's law is applied to these circuits. The calculations for the ground-fault circuit are not shown, as the impedance of the return ground path can vary considerably.

CONDUCTOR WITHSTAND RATING

Up to this point in the text, we have covered in detail how to compute conductor ampacities for branch-circuits, feeders, and service-entrance conductors, with the basis being the connected load and/or volt-amperes per square foot. Then the demand factors and other diversity factors are applied. If we have followed the *Code* rules for calculating conductor size, the conductors will not be damaged under overloaded conditions because the conductors will have the proper ampacity and will be protected by the proper size and type of overcurrent protective device.

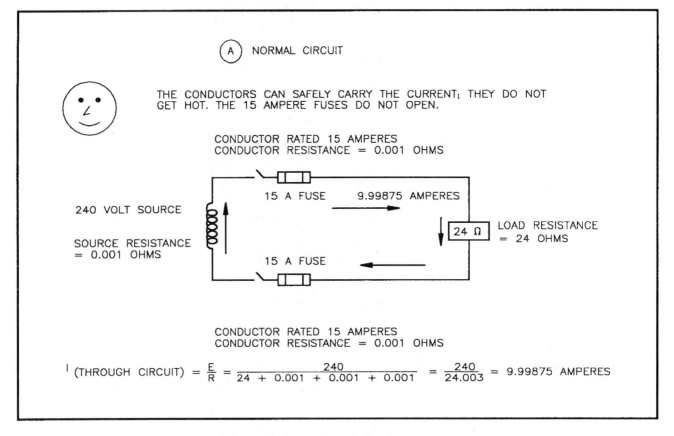

Figure 19-1A   Normal circuit current.

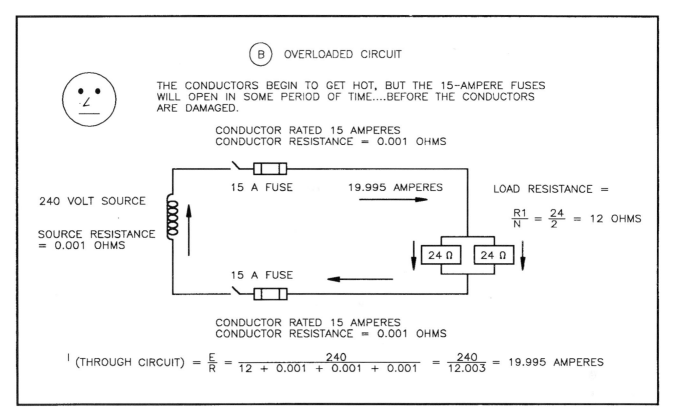

Figure 19-1B   Overloaded circuit current.

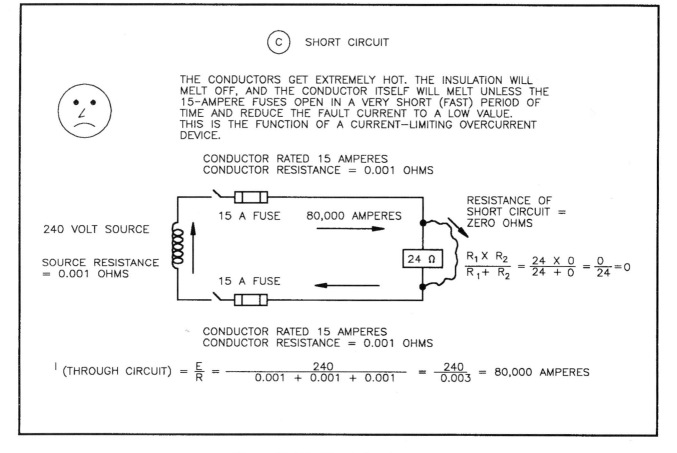

Figure 19-1C   Short-circuit current.

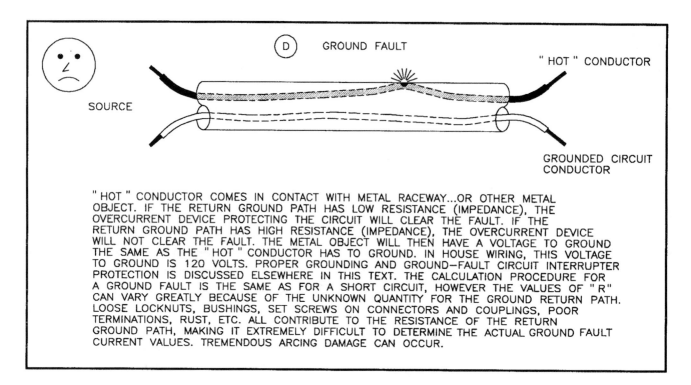

Figure 19-1D   Ground-fault current.

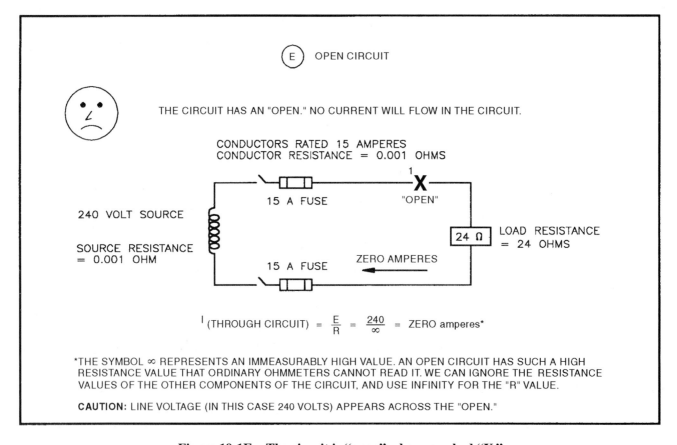

Figure 19-1E   The circuit is "open" where marked "X."

This part of Unit 19 discusses this topic in a reasonable amount of detail. As difficult as the subject may seem, it is extremely important.

NEC® tables such as *Table 310.16* set forth the ampacity ratings for various sizes and insulation types of conductors. These tables consider normal loading conditions.

NEC® 310.10, 310.15(C), and *Annex B* in the *NEC®* also relate to a conductor's current-carrying capability. For those of you having the interest to really dig into this subject, you can find additional information in the ANSI/IEEE Red, Gray, Buff, and Blue books, as well as the Canadian Electrical Code and the International Electrotechnical Commission Standards.

When short circuits and/or ground faults occur, the connected loads are bypassed, thus the term *short circuit.* All that remains in the circuit are the conductors and other electrical devices such as breakers, switches, and motor controllers. The impedance (ac resistance) of these devices is extremely low, so for all practical purposes, the most significant opposition to the flow of current when a fault occurs is the conductor impedance. Fault-current calculations are covered in Unit 18.

The following discussion covers the actual ability of a conductor to maintain its integrity for the time it is called upon to carry fault current, instead of burning off and causing additional electrical problems. It is important that the electrician, electrical contractor, electrical inspector, and consulting engineer have a good understanding of what happens to a conductor under medium- to high-level fault conditions.

The withstand rating of a conductor, such as an equipment grounding conductor, main bonding jumper, or any other current-carrying conductor, reveals that the conductor can withstand a "certain amount of current for a given amount of time." This is the short-time withstand rating of the conductor.

Let us review some key *National Electrical Code®* sections that focus on the importance of equipment and conductor withstand rating, available fault currents, and circuit impedance.

- The interrupting rating is *the highest current at rated voltage that a device is intended to inter-*
rupt under standard test conditions. See NEC® Article 100.

- *Listed or labeled equipment shall be installed and used in accordance with any instructions included in the listing or labeling. See NEC® 110.3(B).*

- **Interrupting Rating.** *Equipment intended to break current at fault levels shall have an interrupting rating sufficient for the system voltage and the current that is available at the line terminals of the equipment. See NEC® 110.9.*

- **Circuit Impedance and Other Characteristics.** *The overcurrent protective devices, the total impedance, the component short-circuit current ratings, and other characteristics of the circuit to be protected shall be selected and coordinated to permit the circuit protective devices to clear a fault without extensive damage to the electrical components of the circuit. This fault shall be assumed to be either between two or more of the circuit conductors or between any circuit conductor and the grounding conductor or enclosing metal raceway. Listed products applied in accordance with their listing shall be considered to meet the requirements of this section. See NEC® 110.10.*

- Service equipment shall be suitable for the short-circuit current available at its supply terminals. See *NEC® 110.9.*

- *Overcurrent protection for conductors and equipment is provided to open the circuit if the current reaches a value that will cause an excessive or dangerous temperature in conductors or conductor insulation. See NEC® 240.1, FPN.*

- *A current-limiting overcurrent protective device is a device that, when interrupting currents in its current-limiting range, will reduce the current flowing in the faulted circuit to a magnitude substantially less than that obtainable in the same circuit if the device were replaced with a solid conductor having comparable impedance. See NEC® 240.2.*

- The required label marking for fuses, one item being the fuse's interrupting rating, is set forth in *NEC® 240.60(C).*

- The required label marking for circuit breakers, one item being the circuit breaker's interrupting rating, is set forth in *NEC® 240.83(C)*.

- The requirements for series-rated equipment are contained in *NEC® 240.86*.

- Ground the electrical system so as to limit the voltage imposed by lightning, line surges, or unintentional contact with higher voltage lines so as to stabilize the voltage to the earth during normal operation. See *NEC® 250.4(A)(1)*.

- Ground conductive electrical conductors and equipment to limit the voltage to ground on these materials. See *NEC® 250.4(A)(2)*.

- Metal piping and other electrically conductive materials that might become energized must be bonded together so as to establish an effective path for fault currents. See *NEC® 250.4(A)(5)*.

- ►*NEC® 250.4(A)(5)* requires that an effective ground-fault current path be established for all electrical equipment, wiring, and other electrically conductive material likely to become energized. This path shall be installed:

  1.   permanently, and
  2.   be of low impedance, and capable of safely carrying the maximum ground fault current likely to be imposed on it.

  The earth shall not be considered as an effective ground-fault current path. See *NEC® 250.4(A)(5)*.◄

- *Bonding shall be provided where necessary to ensure electrical continuity and the capacity to conduct safely any fault current likely to be imposed.* See *NEC® 250.90*.

- *Metal raceways, cable trays, cable armor, cable sheath, enclosures, frames, fittings, and other metal noncurrent-carrying parts that are to serve as grounding conductors with or without the use of supplementary equipment grounding conductors shall be effectively bonded where necessary to ensure electrical continuity and the capacity to conduct safely any fault current likely to be imposed on them. Any nonconductive paint, enamel, or similar coating shall be removed at threads, contact points, and contact surfaces or be connected by means of fittings so*

designed as to make such removal unnecessary. See *NEC® 250.96(A)*.

- The size of the grounding electrode conductor for a grounded or ungrounded system shall not be less than given in *NEC® Table 250.66*. See *NEC® 250.66*.

- The size of copper, aluminum, or copper-clad aluminum equipment grounding conductors shall not be less than given in *NEC® Table 250.122*. See *NEC® 250.122*.

- Attention should be given to the requirement that equipment grounding conductors may need to be sized larger than shown in *NEC® Table 250.122* because of the need to comply with *NEC® 250.4(A)(5)* regarding effective grounding. This would generally be necessary where available fault currents are high, and the possibility of burning off the equipment grounding conductor under fault conditions exists. See *NEC® 250.122*.

- *NEC® 310.10* is rather lengthy, presenting in detail the requirement that *no conductor shall be used in such a manner that its operating temperature will exceed that designated for the type of insulated conductor involved.* This section should be read carefully.

CONDUCTOR HEATING

The value of energy (heat) generated during a fault varies as the square of the root-mean-square (RMS) current multiplied by the time (duration of fault) in seconds. This value is expressed as

$$I^2t$$

The term I^2t is called "ampere squared seconds."

Because watts = I^2R, we could say that the damage that might be expected under severe fault conditions can be related to:

1. the amount of current flowing during the fault.
2. the time in seconds that the fault current flows.
3. the resistance of the fault path.

This relationship is expressed as

$$I^2Rt$$

Because the value of resistance under severe fault conditions is generally extremely low, we can simply think in terms of how much current (I) is flowing and for how long a time (t) it will flow.

It is safe to say that whenever an electrical system is subjected to a high-level short circuit or ground fault, less damage will occur if the fault current can be limited to a low value and if the faulted circuit is cleared as fast as possible.

It is important to understand the time-current characteristics of fuses and circuit breakers in order to minimize equipment damage. Time-current characteristic curves were discussed in Unit 17.

CALCULATING AN INSULATED (167°F [75°C] THERMOPLASTIC) CONDUCTOR'S SHORT-TIME WITHSTAND RATING

Copper conductors can withstand

- one ampere (RMS current)

- for five seconds

- for every 42.25 circular mils (cm) of cross-sectional area.

In the preceding statement both current (how much) and time (how long) are included.

If we wish to provide an equipment grounding conductor for a circuit protected by a 60-ampere overcurrent device, we find in *NEC® Table 250.122* that a 10 AWG copper conductor is the MINIMUM size permitted.

Referring to *NEC® Chapter 9, Table 8,* we find that the cross-sectional area of a 10 AWG conductor is 10,380 circular mils.

This 10 AWG copper conductor has a five-second withstand rating of

$$\frac{10,380 \text{ cm}}{42.25 \text{ cm}} = 246 \text{ amperes}$$

This means that 246 amperes is the maximum amount of current that a 10 AWG copper insulated conductor can carry for five seconds without being damaged. This is the 10 AWG conductor's short-time withstand rating.

Stating this information using the thermal stress (heat) formula, we have

$$\text{Thermal stress} = I^2 t$$

where I = RMS current in amperes and
t = time in seconds

Thus, the 10 AWG copper conductor's withstand rating is:

$$I^2 t = 246^2 \times 5 = 302,580 \text{ A}^2 \text{ seconds}$$

With this basic information, we can easily determine the short-time withstand rating of this 10 AWG copper insulated conductor for other values of time and/or current. The 10 AWG copper conductor's one-second withstand rating is

$$I^2 t = \text{ampere squared seconds}$$

$$I^2 = \frac{\text{ampere squared seconds}}{t}$$

$$I = \sqrt{\frac{\text{ampere squared seconds}}{t}}$$

$$I = \sqrt{\frac{302,580}{1}} = 550 \text{ amperes}$$

The 10 AWG copper conductor's one-cycle withstand rating (given that the approximate opening time of a typical molded-case circuit breaker is one cycle, or $\frac{1}{60}$th of a second, or 0.0167 second) is

$$I = \sqrt{\frac{302,580}{0.0167}} = 4257 \text{ amperes}$$

The 10 AWG copper conductor's one-quarter-cycle withstand rating (given that the typical clearing time for a current-limiting fuse is approximately one-fourth of a cycle, or 0.004 second) is

$$I = \sqrt{\frac{302,580}{0.004}} = 8697 \text{ amperes}$$

Therefore, a conductor can be subjected to large values of fault current if the clearing time is kept very short.

When applying current-limiting overcurrent devices, it is important to use peak let-through charts to determine the apparent RMS let-through current before applying the thermal stress formula. For example, in the case of a 60-ampere Class RK1 current-limiting fuse, the apparent RMS let-through current with an available fault current of 40,000 amperes is approximately 3000 amperes. This fuse will clear in approximately 0.004 second. See Figure 17-27, Figure 19-2, and Figure 19-4.

$I^2 t$ let-through of current-limiting fuse =
$3000 \times 3000 \times 0.004 = 36,000$ ampere squared seconds

Current Limitation Curves

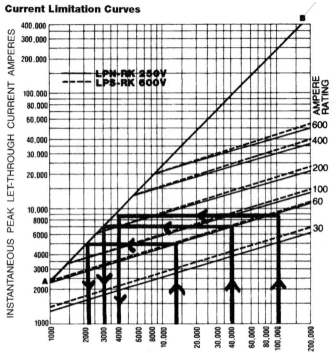

Figure 19-2   Current-limiting effect of Class RK1 fuses. The technique required to use these charts is covered in Unit 17.

Because a current-limiting overcurrent device is used in this example, the 10 AWG copper equipment grounding conductor could be used where the available fault current is 40,000 amperes.

With an available fault current of 100,000 amperes, the apparent RMS let-through current of the 60-ampere Class RK1 current-limiting fuse is approximately 4000 amperes.

$$I^2t \text{ let-through of current-limiting fuse} =$$
$$4000 \times 4000 \times 0.004 = 64,000 \text{ ampere squared seconds}$$

Remember that, as previously discussed, the withstand I^2t rating of a 10 AWG copper conductor is 302,580 ampere squared seconds.

Example:

A 167°F (75°C) thermoplastic insulated Type THW copper conductor can withstand 4200 amperes for one cycle.

a. What is the I^2t withstand rating of the conductor?

b. What is the I^2t let-through value for a non-current-limiting circuit breaker that takes one cycle to open? The available fault current is 40,000 amperes.

c. What is the I^2t let-through value for a current-limiting fuse that opens in 0.004 second when subjected to a fault current of 40,000 amperes? The apparent RMS let-through current is approximately 4600 amperes. Refer to Figure 17-27.

d. Which overcurrent device (b or c) will properly protect the conductor under the 40,000-ampere available fault current?

Answers:

a. $I^2t = 4200 \times 4200 \times 0.016 = 282,240$ ampere squared seconds.

b. $I^2t = 40,000 \times 40,000 \times 0.016 = 25,600,000$ (2.56×10^7) ampere squared seconds.

c. $I^2t = 4600 \times 4600 \times 0.004 = 84,640$ ampere squared seconds.

d. Comparing the I^2t withstand rating of the conductor to the I^2t let-through values of the breaker (b) and fuse (c), the proper choice of protection for the conductor is (c).

The use of peak let-through, current-limiting charts is discussed in Unit 17. Peak let-through charts are available from all manufacturers of current-limiting fuses and current-limiting circuit breakers.

Conductors are particularly vulnerable to damage under fault conditions. Circuit conductors can be heated to a point where the insulation is damaged or completely destroyed. The conductor can actually burn off. In the case of equipment grounding conductors, if the conductor burns off under fault conditions, the equipment can become hot, creating an electrical-shock hazard.

Even if the equipment grounding conductor does not burn off, it can become so hot that it melts the insulation on the other conductors, shorting out the other circuit conductors in the raceway or cable. This results in further fault-current conditions, damage, and hazards.

Of extreme importance are bonding jumpers, particularly the main bonding jumpers in service equipment. These main bonding jumpers must be capable of handling extremely high values of fault current.

CALCULATING A BARE COPPER CONDUCTOR AND/OR ITS BOLTED SHORT-CIRCUIT WITHSTAND RATING

A bare conductor can withstand higher levels of current than an insulated conductor of the same cross-sectional area. Do not exceed

• one ampere (RMS current)
• for five seconds
• for every 29.1 circular mils of cross-sectional area of the conductor.

Because bare equipment grounding conductors are often installed in the same raceway or cable as the insulated circuit conductors, the weakest link of the system would be the insulation on the circuit conductors. Therefore, the conservative approach when considering conductor safe withstand ratings is to apply the ONE AMPERE — FOR FIVE SECONDS — FOR EVERY 42.25 CIRCULAR MILS formula, or simply refer to a conductor short-circuit withstand rating chart, such as Figure 19-3.

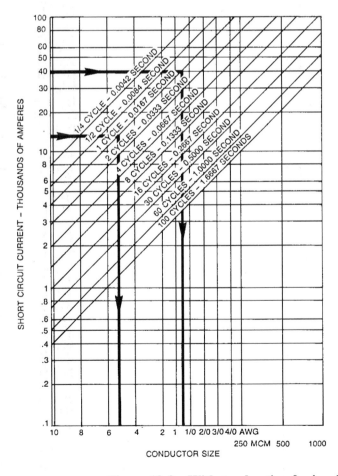

CONDUCTOR–COPPER

INSULATION–THERMOPLASTIC

CURVES BASED ON FORMULA

$$\left[\frac{I}{A}\right]^2 t = 0.0297 \log\left[\frac{T_2 + 234}{T_1 + 234}\right]$$

WHERE

I = SHORT CIRCUIT CURRENT—AMPERES

A = CONDUCTOR AREA—CIRCULAR MILS

t = TIME OF SHORT CIRCUIT—SECONDS

T_1 = MAXIMUM OPERATING
    TEMPERATURE—167°F (75°C)

T_2 = MAXIMUM SHORT CIRCUIT
    TEMPERATURE—150°F (66°C)

Figure 19-3   Withstand rating for insulated copper conductors.

CALCULATING THE MELTING POINT OF A COPPER CONDUCTOR

The melting point of a copper conductor can be calculated by using these values:

- one ampere (RMS current)
- for five seconds
- for every 16.19 circular mils of cross-sectional area of the conductor.

NEC® 250.122(B) states that if the ungrounded conductors are sized larger for any reason, the equipment grounding conductor shall be increased proportionately, based on the cross-sectional area of the ungrounded conductors for that particular branch-circuit or feeder. The need to increase the size of the ungrounded conductor might be for voltage drop reasons, or because high values of fault current are possible. ► In *NEC® 250.4(A)(5),* we find a definition of an effective ground-fault current path: *An intentionally constructed, permanent, low-impedance electrically conductive path designed and intended to carry current under ground-fault conditions from the point of a ground fault on a wiring system to the electrical supply source and that facilitates the operation of the overcurrent protective device or ground fault detectors on high-impedance grounded systems.* *◄ Under fault-current conditions, the possible burning off of an equipment grounding conductor, a bonding jumper, or any conductor that is dependent upon safely carrying fault current until the overcurrent protective device can clear the fault results in a hazard to life, safety, and equipment.

USING CHARTS TO DETERMINE A CONDUCTOR'S SHORT-TIME WITHSTAND RATING

The Insulated Cable Engineers Association, Inc., publishes much data on this subject. The graph in Figure 19-3 shows the withstand rating of 167°F (75°C) thermoplastic insulated copper conductors. Many engineers use these tables for bare grounding conductors because, in most cases, the bare equipment grounding conductor is in the same raceway as the phase conductors. An extremely hot equipment grounding conductor in contact with the phase conductors would damage the insulation on the phase conductors. For example, when nonmetallic conduit is used, the equipment grounding conductor and the phase conductors are in the same raceway.

*Reprinted with permission from NFPA 70-2005.

To use the table, begin on the left side, at the amount of fault current available. Then draw a line to the right, to the time of opening of the overcurrent protective device. Draw a line downward to the bottom of the chart to determine the conductor size.

Example:

A circuit is protected by a 60-ampere overcurrent device. Determine the minimum size equipment grounding conductor for available fault currents of 14,000 amperes and 40,000 amperes. Refer to *NEC® Table 250.122,* the fuse peak let-through chart, and the chart showing allowable short-circuit currents for insulated copper conductors, Figure 19-3.

You can see from Table 19-1 that if the conductor size as determined from the allowable short-

Short-Circuit Currents for Insulated Copper Conductor		
Available Fault Current	**Overcurrent Device**	**Conductor Size**
40,000 amperes	Typical one-cycle breaker	1/0 copper conductor
40,000 amperes	Typical RK1 fuse. Clearing time ¼ cycle or less.	Using a 60-ampere RK1 current-limiting fuse, the apparent RMS let-through current is approximately 3000 amperes. *NEC® Table 250.122* shows a minimum 8 AWG equipment grounding conductor. The allowable short-circuit current chart shows a conductor smaller than a 10 AWG. Therefore, 8 AWG is the minimum size EGC permitted.
14,000 amperes	Typical one-cycle breaker	4 AWG copper conductor
14,000 amperes	Typical RK1 fuse. Clearing time ¼ cycle or less.	Using a 60-ampere RK1 current-limiting fuse, the apparent RMS let-through current is approximately 2200 amperes. *NEC® Table 250.122* shows a minimum 8 AWG equipment grounding conductor. The allowable short-circuit current chart shows a conductor smaller than a 10 AWG. Therefore, 8 AWG is the minimum size EGC permitted.

Table 19-1

circuit current chart is larger than the size given in *NEC® Table 250.122*, then you must install the larger size. Installing a conductor too small to handle the available fault current could result in insulation damage or, in the worst case, the burning off of the equipment grounding conductor, leaving the protected equipment hot.

If the conductor size as determined from the allowable short-circuit current chart is smaller than the size given in *NEC® Table 250.122*, then install an equipment grounding conductor *not smaller* than the minimum size required by *NEC® Table 250.122*.

There is another common situation in which it is necessary to install equipment grounding conductors larger than shown in *NEC® Table 250.122*. When circuit conductors have been increased in size because of a voltage drop, the equipment grounding conductor size shall be increased proportionately.

MAGNETIC FORCES

Magnetic forces acting upon electrical equipment (busbars, contacts, conductors, and so on) are proportional to the square of the PEAK current. This relationship is expressed as I_p^2.

Refer to Figure 19-4 for the case in which there is no fuse in the circuit. The peak current, I_p (available short-circuit current), is indicated as 30,000 amperes. The peak let-through current, I_p, resulting from the current-limiting effect of a fuse is indicated as 10,000 amperes.

Example:

Because magnetic forces are proportional to the square of the peak current, the magnetic forces (stresses) on the electrical equipment subjected to the full 30,000-ampere peak current are nine times that of the 10,000-ampere peak current let-through by the fuse. Stated another way, a current-limiting fuse or circuit breaker that can reduce the available short-circuit peak current from 30,000 amperes to only 10,000 amperes will subject the electrical equipment to only one-ninth the magnetic forces.

Visual signs, indicating that too much current was permitted to flow for too long a time, include conductor insulation burning, melting and bending of busbars, arcing damage, exploded overload elements in motor controllers, and welded contacts in controllers.

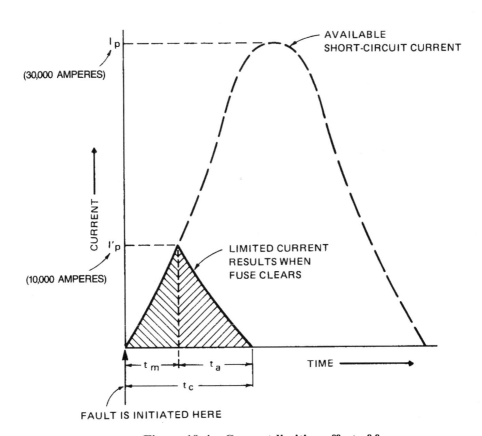

Figure 19-4   Current-limiting effect of fuses.

A current-limiting fuse or current-limiting circuit breaker must be selected carefully. The fuse or breaker must not only have an adequate interrupting rating to clear a fault safely without damage to itself, but it also must be capable of limiting the let-through current (I_p) and the value of I^2t to the withstand rating of the equipment it is to protect.

The graph shown in Figure 19-4 illustrates the current-limiting effect of fuses. The square of the area under the dashed line is energy (I^2t). I_p is the available peak short-circuit current that flows if there is no fuse in the circuit or if the overcurrent device is not a current-limiting type. For a current-limiting fuse, the square of the shaded area of the graph represents the energy (I^2t), and the peak let-through current, I_p. The melting time of the fuse element is t_m, and the square of the shaded area corresponding to this time is the melting energy. The arcing time is shown as t_a; similarly, the square of the shaded area corresponding to this time is the arcing energy. The total clearing time, t_c, is the sum of the melting time and the arcing time. The square of the shaded area for time, t_c, is the total energy to which the circuit is subjected when the fuse has cleared. For the graph in Figure 19-4, the area under the dashed line is six times greater than the shaded area. Because energy is equal to the area squared, then $6 \times 6 = 36$; that is, the circuit is subjected to 36 times as much energy when it is protected by noncurrent-limiting overcurrent devices.

Summary of Conductor Short-Circuit Protection

When selecting equipment grounding conductors, bonding jumpers, or other current-carrying conductors, the amount of short-circuit or ground-fault current available and the clearing time of the overcurrent protective device must be taken into consideration so as to minimize damage to the conductor and associated equipment. The conductor must not become a fuse. The conductor must remain intact under any values of fault current.

As stated earlier, the two important issues when considering conductor withstand rating are:

1. How much current will flow?

2. How long will the current flow?

The choices are

1. **Limit the current.**
   Install current-limiting overcurrent devices that will limit the let-through fault current and will reduce the time it takes to clear the fault. Then refer to *NEC® Table 250.122* to select the minimum size equipment grounding conductor permitted by the *National Electrical Code®* and refer to the other tables such as *NEC® Table 310.16* for the selection of circuit conductors.

2. **Do not limit the current.**
   Install conductors that are large enough to handle the full amount of available fault current for the time it takes a noncurrent-limiting overcurrent device to clear the short circuit or ground fault. Withstand ratings of conductors can be calculated or determined by referring to conductor withstand rating charts.

TAP CONDUCTORS

A tap conductor is defined in *NEC® 240.2*.

1. It is a conductor, other than a service conductor.

2. It has overcurrent protection ahead of its point of supply.

3. The overcurrent protection exceeds the value permitted for similar conductors that are not taps.

Summary.   Branch-circuit conductors must have an ampacity not less than the maximum load served, *NEC® 210.19*.

Feeder conductors must have sufficient ampacity to supply the load served, *NEC® 220.2(A)*.

Service-entrance conductors shall be of sufficient size to carry the loads as calculated, *NEC® 230.42*.

The basic overcurrent protection rule in *NEC® 240.4* is that conductors shall be protected at the conductor's ampacity. There are seven subsections to this section, each pertaining to specific applications.

NEC® 240.21 supports *240.4* in that it states in part: *Overcurrent protection shall be provided in each ungrounded circuit conductor and shall be located where the conductor to be protected receives its supply.* There are many subsections to *240.21*, each pertaining to specific types of conductor taps.

Of particular interest are feeder taps that are not over 10 ft (3 m) and feeder taps that are not over 25 ft (7.5 m). A number of requirements must be met when taps of this sort are made.

There are instances where a smaller conductor must be tapped from a larger conductor. For exam-

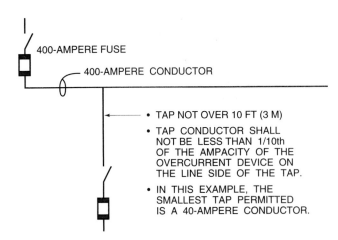

- TAP NOT OVER 10 FT (3 M)
- TAP CONDUCTOR SHALL NOT BE LESS THAN 1/10th OF THE AMPACITY OF THE OVERCURRENT DEVICE ON THE LINE SIDE OF THE TAP.
- IN THIS EXAMPLE, THE SMALLEST TAP PERMITTED IS A 40-AMPERE CONDUCTOR.

Figure 19-5   *NEC® 240.21(B)(1)(4)* **requires that the overcurrent device not be greater than 10 times that of the tap conductor's ampacity.**

ple, Figure 19-5 shows a large feeder protected by a 400-ampere fuse. The minimum size tap in the wireway or gutter shall not be less than one-tenth the ampacity of the 400-ampere fuse. This simply means that the tap conductor must have an ampere rating of not less than 40 amperes.

$$\frac{400}{10} = 40 \text{ amperes}$$

The preceding example of tapping a small conductor from a larger conductor that is protected by an overcurrent device much larger than the small tap conductor is an indication that the *Code* is concerned with the protection of the smaller conductor.

Here again, the electrician must properly size the conductors for the load to be served, take into consideration the possibility of voltage drop, and then check the short-circuit withstand rating of the conductor to be sure that a severe fault will not cause damage to the conductor's insulation or, in the worst case, vaporize the conductor.

Another common example of a tap is for fixture whips that connect a luminaire (fixture) to a branch-circuit. The size of taps to branch-circuits is found in *NEC® Table 210.24*. For instance, for a 20-ampere branch-circuit, the minimum size tap is a 14 AWG copper conductor. The tap conductors in a fixture whip shall be suitable for the temperature to be encountered. In most cases, this will call for a 194°F (90°C) conductor, Type THHN or equivalent.

Important Distinction.   When connecting a receptacle to a 20-ampere branch-circuit, the conductor from a splice in the box to the receptacle terminal must be rated 20-ampere. These short conductors are not taps; they are part of the branch circuit. It would be a violation of the *NEC®* to use 14 AWG conductors with a 15-ampere rating for this connection.

Additional *Code* rules for taps and panelboard protection are found in Unit 12.

REVIEW QUESTIONS

Refer to the *National Electrical Code®* or the working drawings when necessary. Where applicable, responses should be written in complete sentences.

1. Define *withstand rating*. _____

   _____

   _____

   _____

2. Define an *effective grounding path*. _____

   _____

   _____

   _____

3. Define I^2t. _____

_____

_____

_____

4. Define a *current-limiting overcurrent device*. _____

_____

_____

_____

5. Define an *equipment grounding conductor*. _____

_____

_____

_____

6. What table in the *National Electrical Code*® shows the minimum size equipment grounding conductor? _____

For the following questions, show all required calculations.

7. What is the appropriate action if the label on a motor controller or the nameplate on other electrical equipment is marked "Maximum Fuse Size"? Cite the *Code* section that supports your proposed action.

_____

_____

_____

8. What is the maximum current that a 12 AWG, 167°F (75°C) thermoplastic insulated conductor can safely carry for five seconds?

9. What is the maximum current that an 8 AWG, 167°F (75°C) thermoplastic insulated conductor can safely carry for one second?

10. What is the maximum current that a 12 AWG, 167°F (75°C) thermoplastic insulated conductor can safely carry for one-quarter cycle (0.004 second)?

11. Referring to Figure 19-3, if the available fault current is 10,000 amperes and the over-current device has a total clearing time of one cycle, the minimum permitted copper 167°F (75°C) thermoplastic insulated conductor would be a _____ AWG.

12. Referring to Figure 19-2, if the available fault current is 40,000 amperes, a 200-ampere, 600-volt fuse of the type represented by the chart will have an instantaneous peak let-through current of approximately _____ amperes.

13. Referring to Figure 19-2, if the available fault current is 40,000 amperes, a 200-ampere 600-volt fuse of the type represented by the chart will have an apparent RMS let-through current of approximately _____ amperes.

14. A tap conductor that complies with appropriate *Code* requirements has an ampacity of 30 amperes and is 9 ft long. This is a field installation in which the tap conductor leaves the enclosure where the tap is made (a large junction box). The maximum allowable overcurrent protection for the feeder conductor would be _____ amperes.

UNIT 20

Low-Voltage Remote-Control

OBJECTIVES

After studying this unit, the student should be able to

- list the components of a low-voltage remote-control wiring system.
- select the appropriate *NEC®* sections governing the installation of a low-voltage remote-control wiring system.
- demonstrate the correct connections for wiring a low-voltage remote-control system.

LOW-VOLTAGE REMOTE-CONTROL

Conventional general lighting control is used in the major portion of the commercial building. For the drugstore, however, a method known as *low-voltage remote-control* is selected because of the number of switches required and because it is desired to have extensive control of the lighting.

Low-voltage remote-control systems are available ranging from very simple to very complex. These systems can provide control from multiple locations, automatic timing control, handheld (wireless) controls, occupancy sensors (On when motion is detected) and unoccupancy sensors (Off with predetermined time-delay when no one is in the room), photocell control for security (On at dusk, Off at dawn), and programmable controls.

Federally mandated energy codes, such as the ASHRAE 90.1 energy standard for buildings will no doubt lead to more and more low-voltage switching applications for the automatic control of heating, lighting, and cooling.

Many types of systems are available. Some systems have mechanical devices; others have solid-state devices. The purpose of this chapter is to introduce you to low-voltage control. The system installed in the drugstore is a very basic system.

Manufacturers of these products provide a tremendous amount of technical installation literature. Vist Web sites such as http://www.circon.com and http://www.wattstopper.com. Search the Web for "low-voltage systems." You will be amazed at how much information is available on-line.

Relays

A low-voltage remote-control wiring system is relay operated. The relay is controlled by a low-voltage switching system and in turn controls the power circuit connected to it, Figure 20-1. The low-voltage, split-coil relay is the heart of the low-voltage remote-control system, Figure 20-2. When the On coil of the relay is energized, the solenoid mechanism causes the contacts to move into the On position to complete the power circuit. The contacts stay in this position until the Off coil is energized. When this occurs, the contacts are withdrawn and the power circuit is opened. The red wire is the On wire; the black wire is Off, and the blue wire is common to the transformer.

The low-voltage relay is available in two mounting styles. One style of relay is designed to mount through a ½-inch knockout opening, Figure 20-3. For a ¾-inch knockout, a rubber grommet is inserted

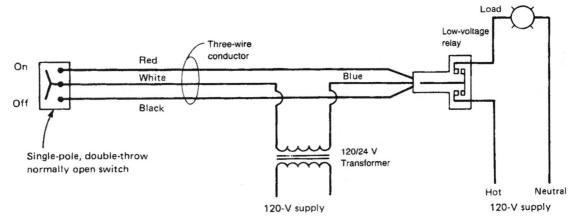

Figure 20-1   Connection diagram for low-voltage remote-control.

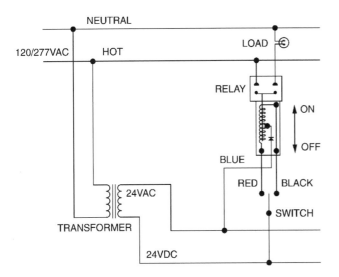

Figure 20-2   A cutaway view of a low-voltage relay showing the internal mechanism as well as the external line voltage and low-voltage leads. Some relays have screw terminals or "quick connects" instead of conductor leads.

to isolate the relay from the metal. This practice should ensure quieter relay operation. The second relay mounting style is the *plug-in relay*. This type of relay is used in an installation where several relays are mounted in one enclosure. The advantage of the plug-in relay is that it plugs directly into a busbar. As a result, it is not necessary to splice the line voltage leads.

Single Switch

The switch used in the low-voltage remote-control system is a normally open, single-pole, double-throw, momentary contact switch, Figure 20-4. This switch is approximately one-third the size of a standard single-pole switch. In general, this type of switch has slide-on terminals for easy connections, Figure 20-5.

To make connections to the low-voltage switches, the white wire is common and is connected to the 24-volt transformer source. The red wire connects

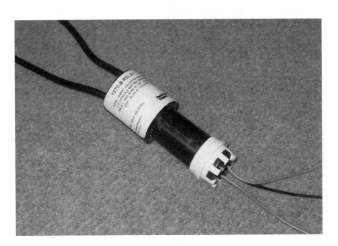

Figure 20-3   Low-voltage relay.

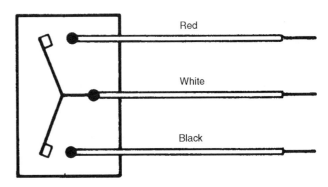

Figure 20-4   Single-pole, double-throw, normally open switch.

Figure 20-5   Low-voltage switch. (*Courtesy* General Electric Wiring Devices.)

Figure 20-6   Eight-switch combination. (*Courtesy* General Electric Wiring Devices.)

to the On circuit and the black wire connects to the Off circuit.

Switches are available in single-, two-, four-, and eight-switch combinations with features such as pilot light, lighted toggle, and being key-operated.

Master Control

It is often desirable to control several circuits from a single location. Up to eight low-voltage switches, Figure 20-6, can be located in the same area that is required by two conventional switches. This eight-switch combination can be mounted on

a 4^{11}⁄$_{16}$ (119 mm) square box using an adapter provided with the switch. Directory strips, identifying the switch's functions, can be prepared and inserted in the switch cover. An eight-switch combination is used in the commercial building drugstore. The connection diagram is shown in Figure 20-7.

If the control requirements are very complex or extensive, a master sequencer, Figure 20-8, can be installed. There is practically no limit to the number of relays that can be controlled, almost instantaneously, with this microprocessor-controlled electronic switch.

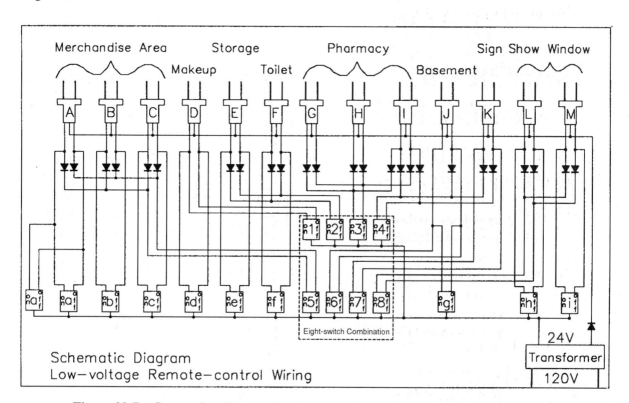

Figure 20-7   Connection diagram for drugstore low-voltage, remote-control wiring.

Figure 20-8   Master sequencer available in three sizes to control 8, 16, or 32 relays.

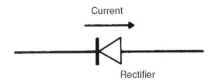

Figure 20-9   Diagram of rectifier showing direction of current.

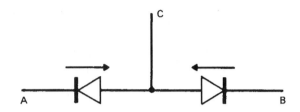

Figure 20-10   Diagram of rectifiers connected in opposition.

Master Control with Rectifiers

Several relays can be operated individually or from a single switch through the use of *rectifiers*. The principle of operation of a master control with rectifiers is based on the fact that a rectifier permits current in only one direction, Figure 20-9.

For example, if two rectifiers are connected as shown in Figure 20-10, current cannot exist from

A to B, or from B to A; however, current can exist from A to C, or from B to C. Thus, if a rectifier is placed in one lead of the low-voltage side of the transformer, and additional rectifiers are used to isolate the switches, then a master switching arrangement is achieved, Figure 20-11. This method of master control is used in the drugstore. Although a switching schedule is included in the specifications,

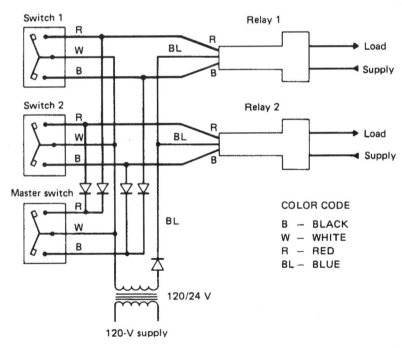

Figure 20-11   Switch 1 and the master switch control relay 1. Switch 2 and the master switch control relay 2. Rectifiers are connected into the leads of the master switch and the transformer. New systems have rectification as part of the motherboard in the main control panel.

the electrician may find it necessary to prepare a connection diagram similar to that shown in Figure 20-7. For the drugstore master control, the transformer, Figure 20-12, and rectifier, Figure 20-13, are located in the low-voltage control panel.

WIRING METHODS

NEC® Article 725 governs the installation of remote-control and signal circuits such as is installed in the drugstore. The provisions of this article apply to remote-control circuits, low-voltage relay switching, low-energy power circuits, and low-voltage circuits.

The drugstore low-voltage wiring is classified as a Class 2 circuit, by *NEC® 725.2*. Because the power source of the circuit is limited (by the definition of a low-voltage circuit), overcurrent protection is not required.

Power source limitations for alternating current Class 2 circuits are found in *NEC® Chapter 9, Tables 11(A)* and *11(B)*.

Figure 20-12   Transformer. (*Courtesy* **Pass & Seymour/Legrand.**)

Figure 20-13   Rectifier. New systems have rectification as part of the motherboard in the main control panel.

The circuit transformer, Figure 20-12, is designed so that in the event of an overload, the output voltage decreases and there is less current output. Any overload can be counteracted by these energy-limiting characteristics through the use of a specially designed transformer core. If the transformer is not self-protected, a thermal device may be used to open the primary side to protect the transformer from overheating. This thermal device resets automatically as soon as the transformer cools. Other transformers are protected with nonresetting fuse links or externally mounted fuses.

Although the *NEC®* does not require that the low-voltage wiring be installed in a raceway, the specifications for the commercial building do contain this requirement. The advantage of using a raceway for the installation is that it provides a means for making future additions at a minimum cost. A disadvantage of this approach is the initial higher construction cost.

Conductors

NEC® 725.55(A) prohibits Class 2 circuits in the same raceway or cable as the wiring for light and power.

Most low-voltage, remote-control systems are wired with 20 AWG conductors, but larger conductors should be used for long runs to minimize the voltage drop. Cables are available with multiple conductors. To simplify connections, the conductor color coding combinations are red-black-white, red-black-yellow-white, red-black-yellow plus an 18 AWG common white, red-black-blue-white, and red-black-yellow-blue-white. To install these color-coded cables correctly throughout an entire installation, the wires are connected to the relays, switches, master controls, and panel like-color to like-color.

The manufacturers of these types of systems provide detailed wiring diagrams. Follow the installation instructions and wiring diagrams exactly without modifications.

Low-Voltage Panelboard

The low-voltage relays in the drugstore installation are to be mounted in an enclosure next to the power panelboard, Figure 20-14. A barrier in this low-voltage panelboard separates the 120-volt power lines from the low-voltage control circuits, in compliance with *NEC® 725.55(B)*.

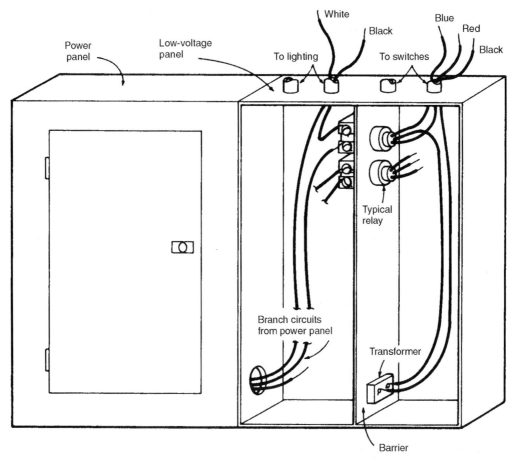

Figure 20-14   Low-voltage panel installation. Prewired panels are available with 12, 24, or 48 relays.

REVIEW QUESTIONS

Refer to the *National Electrical Code*® or the working drawings when necessary. Where applicable, responses should be written in complete sentences.

Refer to Figure 20-1 and respond to the following statements.

1. Describe the action of the low-voltage relay. _____

_____

_____

_____

2. Describe the function of the red conductor. _____

_____

_____

_____

3. Describe the function of the white conductor. _____

_____

_____

_____

4. Describe the function of the black conductor. _____
   _____
   _____
   _____

5. Describe the function of the blue conductor. _____
   _____
   _____
   _____

Respond to the following statements.

6. The *NEC*® article that governs the installation of the low-voltage remote-control system is _____.

7. Complete the following diagram according to the switching schedule that is given at the bottom of the diagram. Indicate the color of the conductors.

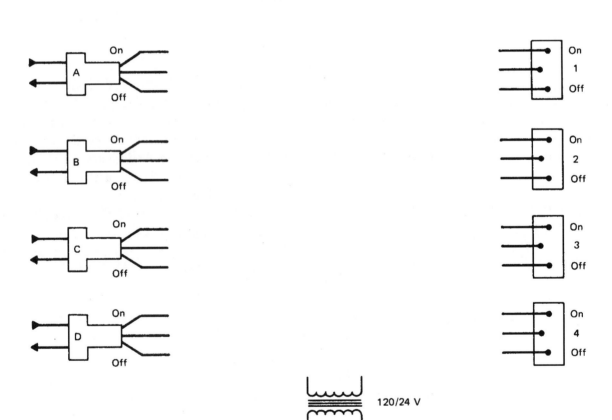

UNIT 21

The Cooling System

OBJECTIVES

After studying this unit, the student should be able to

- list the parts of a cooling system.
- describe the function of each part of the cooling system.
- make the necessary calculations to obtain the sizes of the electrical components.
- read a typical wiring diagram that shows the operation of a cooling unit.

The electrician working on commercial construction is expected to install the wiring of cooling systems and troubleshoot electrical problems in these systems. Therefore, it is recommended that the electrician know the basic theory of refrigeration and the terms associated with it.

REFRIGERATION

Refrigeration is a method of removing energy in the form of heat from an object. When the heat is removed, the object is colder. An energy balance is maintained, which means that the heat must go somewhere. As long as the locations where the heat is discharged and where it is absorbed are remote from each other, it can be said that the space where the heat was absorbed is cooled. The inside of the household refrigerator or freezer is cold to the touch, but this cold cannot be used to cool the kitchen by having the refrigerator door open. Actually, leaving the door open causes the kitchen to become hotter. This situation demonstrates an important principle of mechanical refrigeration: to remove heat energy, it is necessary to add energy or power to it.

Mechanical refrigeration relies primarily on the process of evaporation. This process is responsible for the cool sensation that results when rubbing alcohol is applied to the skin or when gasoline is spilled on the skin. Body heat supplies the energy required to vaporize the alcohol or gasoline. It is the removal of this energy from the body that causes the sensation of cold. In refrigeration systems such as those used in the commercial building, the evaporation process is controlled in a closed system. The purpose of this arrangement is to preserve the refrigerant so that it can be reused many times in what is known as the *refrigerant cycle*. As shown in Figure 21-1, the four main components of the refrigerant cycle are as follows:

- *Evaporator*
  The refrigerant evaporates here as it absorbs energy from the removal of heat.

- *Compressor*
  This device raises the energy level of the refrigerant so that it can be condensed readily to a liquid.

- *Condenser*
  The compressed refrigerant condenses here as the heat is removed.

- *Expansion valve*
  This metering device maintains a sufficient imbalance in the system so that there is a point of low pressure where the refrigerant can expand and evaporate.

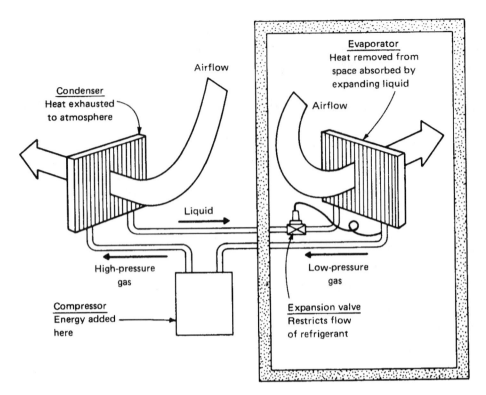

Figure 21-1   The refrigeration cycle.

EVAPORATOR

The evaporator in a commercial installation normally consists of a fin tube coil, as in Figure 21-2, through which the building air is circulated by a motor-driven fan. A typical evaporator unit is shown in Figure 21-3. The evaporator may be located inside or outside the building. In either case,

the function of the evaporator is to remove heat from the interior of the building or the enclosed space. The air is usually circulated through pipes or ductwork to ensure a more even distribution. The window-type air conditioner, however, discharges the air directly from the evaporator coil. In general, the cooling air from the evaporator is recirculated within the space to be cooled and is passed again across the cooling coil. A certain percentage of outside air is added to the circulating air to replace air lost through exhaust systems and fume hoods, or because of the gradual leakage of air through walls, doors, and windows.

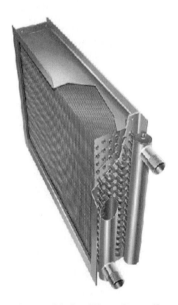

Figure 21-2   Fin tube coils.

Figure 21-3   An evaporator with dual fans.

COMPRESSOR

The compressor (Figure 21-4) serves as a pump to draw the expanded refrigerant gas from the evaporator. In addition, the compressor boosts the pressure of the gas and sends it to the condenser. The compression of the gas is necessary because this process adds the heat necessary to condense the gas

Figure 21-4   A motor-driven reciprocating compressor.

to a liquid. When the temperature of the air or water surrounding the condenser is relatively warm, the gas temperature must be increased to ensure that the temperature around the condenser will liquefy the refrigerant.

Direct-drive compressors are usually used in large installations. For smaller installations, however, the trend is toward the use of hermetically sealed compressors. Due to several built-in electrical characteristics, these hermetic units cannot be used on all installations. (Restrictions on the use of hermetically sealed compressors are covered later in this unit.)

CONDENSER

Condensers are generally available in three types: as air-cooled units, Figure 21-5, as water-cooled units, Figure 21-6, or as evaporative cooling units, Figure 21-7, in which water from a pump is sprayed on air-cooled coils to increase their capacity. The function of the condenser in the refrigerant

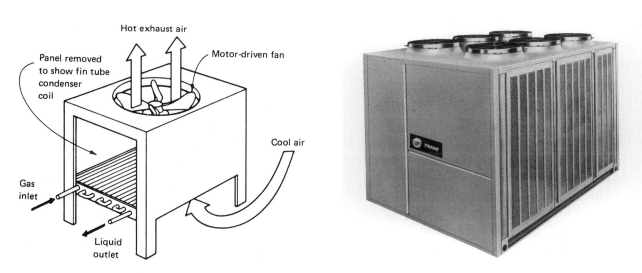

Figure 21-5   An air-cooled condenser.

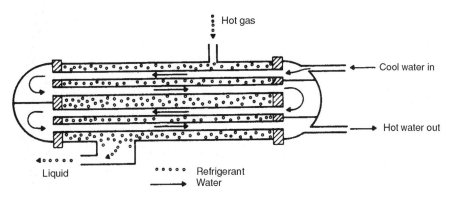

Figure 21-6   A water-cooled condenser.

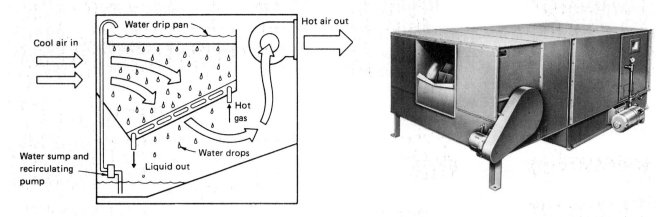

Figure 21-7   An evaporative condenser.

cycle is to remove the heat taken from the evaporator, plus the heat of compression. Thus, it can be seen that keeping the refrigerator door open causes the kitchen to become hotter because the condenser rejects the combined heat load to the condensing medium. In the case of the refrigerator in a residence, the condensing medium is the room air.

Air-cooled condensers use a motor-driven fan to drive air across the condensing coil. Water-cooled condensers require a pump to circulate the water. Once the refrigerant gas is condensed to a liquid state, it is ready to be used again as a coolant.

EXPANSION VALVE

It was stated previously that the refrigerant must evaporate or boil if it is to absorb heat. The process of boiling at ordinary temperatures can occur in the evaporator only if the pressure is reduced. The task of reducing the pressure is simplified because the compressor draws gas away from the evaporator and tends to evacuate it. In addition, a restricted flow of liquid refrigerant is allowed to enter the high side of the evaporator. As a result, the pressure remains fairly low in the evaporator coil so that the liquid refrigerant entering through the restriction flashes into a vapor and fills the evaporator as a boiling refrigerant.

The restriction to the evaporator may be a simple orifice. In commercial systems, however, a form of automatic expansion valve is generally used because it is responsive to changes in the heat load. The expansion valve is located in the liquid refrigerant line at the inlet to the evaporator. The valve controls the rate at which the liquid flows into the evaporator. The liquid flow rate is determined by installing a temperature-sensitive bulb at the outlet of the evaporator to sense the heat gained by the refrigerant as it passes through the evaporator. The bulb is filled with a volatile liquid (which may be similar to the refrigerant). This liquid expands and passes through a capillary tube connected to a spring-loaded diaphragm to cause the expansion valve to meter more or less refrigerant to the coil. The delivery of various amounts of refrigerant compensates for changes in the heat load of the evaporator coil.

HERMETIC COMPRESSORS

The tremendous popularity of mechanical refrigeration for household use in the 1930s and 1940s stimulated the development of a new series of nonexplosive and nontoxic refrigerants. These refrigerants are known as chlorinated hydrofluorides of the halogen family and, at the time of their initial production, were relatively expensive. The expense of these refrigerants was great enough so that it was no longer possible to permit the normal leakage that occurred around the shaft seals of reciprocating, belt-driven units. As a result, the hermetic compressor was developed (Figure 21-8). This unit consists of a motor-compressor completely sealed in a gastight, steel casing. The refrigerant gas is circulated through the compressor and over the motor windings, rotor, bearings, and shaft. The circulation of the expanded gas through the motor helps to cool the motor.

The initial demand for hermetic compressors was for use on residential-type refrigerators. Therefore,

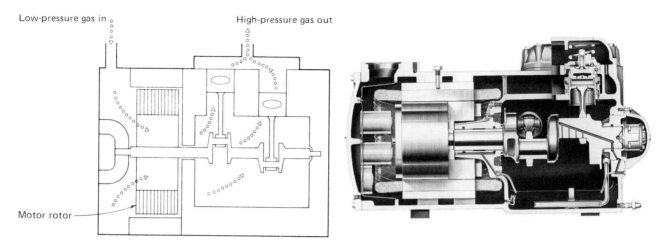

Figure 21-8   The hermetic compressor.

most of the hermetic compressors were constructed for single-phase service. In other words, it was necessary to provide auxiliary winding and starting devices for the compressor installation. Because the refrigerant gas surrounded and filled the motor cavity, it was necessary to remove the centrifugal switch commonly provided to disconnect the starting winding at approximately 85 percent of the full speed. The switch was removed because any arcing in the presence of the refrigerant gas caused the formation of an acid from the hydrocarbons in the gas. This acid attacked and etched the finished surfaces of the shafts, bearings, and valves. The acid also carbonized the organic material used to insulate the motor winding and caused the eventual breakdown of the insulation. To overcome the problem of the switch, the relatively heavy magnetic winding of a relay was connected in series with the main motor winding. The initial heavy inrush of current caused the relay to lift a magnetic core and energize the starting winding. As the motor speed increased, the main winding current decreased and allowed the relay to remove the starting or auxiliary winding from the circuit.

A later refinement of this arrangement was the use of a voltage-sensitive relay that was wound to permit pickup at voltage values greater than the line voltage. The coil of this voltage-sensitive relay was connected across the starting winding. This connection scheme was based on the principle that the rotor of a single-phase induction motor induces in its own starting winding a voltage that is approximately in quadrature (phase) with the main winding voltage and has a value greater than the main voltage. The voltage-sensitive relay broke the circuit to the start-

ing winding and remained in a sealed position until the main winding was de-energized.

Because the major maintenance problems in hermetic compressor systems were due generally to starting relays and capacitors, it was desirable to eliminate as many of these devices as possible. It was soon realized that small- and medium-sized systems could make use of a different form of refrigerant metering device, thus eliminating the automatic expansion valve. Recall that this valve has a positive shut-off characteristic, which means that the refrigerant is restricted when the operation cycle is finished as indicated by the evaporator reaching the design temperature. As a result, when a new refrigeration cycle begins, the compressor must start against a head of pressure. If a small-bore, open capillary tube is substituted for the expansion valve, the refrigerant is still metered to the evaporator coil, but the gas continues to flow after the compressor stops until the system pressure is equalized. Therefore, the motor is required only to start the compressor against the same pressure on each side of the pistons. This ability to decrease the load led to the development of a new series of motors that contained only a running capacitor (no relay was installed). This type of motor furnished sufficient torque to start the unloaded compressor and, at the same time, greatly improved the overall power factor.

COOLING SYSTEM CONTROL

Figure 21-9 shows a wiring diagram that is representative of standard cooling units.

When the selector switch in the heating-cooling control is set to COOL and the fan switch is placed on AUTO, any rise in temperature above the set point causes the cooling contacts (TC) to close and complete two circuits. One circuit through the fan switch energizes the evaporator fan relay (EFR), which, in turn, closes EFR contacts to complete the 208-volt circuit to the evaporator fan motor (EFM).

The second circuit is to the control relay (CR) and causes the CR1 contacts to open. When the CR1 contacts open, the crankcase heater (CH) is de-energized. (This heater is installed to keep the compressor oil warm and dry when the unit is not running.) In addition, CR2 contacts close to com-

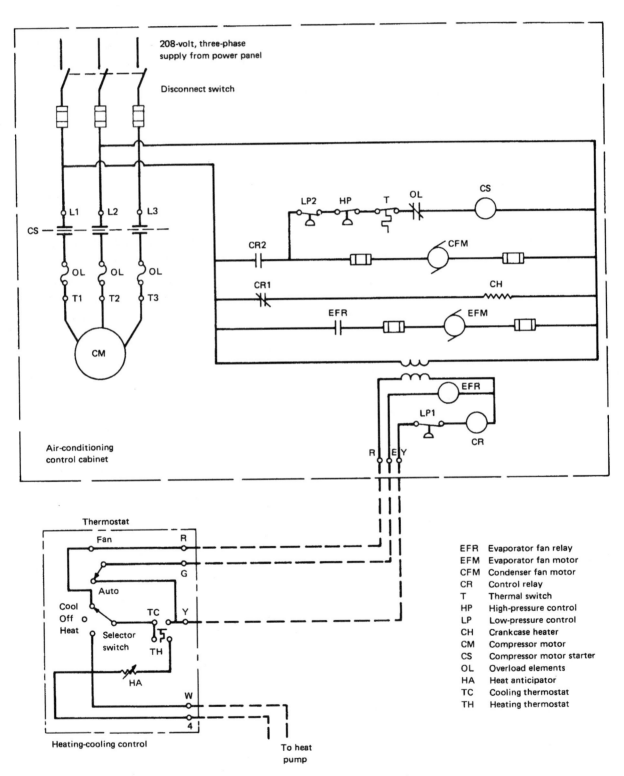

Figure 21-9 A cooling system wiring diagram.

plete a circuit to the condenser fan motor (CFM) and another circuit through low-pressure switch 2 (LP2), the high-pressure switch (HP), the thermal contacts (T), and the overload contacts (OL) to the compressor motor starter coil (CS). When the CR2 contacts close, the three CS contacts in the power circuit to the compressor motor (CM) also close.

Contacts LP1 and LP2 open when the refrigerant pressure drops below a set point. When contacts LP2 open, CS is de-energized. When contacts LP1 open, CR is de-energized, the circuit to CFM opens, and the circuit to CH is completed. The high-pressure switch (HP) contacts open and de-energize the compressor motor starter (CS) when the refrigerant pressure is above the set point. The low-pressure control (LP1) is the normal operating control and the high-pressure control (HP) and LP2 act as safety devices. The T contacts shown in Figure 21-9 are located in the compressor motor and open when the winding temperature of the motor is too high. The OL contacts are controlled by the OL elements installed in the power leads to the motor. The OL elements are sized to correspond to the current draw of the motor. That is, a high current causes the overload elements to overheat and open the OL contacts. As a result, the compressor starter is de-energized and the compressor stops. The evaporator fan motor (EFM) used to circulate air in the store area can be run continuously if the fan switch is turned to FAN. Actually, in many situations, it is recommended that the fan motor run continuously to keep the air in motion.

COOLING SYSTEM INSTALLATION

The owner of the commercial building leases the various office and shop areas on the condition that heat will be furnished to each area. However, tenants agree to pay the cost of operating the refrigeration system to provide cooling in their areas. Only four cooling systems are indicated in the plans for the commercial building because the bakery does not use a cooling system. The cooling equipment for the insurance office, the doctor's office, and the beauty salon are single-package cooling units located on the roof. The compressor, condenser, and evaporator for each of these units are constructed within a single enclosure. The system for the drugstore is a split system with the

compressor and condenser located on the roof and the evaporator located in the basement, as shown in Figure 21-10. Regardless of the type of cooling system installed, a duct system must be provided to connect the evaporator with the area to be cooled. The duct system shown diagrammatically in Figure 21-10 does not represent the actual duct system, which will be installed by another contractor.

The electrician is expected to provide a power circuit to each air-conditioning unit as shown on the plans. In addition, it is necessary to provide wiring to the thermostat in each area and, in the case of the drugstore, wiring must be provided to the evaporator located in the basement.

ELECTRICAL REQUIREMENTS FOR AIR-CONDITIONING AND REFRIGERATION EQUIPMENT

NEC® Article 440 provides the requirements for installing air-conditioning and refrigeration equipment that involves one or more hermetic refrigerant motor-compressors. Where these compressors are not involved, *NEC® 440.3* directs the reader to *NEC® Articles 422, 424,* and *430* and gives other references for special provisions.

For the insurance office, an air-conditioning unit is located on the roof of the commercial building. The data furnished with this air-conditioning unit supply the following information:

Voltage:
208 volts, three phase, three wire, 60 hertz

Hermetic refrigerant compressor-motor:
Rated-load amperes 20.2 at 208 volts three phase

Evaporator motor:
Full-load amperes 3.2 at 208 volts single phase

Condenser motor:
Full-load amperes 3.2 at 208 volts single phase

Minimum circuit ampacity:[1]
31.65 amperes

Maximum overcurrent protection:[2]
50 amperes, time-delay fuse

Locked-rotor current:
176 amperes

---

[1] $(20.2 \times 1.25) + 3.2 + 3.2 = 31.65$ amperes.
[2] $(20.2 \times 2.25) + 3.2 + 3.2 = 51.85$ amperes.
The next lower standard size OCPD from *NEC® 240.6* is 50 amperes.

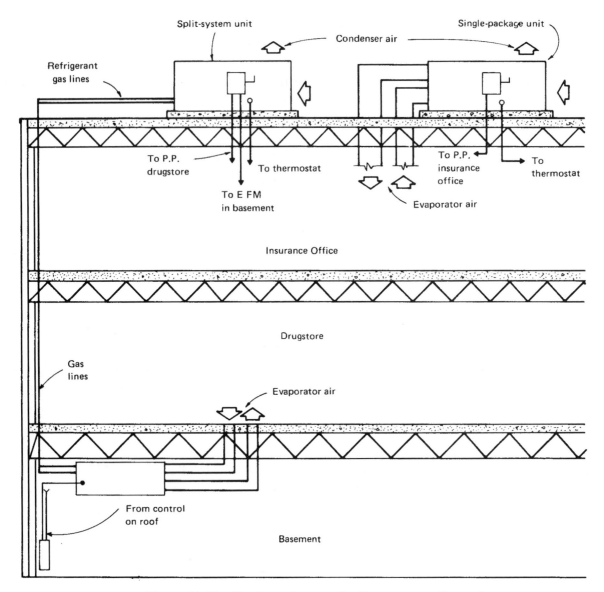

Figure 21-10   Single-package and split-system cooling units.

As discussed elsewhere in this text, when the nameplate on equipment specifies fuses, then only fuses are permitted to be used as the branch-circuit overcurrent protection. Circuit breakers would not be permitted. See *NEC® 110.3(B)*. This has been covered in Unit 17.

Air-conditioning and refrigeration equipment is required by *NEC® 440.4(B)* to be marked with a *visible nameplate marked with the maker's name, the rating in volts, frequency and number of phases, minimum supply circuit conductor ampacity, and the maximum rating of the branch-circuit, short-circuit, and ground-fault protective device, and the short-circuit current rating of the motor controllers or industrial control panel*. The exceptions to the

short-circuit current rating requirement are multi-motor and combination-load equipment in one- and two-family dwellings, cord- and plug-connect equipment, or equipment supplied by a branch-circuit rated at 60 amperes or less.

SPECIAL TERMINOLOGY

Understanding the following definitions is important when installing air conditioners, heat pumps, and other equipment using hermetic refrigerant motor-compressors. These terms are found throughout *NEC® Article 440* and *UL Standard 1995*.

Rated Load-Current (RLA or RLC): The RLA is determined by the manufacturer of the hermetic refrigerant motor-compressor through testing at

---

*Reprinted with permission from NFPA 70-2005.

rated refrigerant pressure, temperature conditions, and voltage.

In most instances, the RLA is at least equal to 64.1 percent of the hermetic refrigerant motor-compressor's maximum continuous current (MCC).

Example:

The nameplate on an air-conditioning unit is marked:

Compressor RLA: 17.8 amperes

Branch-Circuit Selection Current (BCSC): Some hermetic refrigerant motor-compressors are designed to operate continuously at currents greater than 156 percent of the RLA. In such cases, the unit's nameplate is marked with "Branch-Circuit Selection Current." The BCSC will be no less at least 64.1 percent of the MCC rating of the hermetic refrigerant motor-compressor.

Note: 156% and 64.1% have an inverse relationship:
$$1/_{1.56} = 0.641 \text{ and } 1/_{0.641} = 1.56$$

Example:

The MCC of a hermetic refrigerant motor-compressor is 31 amperes. The BCSC will be no less than:

$$31 \times 0.641 = 19.9 \text{ amperes}$$

Because the BCSC value, when marked, is always equal to or greater than the unit's nameplate marked RLA value, the manufacturer of the air-conditioning unit must use the BCSC value, instead of the RLA value, to determine the minimum circuit ampacity (MCA), and the maximum overcurrent protection (MOP). For installation of individual hermetic refrigerant motor-compressors, where the electrician must select the conductors, the controller, the disconnecting means, and the short-circuit and ground-fault protection, the electrician must use the BCSC, if given, instead of the RLA; see *NEC® 440.4(C)*.

Maximum Continuous Current (MCC): The MCC is determined by the manufacturer of the hermetic refrigerant motor-compressor under specific test conditions. The MCC is needed to properly design the end-use product. The installing electrician is not directly involved with the MCC.

The MCC is not on the nameplate of the packaged air-conditioning unit. The MCC is established by the *NEC®* as being no greater than 156 percent of the RLA or the BCSC, for the hermetic refrigerant motor-compressor.

Except in special conditions, the overload protective system must operate for current in excess of 156 percent.

Example:

A hermetic refrigerant motor-compressor is marked:

Maximum Continuous Current: 31 amperes

Minimum Circuit Ampacity (MCA): The manufacturer of an air-conditioning unit is required to mark the nameplate with this value. This is what the electrician needs to know. The manufacturer determines the MCA by multiplying the RLA, or the BCSC, of the hermetic refrigerant motor-compressor by 125 percent. The current ratings of all other concurrent loads, such as fan motors, transformers, relay coils, and so on, are then added to this value.

Example:

An air-conditioning unit's nameplate is marked:

Minimum Circuit Ampacity: 26.4 amperes

This value was derived as follows:

BCSC of 19.9×1.25	=	24.875 amperes
plus ¼-hp fan motor	@	1.5 amperes
Therefore, the MCA	=	26.4 amperes

The electrician must install the conductors, the disconnect switch, and the overcurrent protection based upon the MCA value. No further calculations are needed; the manufacturer of the equipment has done it all.

Maximum Overcurrent Protective (MOP): The electrician should always check the nameplate of an air-conditioning unit for this information.

The manufacturer is required to mark this value on the nameplate. This value is determined by multiplying the RLA, or the BCSC, of the hermetic refrigerant motor-compressor by 225 percent, then adding all concurrent loads such as electric heaters, motors, and so on.

Example:

The nameplate of an air-conditioning unit is marked as follows:

Maximum Time-Delay Fuse: 45 amperes

This value was derived as follows:

BCSC of 19.9×2.25	=	44.775 amperes
plus ¼-hp fan motor @		1.5 amperes
Therefore, the MOP	=	46.275 amperes

Because 46.275 is the maximum, the next lower standard size time-delay fuse (that is, 45 amperes) must be used. See *NEC® 240.6* and *Article 440 Part III.*

The electrician installs branch-circuit overcurrent devices with a rating not to exceed the MOP, as marked on the nameplate of the air-conditioning unit. The electrician makes no further calculations; all calculations have been done by the manufacturer of the air-conditioning unit.

Overcurrent Protection Device Selection: How does an electrician know if fuses or circuit breakers are to be used for the branch-circuit overcurrent device? The electrician reads the nameplate and the instructions carefully. *NEC® 110.3(B)* requires that *listed or labeled equipment be used or installed in accordance with any instructions included in the listing or labeling.*

Example:

If the nameplate of an air-conditioning unit reads "Maximum Size Time-Delay Fuse: 45 amperes," fuses are required for the branch-circuit protection. To install a circuit breaker would be a violation of *NEC® 110.3(B).*

If the nameplate reads "Maximum Fuse or HACR Type Breaker . . . ," then either fuses or an HACR-type circuit breaker is permitted. See Figure 17-33, Figure 17-34, and Figure 17-35.

Disconnecting Means Rating: The *Code* rules for determining the size of the disconnecting means required for HVAC equipment is covered in *NEC® Article 440 Part II.*

The horsepower rating of the disconnecting means must be at least equal to the sum of all the individual loads within the equipment, at rated load conditions and at locked-rotor conditions. See *NEC® 440.12(B)(1).*

The ampere rating of the disconnecting means must be at least 115 percent of the sum of all the individual loads within the equipment. Refer to *NEC® 440.12(A)(1)* and *440.12(B)(2)* for details.

Because disconnect switches are horsepower rated, it is sometimes necessary to convert the locked-rotor current information found on the nameplate of air-conditioning equipment to an equivalent horsepower rating; *NEC® Article 440 Part II* sets forth the conversion procedure.

Air-Conditioning and Refrigeration Equipment Disconnecting Means

- The disconnecting means for an individual hermetic motor-compressor must not be less than 115 percent of the rated load current or the branch-circuit selection current, whichever is greater. See *NEC® 440.12(A)(1).*

- The disconnecting means for an individual hermetic motor-compressor shall have a horsepower rating equivalent to the horsepower rating shown in *NEC® Table 430.251(A)* (Table 21-1) and *Table 430.251(B)* (Table 21-2). This table is a conversion table to convert locked-rotor current to an equivalent horsepower. See *NEC® 440.12(A)(2).*

- For equipment that has at least one hermetic motor-compressor and for other loads, such as fans, heaters, solenoids, and coils, the disconnecting means must be not less than 115 percent of the sum of the currents of all of the components. See *NEC® 440.12(B)(2).*

Single-Phase, Locked-Rotor Currents for Selection of Disconnect Means See *NEC® Table 430.251(A)*			
Rated Horsepower	Maximum Locked-Rotor Current (Amperes, Single Phase)		
	115 Volts	208 Volts	230 Volts
½	58.8	32.5	29.4
¾	82.8	45.8	41.4
1	96	53	48
1½	120	66	60
2	144	80	72
3	204	113	102
5	336	186	168
7½	480	265	240
10	600	332	300

For use only with *NEC® 430.110, 440.12, 440.41* and *455.8(C).*

Table 21-1

Table 430.251(B) Conversion Table of Polyphase Design B, C, and D Maximum Locked-Rotor Currents for Selection of Disconnecting Means and Controllers as Determined from Horsepower and Voltage Rating and Design Letter
For use only with 430.110, 440.12, 440.41 and 455.8(C).

Rated Horsepower	Maximum Motor Locked-Rotor Current in Amperes, Two- and Three-Phase, Design B, C, and D*					
	115 Volts	200 Volts	208 Volts	230 Volts	460 Volts	575 Volts
	B, C, D	B, C, D	B, C, D	B, C, D	B, C, D	B, C, D
½	40	23	22.1	20	10	8
¾	50	28.8	27.6	25	12.5	10
1	60	34.5	33	30	15	12
1½	80	46	44	40	20	16
2	100	57.5	55	50	25	20
3	—	73.6	71	64	32	25.6
5	—	105.8	102	92	46	36.8
7½	—	146	140	127	63.5	50.8
10	—	186.3	179	162	81	64.8
15	—	267	257	232	116	93
20	—	334	321	290	145	116
25	—	420	404	365	183	146
30	—	500	481	435	218	174
40	—	667	641	580	290	232
50	—	834	802	725	363	290
60	—	1001	962	870	435	348
75	—	1248	1200	1085	543	434
100	—	1668	1603	1450	725	580
125	—	2087	2007	1815	908	726
150	—	2496	2400	2170	1085	868
200	—	3335	3207	2900	1450	1160
250	—	—	—	—	1825	1460
300	—	—	—	—	2200	1760
350	—	—	—	—	2550	2040
400	—	—	—	—	2900	2320
450	—	—	—	—	3250	2600
500	—	—	—	—	3625	2900

*Design A motors are not limited to a maximum starting current or locked rotor current.

Table 21-2

Reprinted with permission from NFPA 70-2005

- For equipment that has at least one hermetic motor-compressor and for other loads, such as fans, heaters, solenoids, and coils, the disconnecting means rating is determined by adding all the individual currents. This total is then considered to be a single motor for the purpose of determining the disconnecting means horsepower rating. See *NEC® 440.12(B)(1)*.

In our example, for simplicity, we will consider the minimum circuit ampacity to be the full-load current. This is 31.65 amperes. Checking *NEC® Table 430.250* (Table 21-3) we find that a full-load current rating of 31.65 amperes falls between a 10-horsepower and a 15-horsepower motor. Therefore, we must consider our example to be equivalent to a 15-horsepower motor.

NEC® Table 430.248 (Table 21-4) lists the full-load current ratings for single-phase motors.

- The total locked-rotor current must also be checked to be sure that the horsepower rating of the disconnecting means is capable of safely disconnecting the full amount of locked-rotor current, see *NEC® 440.12(B)(I)*.

In our example, the locked-rotor current equals 182 amperes. Checking *NEC® Table 430.251(B)* (Table 21-2) in the 208-volt column for the conversion of locked-rotor current to horsepower, we find that an LRA of 182

Table 430.250 Full-Load Current, Three-Phase Alternating-Current Motors
The following values of full-load currents are typical for motors running at speeds usual for belted motors and motors with normal torque characteristics.
  The voltages listed are rated motor voltages. The currents listed shall be permitted for system voltage ranges of 110 to 120, 220 to 240, 440 to 480, and 550 to 600 volts.

Horsepower	Induction-Type Squirrel Cage and Wound Rotor (Amperes)							Synchronous-Type Unity Power Factor* (Amperes)			
	115 Volts	200 Volts	208 Volts	230 Volts	460 Volts	575 Volts	2300 Volts	230 Volts	460 Volts	575 Volts	2300 Volts
½	4.4	2.5	2.4	2.2	1.1	0.9	—	—	—	—	—
¾	6.4	3.7	3.5	3.2	1.6	1.3	—	—	—	—	—
1	8.4	4.8	4.6	4.2	2.1	1.7	—	—	—	—	—
1½	12.0	6.9	6.6	6.0	3.0	2.4	—	—	—	—	—
2	13.6	7.8	7.5	6.8	3.4	2.7	—	—	—	—	—
3	—	11.0	10.6	9.6	4.8	3.9	—	—	—	—	—
5	—	17.5	16.7	15.2	7.6	6.1	—	—	—	—	—
7½	—	25.3	24.2	22	11	9	—	—	—	—	—
10	—	32.2	30.8	28	14	11	—	—	—	—	—
15	—	48.3	46.2	42	21	17	—	—	—	—	—
20	—	62.1	59.4	54	27	22	—	—	—	—	—
25	—	78.2	74.8	68	34	27	—	53	26	21	—
30	—	92	88	80	40	32	—	63	32	26	—
40	—	120	114	104	52	41	—	83	41	33	—
50	—	150	143	130	65	52	—	104	52	42	—
60	—	177	169	154	77	62	16	123	61	49	12
75	—	221	211	192	96	77	20	155	78	62	15
100	—	285	273	248	124	99	26	202	101	81	20
125	—	359	343	312	156	125	31	253	126	101	25
150	—	414	396	360	180	144	37	302	151	121	30
200		552	528	480	240	192	49	400	201	161	40
250	—	—	—	—	302	242	60	—	—	—	—
300	—	—	—	—	361	289	72	—	—	—	—
350	—	—	—	—	414	336	83	—	—	—	—
400	—	—	—	—	477	382	95	—	—	—	—
450	—	—	—	—	515	412	103	—	—	—	—
500	—	—	—	—	590	472	118	—	—	—	—

*For 90 and 80 percent power factor, the figures shall be multiplied by 1.1 and 1.25, respectively.

Table 21-3
Reprinted with permission from NFPA 70-2005

amperes falls between a 10-horsepower and a 15-horsepower motor. Therefore, we again consider our example to be equivalent to a 15-horsepower motor, see *NEC® 440.12(A)* and *440.12(B)*.

  If one of the selection methods (FLA or LRA) indicates a larger size disconnect switch than the other method does, then the larger switch shall be installed. See *NEC® 440.12(A)* and *440.12(B)*.

- The disconnecting means must be within sight of and readily accessible from the air-conditioning and refrigeration equipment. See *NEC® 440.14.*

- The disconnecting means may be mounted inside the equipment, or it may be mounted on the equipment. Very large equipment often has the disconnect switch mounted inside the enclosure. See *NEC® 440.14.*

Air-Conditioning and Refrigeration Equipment Branch-Circuit Conductors

- For individual motor-compressor equipment, the branch-circuit conductors must have an ampacity of not less than 125 percent of the rated load current or branch-circuit selection current, whichever is greater. See *NEC® 440.32.*

- For motor-compressor(s), plus other loads such as a typical air-conditioning unit that contains a motor-compressor, fan, heater, coils, and so on, add all the individual current values, plus 25 percent of that of the largest motor or motor-compressor. See *NEC® 440.33.*

Table 430.248 Full-Load Currents in Amperes, Single-Phase Alternating-Current Motors

The following values of full-load currents are for motors running at usual speeds and motors with normal torque characteristics. The voltages listed are rated motor voltages. The currents listed shall be permitted for system voltage ranges of 110 to 120 and 220 to 240 volts.

Horsepower	115 Volts	200 Volts	208 Volts	230 Volts
1/6	4.4	2.5	2.4	2.2
1/4	5.8	3.3	3.2	2.9
1/3	7.2	4.1	4.0	3.6
1/2	9.8	5.6	5.4	4.9
3/4	13.8	7.9	7.6	6.9
1	16	9.2	8.8	8.0
1 1/2	20	11.5	11.0	10
2	24	13.8	13.2	12
3	34	19.6	18.7	17
5	56	32.2	30.8	28
7 1/2	80	46.0	44.0	40
10	100	57.5	55.0	50

Table 21-4

Reprinted with permission from NFPA 70-2005

One can readily see the difficulty encountered when attempting to calculate the actual currents flowing in phases A, B, and C for a typical three-phase unit, as shown in Figure 21-11.

Rather than attempting to add the currents vectorally, for simplicity, one generally uses the arithmetic sum of the current. However, the student must realize that an ammeter reading taken while the air-conditioning unit is operating will not be the same as the calculated current values.

For the air-conditioning unit in the insurance

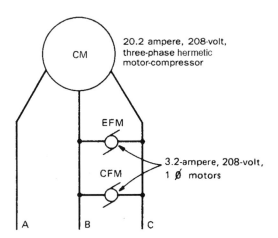

Figure 21-11   Load connections.

office, the calculation is:

$$20.2 + 3.2 + 3.2 = 26.60 \text{ amperes}$$
$$\text{plus } 25\% \text{ of } 20.2 = \underline{5.05} \text{ amperes}$$
$$\text{Total} = 31.65 \text{ amperes}$$

The minimum ampacity for the conductors serving the preceding air-conditioning unit is 31.65 amperes, with the preceding calculation done.

However, the nameplate of the air-conditioning unit calls for a minimum circuit ampacity of 31.65 amperes. Referring to *NEC® Table 310.16*, it is determined that an 8 AWG, Type THHN conductor has an allowable ampacity of 40 amperes.

NEC® 440.35 references *NEC® 440.4(B)*, which sets forth the requirement that the manufacturers mark the equipment with the minimum supply circuit ampacity and the maximum branch-circuit overcurrent protective device ampere rating.

Air-Conditioning and Refrigeration Equipment Branch-Circuit, Short-Circuit, and Ground-Fault Protection

- For individual units, the branch-circuit overcurrent protection shall not exceed 175 percent of the rated load current or the branch-circuit selection current, whichever is greater. But if the 175 percent sizing will not allow the equipment to start, then *NEC® 440.22(A)* permits a further increase in size not to exceed 225 percent.

- Where the circuit has more than one hermetic refrigerant motor-compressor, or one hermetic refrigerant motor-compressor and other motors or other loads, *NEC® 440.22(B)* references *430.53*, which addresses the situation where more than one motor is fed by a single circuit.

The branch-circuit overcurrent protection must also meet the requirements of *NEC® 440.22(B)*. Simply stated, this means that overcurrent protection must be provided that does not exceed the protection selected for the motor-compressor that draws the largest amount of current, plus the sum of the rated load current of branch-circuit selection currents of other motor-compressors, plus the ampere ratings of any other loads.

In our example:

$$(20.2 \times 1.75) + 3.2 + 3.2 = 41.75 \text{ amperes}$$

It is important to know that *NEC®* 440.22(C) states that the overcurrent protection shall not exceed the value indicated on the nameplate as determined by the manufacturer of the unit.

Underwriters Laboratories *Standard No. 1995* allows the manufacturer of hermetic motor-compressor equipment to determine the maximum size overcurrent protection as follows:

$$
\begin{aligned}
\text{Largest load} \times 2.25 = \\
20.2 \times 2.25 = 45.45 \text{ amperes} \\
\text{Plus other loads} \quad 3.20 \text{ amperes} \\
\underline{3.20 \text{ amperes}} \\
51.85 \text{ amperes} \\
\text{maximum}
\end{aligned}
$$

This value is "rounded down" to the next lower standard fuse rating, which is 50-ampere (*NEC®* 240.6). This size requires a 60-ampere disconnect switch. Because the 50-ampere value is maximum, the use of a lower-rated time-delay fuse, such as a 45-ampere or 40-ampere fuse, is permitted. The question is, will the lower ratings open unnecessarily on repetitive start-ups of the air-conditioning unit? The ampere rating of the maximum size fuse or circuit breaker as determined by the manufacturer is generally the size to install.

The nameplate for the air-conditioning unit supplying the cooling for the drugstore calls for maximum 50-ampere, dual-element, time-delay fuses. This value must not be exceeded.

NEC® 110.3(B) requires that *Listed or labeled equipment shall be installed and used in accordance with any instructions included in the listing or labeling.* It is this section that prohibits the use of a circuit breaker to protect this particular air-conditioning unit. If the feeder or branch-circuit originates from a circuit-breaker panel, then the requirements of this section can be met by installing a fusible disconnect switch near, on, or within the air-conditioning equipment.

Air-Conditioning and Refrigeration Equipment Motor-Compressor and Branch-Circuit Overload Protection

This topic is covered in *NEC®* Article 440, Part VI.

The manufacturer of this type of equipment provides the overload protection for the various internal components that make up the complete air-conditioning unit. The manufacturer submits the equipment to Underwriters Laboratories for testing and listing of the product. Overload protection of the internal components is covered in the UL Standards. This protection can be in the form of fuses, thermal protectors, certain types of inverse-time circuit breakers, overload relays, and so forth.

NEC® 440.52 discusses overload protection for motor-compressor appliances. This section requires that the motor-compressor be protected from overloads and failure to start by:

- a separate overload relay that is responsive to current flow and trips at not more than 140 percent of the rated load current, or

- an integral thermal protector arranged so that its action interrupts the current flow to the motor-compressor, or

- a time-delay fuse or inverse-time circuit breaker rated at not more than 125 percent of the rated load current.

It is rare for the electrician to have to provide additional overload protection for these components. The electrician's job is to make sure that he or she supplies the equipment with the proper size conductors, the proper size and type of branch-circuit overcurrent protection, and the proper horsepower-rated disconnecting means.

Tons versus Amperes

It is possible to estimate the current required for air-conditioning equipment, heat pumps, or other hermetic motor-compressor-type loads, based on the equipment tons, but the procedure is complex and is not necessary. ▶In conformance to *NEC®* 440.4(B), all units are marked with the *maker's name, the rating in volts, frequency and number of phases, minimum supply circuit conductor ampacity, the maximum rating of the branch-circuit short circuit and ground-fault protective device, and the short circuit current rating of the motor controllers or industrial control panel.*◀

---

*Reprinted with permission from NFPA 70-2005.

REVIEW QUESTIONS

Refer to the *National Electrical Code*® or the working drawings when necessary. Where applicable, responses should be written in complete sentences.

Complete the following sentences by writing a component used in a refrigeration system.

1. The _____ raises the energy level of the refrigerant.

2. The heat is removed from the refrigerant in the _____.

3. The refrigerant absorbs heat energy in the _____.

4. The _____ is a metering device.

5. In the typical residential heating/cooling system, the evaporator would be located _____.

6. In the typical residential heating/cooling system, the condenser would be located _____.

Respond to the following statements.

7. Name three types of condensers. (1) _____ , (2) _____ and (3) _____

8. A 230-volt, single-phase, hermetic compressor with an FLA of 21 and an LRA of 250 requires a disconnect switch with a rating of _____ horsepower.

9. A motor-compressor unit with a rating of 32 amperes is protected from overloads by a separate overload relay selected to trip at not more than _____ ampere(s).

10. Which *NEC*® section permits the disconnect switch for a large rooftop air conditioner to be mounted inside the unit?

_____

11. The nameplate of an air-conditioning unit specifies 85 amperes as the minimum ampacity for the supply conductors. Select the conductor size and type.

_____

12. Where the nameplate of an air-conditioning unit specifies the minimum ampacity required for the supply conductors, should the electrician multiply this ampacity by 1.25 to determine the correct size conductors? Explain.

_____

_____

13. Which current rating of an air-conditioning unit is greater? Check the correct answer.

   a. _____ rated load current

   b. _____ branch-circuit selection current

APPENDIX

ELECTRICAL SPECIFICATIONS

General Clauses and Conditions

The General Clauses and Conditions and the Supplementary General Conditions are hereby made a part of the Electrical Specifications.

Scope

The electrical contractor shall furnish all labor and material to install the electrical equipment shown on the drawings herein specified, or both. The contractor shall secure all necessary permits and licenses required and in accepting the contract agrees to have all equipment in working order at the completion of the project.

Materials

All materials used shall be new, shall bear the label of a Nationally Recognized Testing Laboratory (NRTL), and shall meet the requirements of the drawings and specifications.

Workmanship

All electrical work shall be in accordance with the requirements of the *National Electrical Code,*® shall be executed in a workmanlike manner, and shall present a neat and symmetrical appearance when completed.

Show Drawings

The contractor shall submit for approval descriptive literature for all equipment installed as part of this contract.

Motors

All motors shall be installed by the contractor furnishing the motor. All wiring to the motors and the connection of motors will be completed by the electrical contractor. All control wiring will be the responsibility of the contractor furnishing the equipment to be controlled.

Wiring

The wiring shall meet the following requirements:

a. The minimum size conductors will be 12 AWG.

b. In the doctor's office, all branch-circuit conductors will be installed in electrical metallic tubing. In all other areas of the building, branch-circuit conductors may be installed in either electrical metallic tubing or electrical nonmetallic tubing.

c. Three spare trade size ¾ conduits shall be installed from each flush-mounted panel into the ceiling joist space above the panel.

d. In the doctor's office, branch-circuit raceways shall be electrical metallic tubing. In all other areas of the building, raceways may be either electrical metallic tubing or electrical nonmetallic conduit.

e. Feeders shall be installed in either rigid metal conduit or rigid nonmetallic conduit. The supply to the boiler shall be rigid metal conduit or Schedule 80 rigid nonmetallic conduit.

f. Conductors shall have a 194°F (90°C) rating, Types THHN, THHW, THWN, or XHHW. All conductors shall be copper.

Switches

Switches shall be ac general-use, snap switches, specification grade, 20 amperes, 120–277 volts, either single-pole, three-way, or four-way, as shown on the plans.

Time Clock

A time clock shall be installed to control the lighting in the front and rear entries. The clock will be connected to Panel EM circuit No. 1. The clock will be 120 volts, one circuit, with astronomic control and a spring-wound carry-over mechanism.

Receptacles

The electrical contractor shall furnish and install, as indicated in blueprint E-4 (Electrical Symbol Schedule), receptacles meeting NEMA standards and listed by Underwriters Laboratories, Inc.

Panelboards

The electrical contractor shall furnish panelboards, as shown on the plans and detailed in the panelboard schedules. Panelboards shall be listed by a Nationally Recognized Testing Laboratory. All interiors will have 225-ampere bus with lugs, rated at 167°F (75°C), for feed conductors. Boxes will be of galvanized sheet steel and will provide wiring gutters, as required by the *National Electrical Code.®* Fronts will be suitable for either flush or surface installation and shall be equipped with a keyed lock and a directory card holder. All panelboards shall have an equipment grounding bus bonded to the cabinet. See Sheet E4 for a listing of the required poles and overcurrent devices.

Molded-Case Circuit Breakers

Molded-case circuit breakers shall be installed in branch-circuit panelboards as indicated on the panelboard schedules following the Specifications. Each breaker shall provide inverse time-delay under overload conditions and magnetic tripping for short circuits. The breaker operating mechanism shall be trip-free and multipole units will open all poles if one pole trips. Breakers shall have sufficient interrupting rating to interrupt 10,000 RMS symmetrical amperes.

Motor-Generator

The electrical contractor will furnish and install a motor-generator plant capable of delivering 12 kilovolt-amperes at 208Y/120 volts, three phase. Motor shall be for use with diesel fuel, liquid cooled, complete with 12-volt batteries and battery charger, mounted on antivibration mounts with all necessary accessories, including mufflers, exhaust piping, fuel tanks, fuel lines, and remote derangement annunciator.

Transfer Switch

The electrical contractor will furnish and install a complete listed automatic load transfer switch capable of handling 12 kilovolt-amperes at 208Y/120 volts, three phase. The switch control will sense a loss of power on any phase and signal the motor-generator to start. When emergency power reaches 80 percent of voltage and frequency, the switch will automatically transfer to the generator source. When the normal power has been restored for a minimum of five minutes, the switch will reconnect the load to the regular power and shut off the motor-generator. Switch shall be in a NEMA 1 enclosure.

Luminaires (Fixtures)

The electrical contractor will furnish and install luminaires (fixtures), as described in the schedule shown in the plans. Luminaires (fixtures) shall be complete with diffusers and lamps where indicated. Ballast, when required, shall be Class P and have a sound rating of A. Each ballast is to be individually fused; fuse size and type shall be as recommended by the ballast manufacturer. All rapid-start ballasts shall be Class E.

Service Entrance and Equipment

The electrical contractor will furnish and install a service entrance as shown on the Electrical Power Distribution Riser Diagram, which is found in the plans. Transformers and all primary service will be installed by the power company. The electrical contractor will provide a concrete pad and conduit rough-in as required by the power company.

The electrical contractor will install a ground field adjacent to the transformer pad consisting of three 8-foot-long copperweld rods connected to the grounded conductor. In addition, a connection shall be made to the main water pipe.

The electrical contractor will furnish and install a service lateral consisting of three sets of three 500-kcmil and one 3/0 AWG conductors, each set in a trade size 3 rigid metallic or nonmetallic conduit.

The electrical contractor will furnish and install service-entrance equipment, as shown in the plans and detailed herein. The equipment will consist of nine switches, and the metering equipment will be fabricated in three type NEMA 1 sections. A continuous neutral bus will be furnished for the length of the equipment and shall be isolated except for the main bonding jumper to the grounding bus, which shall also be connected to each section of the service-entrance equipment and to the water main. The switchboard shall be braced for 50,000 RMS symmetrical amperes.

The metering will be located in one section and shall consist of 7 meters. Five of these meters shall be for the occupants of the building and two meters shall serve the owner's equipment.

The switches shall be as follows:

1. Bolted pressure switch, three-pole, 600 amperes with three 600-ampere fuses.

2. Bolted pressure switch, three-pole, 800 amperes with three 700-ampere fuses.

3. Quick make-quick break switch, three-pole, 100 amperes, with three 100-ampere fuses.

4. Quick make-quick break switch, three-pole, 200 amperes with three 200-ampere fuses.

5. Quick make-quick break switch, three-pole, 200 amperes with three 150-ampere fuses.

6. Quick make-quick break switch, two-pole, 200-ampere switch with two 125-ampere fuses.

7. Quick make-quick break switch, three-pole, 100-ampere with three 90-ampere fuses.

8. Quick make-quick break switch, three-pole, 100-ampere with three 60-ampere fuses.

9. Quick make-quick break switch, three-pole, 100-ampere switch.

The bolted pressure switches shall be knife-type switches constructed with a mechanism that automatically applies a high pressure to the blade when the switch is closed. Switch shall be rated to interrupt 200,000 symmetrical RMS amperes when used with current-limiting fuses having an equal rating.

The quick make-quick break switches shall be constructed with a device that assists the operator in opening or closing the switch to minimize arcing. Switches shall be rated to interrupt 200,000 symmetrical RMS amperes when used with fuses having an equal rating. They shall have rejection type R fuse clips for Class R fuses.

Fuses

a. Fuses 601 amperes and larger shall have an interrupting rating of 200,000 symmetrical RMS amperes. They shall provide time-delay of not less than 4 seconds at 500 percent of their ampere rating. They shall be current limiting and of silver-sand Class L construction.

b. Fuses 600 amperes and less shall have an interrupting rating of 200,000 symmetrical RMS amperes. They shall be of the rejection Class RK1 type, having a time delay of not less than 10 seconds at 500 percent of their ampere rating.

c. Plug fuses shall be dual-element, time-delay type with Type S base.

d. All fuses shall be so selected to ensure positive selective coordination as indicated on the schedule.

e. All lighting ballasts (fluorescent, mercury vapor, or others) shall be protected on the supply side with an appropriate fuse in a suitable approved fuseholder mounted within or on the fixture. Fuses shall be sized as recommended by the ballast manufacturer.

f. Spare fuses shall be provided in the amount of 20 percent of each size and type installed, but in no case shall less than three spares of a specific size and type be supplied. These spare fuses shall be delivered to the owner at the time of acceptance of the project, and shall be placed in a spare fuse cabinet mounted on the wall adjacent to, or located in, the switchboard.

g. Fuse identification labels, showing the size and type of fuses installed, shall be placed inside the cover of each switch.

Low-Voltage Remote-Control Switching

A low-voltage, remote-control switch system shall be installed in the drugstore as shown on the plans and detailed herein. All components shall be specification grade and constructed to operate on 24-volt control power. The transformer shall be a 120/24-volt, energy-limiting type for use on a Class II signal system.

Cabinet. A metal cabinet matching the panelboard cabinets shall be installed for the installation of relays and other components. A barrier will separate the control section from the power wiring.

Relay.   A 24-volt ac, split-coil design relay rated to control 20 amperes of tungsten or fluorescent lamp loads shall be provided.

Switches.   The switches shall be complete with wall plate and mounting bracket. They shall be normally open, single-pole, double-throw, momentary contact switches with on–off identification.

Rectifiers.   A heavy-duty silicon rectifier with 7.5-ampere continuous duty rating shall be provided.

Wire.   The wiring shall be in two- or three-conductor, color-coded, 20 AWG wire.

Rubber grommet.   An adapter will be installed on all relays to isolate the relay from the metal cabinet to reduce noise.

Switching schedule.   Connections will be made to accomplish the lighting control as shown in the switching schedule, Table A-1.

HEATING AND AIR-CONDITIONING SPECIFICATIONS

Only those sections that pertain to the electrical work are listed here.

Boiler

The heating contractor will furnish and install a 200-kilowatt electric hot-water heating boiler completely equipped with safety, operating, and sequencing controls for 208-volt, three-phase electric power.

Hot-Water Circulating Pumps

The heating contractor will furnish and install five circulating pumps. Each pump will serve a separate rental area and be controlled from a thermostat located in that area as indicated on the electrical plans. Pumps will be 1/6 horsepower, 120 volts, single phase. Overload protection will be provided by a manual motor starter.

TABLE A-1
Drugstore Low-Voltage, Remote-Control Switching Schedule

Switch	Relay	Area Served	Branch Circuit #
RCa	A	Main area lighting	1
RCb	B	" " "	3
RCc	C	" " "	3
RCd	D	Makeup area	5
RCe	E	Storage	5
RCf	F	Toilet	5
	G	Pharmacy	10
	H	"	10
	I	"	10
RCg	J	Stairway and Basement	9
RCh, RCi	L,M	Show window	11,13
	K	Sign	15
RCM-1	D		
RCM-2	E,F		
RCM-3	H,I,J		
RCM-4	I,J,K		
RCM-5	A,B,C		
RCM-6	J		
RCM-7	K		
RCM-8	L,M		

Air-Conditioning Equipment

Air-conditioning equipment will be furnished and installed by the heating contractor in four of the rental areas indicated as follows:

Drugstore.   A split system packaged unit will be installed with a rooftop compressor-condenser and a remote evaporator located in the basement. The electrical characteristics of this system are:

Voltage:
208 volts, three-phase, 3 wire, 60 hertz
Hermetic refrigerant compressor-motor:
Rated load current 20.2 amperes, 208 volts, three-phase
Evaporator motor:
Full load current 3.2 amperes, 208 volts, single-phase
Condenser motor:
Full load current 3.2 amperes, 208 volts, single-phase
Minimum circuit ampacity:
31.65 amperes
Maximum overcurrent protection:
50 amperes, time-delay fuse

Insurance Office.   A single package unit will be installed on the roof with electrical data identical to that of the unit specified for the drugstore.

Beauty Salon.   A single package unit will be installed on the roof and will have the following electrical characteristics:
Voltage:
208 volts, three-phase, 3 wire, 60 hertz
Hermetic refrigerant compressor-motor:
Rated load amperes 14.1, 208 volts, three-phase
Condenser-evaporator motor:
Full load amperes 3.3, 208 volts, single-phase
Minimum circuit ampacity:
20.92 amperes
Maximum overcurrent protection:
35 amperes, time-delay fuse

Doctor's Office.   A single package unit will be installed on the roof and will have the following electrical characteristics:
Voltage:
208 volts, single-phase, 2 wire, 60 hertz
Hermetic refrigerant compressor-motor:
Rated load amperes 16.8 at 208 volts, single-phase
Condenser-evaporator motor:
Full load amperes 3.7 at 208 volts, single-phase
Minimum circuit ampacity:
24.7 amperes
Maximum overcurrent protection:
40 amperes, time-delay fuse

Heating Control

In cooled areas, the heating and cooling will be controlled by a combination thermostat located in the proper area as shown on the electrical plans.

PLUMBING SPECIFICATIONS

Only those sections that pertain to the electrical work are listed here.

Motors

All motors will be installed by the contractor furnishing the motor. All electrical power wiring to the motors and the connection of the motors shall be made by the electrical contractor. All control wiring and control devices will be the responsibility of the contractor furnishing the equipment to be controlled.

Sump Pump

The plumbing contractor will furnish and install an electric motor-driven, fully automatic sump pump. Motor will be ½ horsepower, 120 volts, single phase. Overload protection will be provided by a manual motor starter.

HOW TO USE TABLES A-2 AND A-3 TO CALCULATE VOLTAGE LOSS

1. Multiply length of circuit in feet (one way) by the current (in amperes) by the number shown in the table for the kind of current (3-phase, 1-phase, power factor) and the size of conductor.

2. Put a decimal in front of the last 6 digits. This is the approximate voltage drop to be expected in the circuit.

3. If you have conductors in parallel, one extra step is needed: for two conductors in parallel per phase, divide answer in Step 2 by 2; for three conductors in parallel per phase, divide answer in Step 2 by 3; for four conductors in parallel per phase, divide answer in Step 2 by 4.

Example: The 208Y/120-volt commercial building service is made up of three 500-kcmil copper conductors per phase. The length of the service conductors between the switchboard and the transformer secondary is approximately 25 feet. The wiring method is steel conduit. Find the approximate voltage drop when the load is 1000 amperes, 90 percent power factor.

1. $25 \times 1000 \times 81 = 2{,}025{,}000$

2. $2{,}025{,}000$

3. $\dfrac{2{,}025{,}000}{3} = 0.675$ volts loss

HOW TO USE TABLES A-2 AND A-3 TO SELECT CONDUCTOR SIZE

1. Multiply length of circuit in feet (one way) by the current (in amperes).

2. Multiply the permitted voltage loss by 1,000,000.

3. Divide Step 2 by Step 1.

4. Look in the column that shows the type of conductor, current, and power factor for the number found in Step 3 that is equal to or less than that number.

Example: Find the size copper conductors needed to limit the voltage loss on a 180-foot run. The wiring method is steel conduit. The system is 3-phase. The load is 40 amperes, 80 percent power factor.

1. $180 \times 40 = 7200$

2. $5.5 \times 1{,}000{,}000 = 5{,}500{,}000$

3. $\dfrac{5{,}500{,}000}{7200} = 764$

4. In the table, in the proper column, find the number 764 or less. The lesser number is 745. Go to the left column and find the minimum conductor size to be 6 AWG.

TABLE A-2
Aluminum Conductors* — Ratings and Volt Loss**

Conduit	Wire Size	Ampacity			Direct Current	Volt Loss									
		Type T, TW, UF (60°C Wire)	Type RHW, THHW (wet), THW, THWN, XHHW (wet), USE (75°C Wire)	Type RHH, THHW, THHN (dry), XHHW (dry) (90°C Wire) RHW-2, THW-2 (90°C wet or dry)		Three Phase (60 Cycle, Lagging Power Factor.)					Single Phase (60 Cycle, Lagging Power Factor.)				
						100%	90%	80%	70%	60%	100%	90%	80%	70%	60%
Steel Conduit	12	20*	20*	25*	6360	5542	5039	4504	3963	3419	6400	5819	5201	4577	3948
	10	25	30*	35*	4000	3464	3165	2836	2502	2165	4000	3654	3275	2889	2500
	8	30	40	45	2520	2251	2075	1868	1656	1441	2600	2396	2158	1912	1663
	6	40	50	60	1616	1402	1310	1188	1061	930	1620	1513	1372	1225	1074
	4	55	65	75	1016	883	840	769	692	613	1020	970	888	799	708
	3	65	75	85	796	692	668	615	557	497	800	771	710	644	574
	2	75	90	110	638	554	541	502	458	411	640	625	580	529	475
	1	85	100	115	506	433	432	405	373	338	500	499	468	431	391
	0	100	120	135	402	346	353	334	310	284	400	407	386	358	328
	00	115	135	150	318	277	290	277	260	241	320	335	320	301	278
	000	130	155	175	252	225	241	234	221	207	260	279	270	256	239
	0000	155	180	205	200	173	194	191	184	174	200	224	221	212	201
	250	170	205	230	169	148	173	173	168	161	172	200	200	194	186
	300	190	230	255	141	124	150	152	150	145	144	174	176	173	168
	350	210	250	280	121	109	135	139	138	134	126	156	160	159	155
	400	225	270	305	106	95	122	127	127	125	110	141	146	146	144
	500	260	310	350	85	77	106	112	113	113	90	122	129	131	130
	600	285	340	385	71	65	95	102	105	106	76	110	118	121	122
	750	320	385	435	56	53	84	92	96	98	62	97	107	111	114
	1000	375	445	500	42	43	73	82	87	89	50	85	95	100	103
Non-Magnetic Conduit (Lead Covered Cables Or Installation In Fibre Or Other Non-Magnetic Conduit. Etc.)	12	20*	20*	25*	6360	5542	5029	4490	3946	3400	6400	5807	5184	4557	3926
	10	25	30*	35*	4000	3464	3155	2823	2486	2147	4000	3643	3260	2871	2480
	8	30	40	45	2520	2251	2065	1855	1640	1423	2600	2385	2142	1894	1643
	6	40	50	60	1616	1402	1301	1175	1045	912	1620	1502	1357	1206	1053
	4	55	65	75	1016	883	831	756	677	596	1020	959	873	782	688
	3	65	75	85	796	692	659	603	543	480	800	760	696	627	555
	2	75	90	100	638	554	532	490	443	394	640	615	566	512	456
	1	85	100	115	506	433	424	394	360	323	500	490	455	415	373
	0	100	120	135	402	346	344	322	296	268	400	398	372	342	310
	00	115	135	150	318	277	281	266	247	225	320	325	307	285	260
	000	130	155	175	252	225	234	223	209	193	260	270	258	241	223
	0000	155	180	205	200	173	186	181	171	160	200	215	209	198	185
	250	170	205	230	169	147	163	160	153	145	170	188	185	177	167
	300	190	230	255	141	122	141	140	136	130	142	163	162	157	150
	350	210	250	280	121	105	125	125	123	118	122	144	145	142	137
	400	225	270	305	106	93	114	116	114	111	108	132	134	132	128
	500	260	310	350	85	74	96	100	100	98	86	111	115	115	114
	600	285	340	385	71	62	85	90	91	91	72	98	104	106	105
	750	320	385	435	56	50	73	79	82	82	58	85	92	94	95
	1000	375	445	500	42	39	63	70	73	75	46	73	81	85	86

*The overcurrent protection for conductor types marked with an (*) shall not exceed 15-ampere for 12 AWG, or 25-ampere for 10 AWG after correction factors for ambient temperature and adjustment factors for the number of conductor have been applied.

**Figures are L-L for both single-phase and three-phase. Three-phase figures are average for the three phases.

Table *courtesy of* Cooper Bussman, Inc.

TABLE A-3
Copper Conductors* — Ratings and Volt Loss**

Conduit	Wire Size	Ampacity Type TW, UF (60°C Wire)	Type RHW, THHW (wet), THW, THWN, XHHW (wet), USE (75°C Wire)	Type RHH, THHW, THHN (dry), XHHW (dry) (90°C Wire) RHW-2, THW-2 (90°C wet or dry)	Direct Current	Volt Loss Three Phase (60 Cycle, Lagging Power Factor.) 100%	90%	80%	70%	60%	Single Phase (60 Cycle, Lagging Power Factor.) 100%	90%	80%	70%	60%
Steel Conduit	14	20*	20*	25*	6140	5369	4887	4371	3848	3322	6200	5643	5047	4444	3836
	12	25*	25*	30*	3860	3464	3169	2841	2508	2172	4000	3659	3281	2897	2508
	10	30	35*	40*	2420	2078	1918	1728	1532	1334	2400	2214	1995	1769	1540
	8	40	50	55	1528	1350	1264	1148	1026	900	1560	1460	1326	1184	1040
	6	55	65	75	982	848	812	745	673	597	980	937	860	777	690
	4	70	85	95	616	536	528	491	450	405	620	610	568	519	468
	3	85	100	110	490	433	434	407	376	341	500	501	470	434	394
	2	95	115	130	388	346	354	336	312	286	400	409	388	361	331
	1	110	130	150	308	277	292	280	264	245	320	337	324	305	283
	0	125	150	170	244	207	228	223	213	200	240	263	258	246	232
	00	145	175	195	193	173	196	194	188	178	200	227	224	217	206
	000	165	200	225	153	136	162	163	160	154	158	187	188	184	178
	0000	195	230	260	122	109	136	140	139	136	126	157	162	161	157
	250	215	255	290	103	93	123	128	129	128	108	142	148	149	148
	300	240	285	320	86	77	108	115	117	117	90	125	133	135	135
	350	260	310	350	73	67	98	106	109	109	78	113	122	126	126
	400	280	335	380	64	60	91	99	103	104	70	105	114	118	120
	500	320	380	430	52	50	81	90	94	96	58	94	104	109	111
	600	355	420	475	43	43	75	84	89	92	50	86	97	103	106
	750	400	475	535	34	36	68	78	84	88	42	79	91	97	102
	1000	455	545	615	26	31	62	72	78	82	36	72	84	90	95
Non-Magnetic Conduit (Lead Covered Cables Or Installation In Fibre Or Other Non-Magnetic Conduit, Etc.)	14	20*	20*	25*	6140	5369	4876	4355	3830	3301	6200	5630	5029	4422	3812
	12	25*	25*	30*	3860	3464	3158	2827	2491	2153	4000	3647	3264	2877	2486
	10	30	35*	40*	2420	2078	1908	1714	1516	1316	2400	2203	1980	1751	1520
	8	40	50	55	1528	1350	1255	1134	1010	882	1560	1449	1310	1166	1019
	6	55	65	75	982	848	802	731	657	579	980	926	845	758	669
	4	70	85	95	616	536	519	479	435	388	620	599	553	502	448
	3	85	100	110	470	433	425	395	361	324	500	490	456	417	375
	2	95	115	130	388	329	330	310	286	259	380	381	358	330	300
	1	110	130	150	308	259	268	255	238	219	300	310	295	275	253
	0	125	150	170	244	207	220	212	199	185	240	254	244	230	214
	00	145	175	195	193	173	188	183	174	163	200	217	211	201	188
	000	165	200	225	153	133	151	150	145	138	154	175	173	167	159
	0000	195	230	260	122	107	127	128	125	121	124	147	148	145	140
	250	215	255	290	103	90	112	114	113	110	104	129	132	131	128
	300	240	285	320	86	76	99	103	104	102	88	114	119	120	118
	350	260	310	350	73	65	89	94	95	94	76	103	108	110	109
	400	280	335	380	64	57	81	87	89	89	66	94	100	103	103
	500	320	380	430	52	46	71	77	80	82	54	82	90	93	94
	600	355	420	475	43	39	65	72	76	77	46	75	83	87	90
	750	400	475	535	34	32	58	65	70	72	38	67	76	80	83
	1000	455	545	615	26	25	51	59	63	66	30	59	68	73	77

*The overcurrent protection for conductor types marked with an (*) shall not exceed 15-ampere for 14 AWG, or 20-ampere for 12 AWG, and 30-ampere for 10 AWG after correction factors for ambient temperature and adjustment factors for the number of conductor have been applied.

**Figures are L-L for both single-phase and three-phase. Three-phase figures are average for the three phases.

Table *courtesy of* Cooper Bussman, Inc.

TABLE A-4
Bakery – Loading Schedule

	Count	VA/Unit	*NEC*®	Connected	Calculated	Balanced	Nonlinear
General Lighting:							
Style C luminaire (fixture)	15	87		1305			1305
Style B luminaire (fixture)	4	87		348			348
Style E luminaire (fixture)	2	144		288			288
Style N luminaire (fixture)	2	60		120			
Totals:				2061	2061		
Storage:							
NEC® 220.12	1200	0.25	300				
Style L luminaire (fixture)	12	87		1044			1044
Totals:			300	1044	1044		
Show Window:							
NEC® 220.12	14	200	2800				
Style C luminaire (fixture)	3	87		261			261
Receptacle outlets	4	650		2600			
Totals:			2800	2861	2861		
Other Loads:							
Receptacle outlets	21	180		3780			
Sign outlet	1	1200		1200			1200
Bake oven outlet	1	16,000		16,000		16,000	
Totals:				20,980	20,980		
Motors and Appliances:	**Amperes**						
Exhaust fan	2.9	120		348			
Multi-mixer	3.96	360		1426		1426	
Multi-mixer	7.48	360		2693		2693	
Dough divider	2.2	360		792		792	
Doughnut machine	2.2	360		792		792	
Doughnut heater		2000		2000		2000	
Dishwasher	23.9	360		8604		8604	
Disposer	7.44	360		2678		2678	
Totals:				19,333	19,333		
TOTAL LOADS:					46,279	34,985	4446

TABLE A-5
Bakery – Panelboard Worksheet

Phase	Circuit Number	Load/Area Served	Calculated Volt-Amperes	Calculated Amperes	Load Type	Load Modifier	OCPD Selection Amperes	OCPD Rating	Minimum Conductor Size AWG	Minimum Ampacity	Ambient Temp. C°	Correction Factor	Current-Carrying Conductors	Adjustment Factor	Minimum Allowable Ampacity	Conductor Size AWG	Allowable Ampacity	Ampacity
A	1	15 Style C, Bake Area Exhaust fan	1305	11	C	1.25	17	20	12	18	30	1		1	17	12	30	30
A			348	2.9	C	1.25												
A	2	4 Rec., Sales Area	720	6	R	1	6	20	12	21	30	1		1	21	12	30	30
B	3	4 Style B, 3 Style C	609	5.1	C	1.25	7	20	12	16	30	1		1	7	12	30	30
B	4	2 Style N, 1 E, Toilet	264	2.2	C	1.25	3	20	12	16	30	1		1	3	12	30	30
C	5	Sign Outlet	1200	10	C	1.25	13	20	12	16	30	1		1	13	12	30	30
C	6	12 Style L, 1 E	1188	9.9	C	1.25	13	20	12	16	30	1		1	13	12	30	30
A	7	Multi-mixer	1426	4														
B		Multi-mixer	2693	7.5	R	1	14	20	12	20	30	1		1	20	12	30	30
C		Dough Divider	792	2.2														
A	8	Motor	792	2.2														
B		Doughnut Machine	2000	5.6	R	1	8	20	12	20	30	1		1	20	12	30	30
C																		
A	9	Recp. Show Window	1300	11	R	1	11	20	12	16	30	1		1	21	12	30	30
C	10	Oven	16000	44	C	1.25	56	60	4	56	30	1		1	56	4	95	95
B	11	Recp. Show Window	1300	11	R	1	11	20	12	21	30	1		1	21	12	30	30
C	12	Dishwasher	8604	24	R	1	24	50	8	50	30	1		1	50	8	55	55
A	13	Recp. Basement North	900	7.5	R	1	8	20	12	21	30	1		1	21	12	30	30
C	14	Disposer	2678	22	R	1	22	50	8	50	30	1		1	50	8	55	55
A	15	Recp. North Wall Bake	540	4.5	R	1	5	20	12	21	30	1		1	21	12	30	30
B	17	Recp. Basement South	720	6	R	1	6	20	12	21	30	1		1	21	12	30	30
C	19	Recp. S & W Walls Bake	900	7.5	R	1	8	20	12	21	30	1		1	21	12	30	30

TABLE A-6
Bakery – Load Summary

Connection Summary:	VA Total
Connected Load Phase A	4213
Connected Load Phase B	2893
Connected Load Phase C	4188
Balanced Loads	34,985
Connection Total	46,279
Loads Summary:	**VA Total**
Continuous	22,566
Noncontinuous + Receptacle	6380
Highest Motor	8604
Other Motors	8729
Loads Summary Total	46,279

TABLE A-7
Bakery – Feeder Selection

Load Summary, 208Y/120 Volt, 3 Phase, 4 Wire	Calculated	OCPD
Continuous Load	22,566	28,208
Noncontinuous + Receptacle Load	6380	6380
Highest Motor Load	8604	10,755
Other Motor Load	8729	8729
Growth	11,570	14,463
Calculated Load & OCPD Load	57,849	68,535
OCPD Selection	**Input**	**Output**
OCPD Load Volt-Amperes & Amperes	68,535	190
OCPD Rating		200
Minimum Conductor Size		3/0 AWG
Minimum Ampacity		176
Phase Conductor Selection	**Input**	**Output**
Ambient Temperature & Correction Factor	28°C	1
Current-Carrying Conductors & Adjustment Factor	3	1
Derating Factor		1
Minimum Allowable Ampacity		176
Conductor Size		3/0 AWG
Conductor Type, Allowable Ampacity	THHN	225
Ampacity		225
Voltage Drop, 0.9 pf, per 100 ft		2.6
Neutral Conductor Selection	**Input**	**Output**
OCPD Load Volt-Amperes	68,535	
Balanced Load Volt-Amperes	−34,985	
Nonlinear Load Volt-Amperes	4446	
Total Load Volt-Amperes & Amperes	37,996	106
Minimum Neutral Size		2 AWG
Minimum Allowable Ampacity		106
Neutral Conductor Type & Size	THHN	2 AWG
Allowable Ampacity & Ampacity	130	130
Raceway Size Determination	**Input**	**Output**
Feeder Conductors Size & Total Area	3/0 AWG	0.8037
Neutral Conductor Size & Area	2 AWG	0.1158
Grounding Conductor Size & Area		0
Total Conductor Area		0.9195
Raceway Type & Trade Size	RMC	2

TABLE A-8
Beauty Salon – Loading Schedule

	Count	VA/Unit	NEC®	Connected	Calculated	Balanced	Nonlinear
General Lighting:							
NEC® 220.2	480	3	1440				
Style A luminaire (fixture)	5	87		435			
Style M luminaire (fixture)	9	50		450			
Style E luminaire (fixture)	5	144		720			720
Totals:			1440	1605	1605		
Other Loads:							
Receptacle outlets	6	180		1080			
Receptacle outlets	3	1500		4500			
Roof receptacle	1	1500		1500			
Washer/dryer	1	4000		4000		4000	
Water heater	1	3800		3800		3800	
Totals:				14,880	14,880		
Motors and Appliances:	**Amperes**						
Air Conditioning:							
Compressor	14.1	360		5076		5076	
Cond-Evap motor	3.3	208		686		686	
Totals:				5762	5762		
TOTAL LOADS:					22,247	13,562	1155

TABLE A-9
Beauty Salon – Panelboard Worksheet

Phase	Circuit Number	Load/Area Served	Calculated Volt-Amperes	Calculated Amperes	Load Type	Load Modifier	OCPD Selection Amperes	OCPD Rating	Minimum Conductor Size AWG	Minimum Ampacity	Ambient Temp. °C	Correction Factor	Current-Carrying Conductors	Adjustment Factor	Minimum Allowable Ampacity	Conductor Size AWG	Allowable Ampacity	Ampacity
A	1	General-use receptacles	1080	9	R	1	9	20	12	21	30	1		1	21	12	30	30
A	2	Receptacle Station A	1500	13	N	1	13	20	12	16	30	1		1	13	12	30	30
B	3	9 Style M Station Spots	450	3.8	C	1.25	5	20	12	16	30	1		1	5	12	30	30
B	4	Receptacle Station B	1500	13	N	1	13	20	12	16	30	1		1	13	12	30	30
C	5	Evaporator/Condenser		3.3		1												
A		Cooling System	5762	14	H	2.25	35	35	10	21	38	0.9	5	0.8	35	10	40	29
B																		
C	6	Receptacle Station C	1500	13	N	1	13	20	12	16	30	1		1	13	12	30	30
C	7	4 Style A Station Area	348	2.9	C	1.25	4	20	12	16	30	1		1	4	12	30	30
A	8	5 Style E, 1 A Waiting Area	807	6.7	C	1.25	9	20	12	16	30	1		1	9	12	30	30
A	9	Water Heater	3800	18	M	1.25	23	25	12	23	30	1	5	0.8	23	12	30	24
B																		
B	10	Washer/Dryer	4000	19	C	1.25	25	25	12	25	30	1	5	0.8	25	12	30	24
C																		
C	11	Roof Receptacle	1500	13	N	1	13	20	12	16	38	0.9	5	0.8	18	12	30	21

TABLE A-10
Beauty Salon – Load Summary

Connection Summary:	VA Total
Connected Load Phase A	3387
Connected Load Phase B	1950
Connected Load Phase C	3348
Balanced Loads	13,562
Connection Total	22,247
Loads Summary:	**VA Total**
Continuous	5605
Noncontinuous + Receptacle	7080
Highest Motor	5762
Other Motors	3800
Loads Summary Total	22,247

TABLE A-11
Beauty Salon – Feeder Selection

Load Summary, 208Y/120 Volt, 3 Phase, 4 Wire	Calculated	OCPD
Continuous Load	5605	7006
Noncontinuous + Receptacle Load	7080	7080
Highest Motor Load	5762	7203
Other Motor Load	3800	3800
Growth	5562	6953
Calculated Load & OCPD Load	27,809	32,042
OCPD Selection	**Input**	**Output**
OCPD Load Volt-Amperes & Amperes	32,042	89
OCPD Rating		90
Minimum Conductor Size		2 AWG
Minimum Ampacity		81
Phase Conductor Selection	**Input**	**Output**
Ambient Temperature & Correction Factor	28°C	1
Current-Carrying Conductors & Adjustment Factor	3	1
Derating Factor		1
Minimum Allowable Ampacity		81
Conductor Size		2 AWG
Conductor Type, Allowable Ampacity	THHN	130
Ampacity		130
Voltage Drop, 0.9 pf, per 100 ft		2.73
Neutral Conductor Selection	**Input**	**Output**
OCPD Load Volt-Amperes	32,042	
Balanced Load Volt-Amperes	−13,562	
Nonlinear Load Volt-Amperes	1155	
Total Load Volt-Amperes & Amperes	19,635	55
Minimum Neutral Size		6 AWG
Minimum Allowable Ampacity		55
Neutral Conductor Type & Size	THHN	6 AWG
Allowable Ampacity & Ampacity	75	75
Raceway Size Determination	**Input**	**Output**
Feeder Conductors Size & Total Area	2 AWG	0.3474
Neutral Conductor Size & Area	6 AWG	0.0507
Grounding Conductor Size & Area		0
Total Conductor Area		0.3981
Raceway Type & Trade Size	RMC	1¼

TABLE A-12
Doctor's Office – Loading Schedule

	Count	VA/Unit	NEC®	Connected	Calculated	Balanced	Nonlinear
General Lighting:							
NEC® 220.12	457	3	1600				
Style F luminaire (fixture)	3	132		396			396
Style G luminaire (fixture)	2	132		264			264
Style J luminaire (fixture)	5	150		750			
Style N luminaire (fixture)	2	60		120			
Totals:			1600	1530	1600		
Other Loads:							
Receptacle outlets	10	180		1800			
Equipment outlet	1	2000		2000			
Roof receptacle	1	1500		1500			
Water heater	1	3800		3800		3800	
Totals:				9100	9100		
Motors and Appliances:	**Amperes**						
Air Conditioning:							
Compressor	16.8	208		3494		3494	
Cond-Evap motor	3.7	208		770		770	
Totals:				4264	4264		
TOTAL LOADS:					14,964	8064	660

TABLE A-13
Doctor's Office – Branch-Circuit Selection

Phase	Circuit Number	Load/Area Served	Calculated Volt-Amperes	Calculated Amperes	Load Type	Load Modifier	OCPD Selection Amperes	OCPD Rating	Minimum Conductor Size AWG	Minimum Ampacity	Ambient Temp. °C	Correction Factor	Current-Carrying Conductors	Adjustment Factor	Minimum Allowable Ampacity	Conductor Size AWG	Allowable Ampacity	Ampacity
B	1	5 Style J, 2 N, Waiting Room	820	7.3	C	1.25	9	20	12	16	30	1		1	9	12	30	30
B	2	5 Receptacles, Waiting Rm	720	6	R	1	6	20	12	21	30	1		1	21	12	30	30
C	3	2 Style G, 3 F, Examining Rm	660	5.5	C	1.25	7	20	12	16	30	1		1	7	12	30	30
C	4	4 Receptacles, Examining Rm	720	6	R	1	6	20	12	21	30	1		1	21	12	30	30
B	5	Equipment Recp., Exam Rm	2000	17	N	1	17	20	12	16	30	1		1	17	12	30	30
B	6	2 Receptacles, Laboratory	360	3	R	1	3	20	12	21	30	1		1	21	12	30	30
C / B	7	Cooling System	4264	21	H	2.25	46	45	10	26	38	0.9	4	0.8	29	10	40	29
C	8	Roof Receptacle	1500	13	N	1	13	20	12	16	38	0.9	4	0.8	18	12	30	21
C / B	9	Water Heater	3800	32	C	1.25	40	40	8	40	30	1		1	40	8	55	55

TABLE A-14
Doctor's Office – Load Summary

Connection Summary:	VA Total
Connected Load Phase A	XXX
Connected Load Phase B	3950
Connected Load Phase C	2880
Balanced Loads	8064
Connection Total	14,894
Loads Summary:	**VA Total**
Continuous	5330
Noncontinuous + Receptacle	5300
Highest Motor	4264
Other Motors	0
Loads Summary Total	14,894

TABLE A-15
Doctor's Office – Feeder Selection

Load Summary, 208Y/120 Volt, 1 Phase, 3 Wire	Calculated	OCPD
Continuous Load	5330	6663
Noncontinuous + Receptacle Load	5330	5330
Highest Motor Load	4264	5330
Other Motor Load	0	0
Growth	3724	4655
Calculated Load & OCPD Load	18,618	21,948
OCPD Selection	**Input**	**Output**
OCPD Load Volt-Amperes & Amperes	21,948	106
OCPD Rating		110
Minimum Conductor Size		2 AWG
Minimum Ampacity		101
Phase Conductor Selection	**Input**	**Output**
Ambient Temperature & Correction Factor	28°C	1
Current-Carrying Conductors & Adjustment Factor	3	1
Derating Factor		1
Minimum Allowable Ampacity		101
Conductor Size		2 AWG
Conductor Type, Allowable Ampacity	THHN	130
Ampacity		130
Voltage Drop, 0.9 pf, per 100 ft		3.66
Neutral Conductor Selection	**Input**	**Output**
OCPD Load Volt-Amperes	21,948	
Balanced Load Volt-Amperes	0	
Nonlinear Load Volt-Amperes	660	
Neutral Load Volt-Amperes & Amperes	22,608	109
Minimum Neutral Size		2 AWG
Minimum Allowable Ampacity		109
Neutral Conductor Type & Size	THHN	2 AWG
Allowable Ampacity & Ampacity	130	130
Raceway Size Determination	**Input**	**Output**
Feeder Conductors Size & Total Area	2 AWG	0.2316
Neutral Conductor Size & Area	2 AWG	0.1158
Grounding Conductor Size & Area		0
Total Conductor Area		0.3474
Raceway Type & Trade Size	RMC	1

TABLE A-16
Drugstore – Loading Schedule

	Count	VA/Unit	NEC®	Connected	Calculated	Balanced	Nonlinear
General Lighting:							
NEC® 220.12	1395	3	4185				
Style I luminaire (fixture)	27	87		2349			2349
Style E luminaire (fixture)	4	144		576			576
Style D luminaire (fixture)	15	74		1110			1110
Style N luminaire (fixture)	2	60		120			
Totals:			4185	4155	4185		
Storage:							
NEC® 220.12	1012	0.25	253				
Style L luminaire (fixture)	9	87		783			783
Style E luminaire (fixture)	1	144		144			144
Totals:			253	927	927		
Show Window:							
NEC® 220.14(G)	16	200	3200				
Receptacle outlets	3	500		1500			
Lighting track	8	150		1200			
Totals:			3200	2700	3200		
Other Loads:							
Receptacle outlets	21	180		3780			
Roof receptacle	1	1500		1500			
Sign outlet	1	1200		1200			
Totals:				6480	6480		
Motors and Appliances:	**Amperes**						
Air Conditioning:							
Compressor	20.2	360		7272		7272	
Condenser	3.2	208		666		666	
Evaporator	3.2	208		666		666	
Totals:				8604	8604		
TOTAL LOADS:					23,396	8604	4962

TABLE A-17
Drugstore – Panelboard Worksheet

Phase	Circuit Number	Load/Area Served	Calculated Volt-Amperes	Calculated Amperes	Load Type	Load Modifier	OCPD Selection Amperes	OCPD Rating	Minimum Conductor Size AWG	Minimum Ampacity	Ambient Temp. °C	Correction Factor	Current-Carrying Conductors	Adjustment Factor	Minimum Allowable Ampacity	Conductor Size AWG	Allowable Ampacity	Ampacity
A	1	9 Style I,	783	6.5	C	1.25	9	20	12	16	30	1		1	16	12	30	30
A	2	4 Rec., Merchandise	720	6	R	1	6	20	12	21	30	1		1	21	12	30	30
B	3	18 Style I,	1566	13	C	1.25	17	20	12	17	30	1		1	17	12	30	30
B	4	3 Rec., Toilet Area	540	4.5	R	1	5	20	12	21	30	1		1	21	12	30	30
C	5	3 Style E, 2 N	552	4.6	C	1.25	6	20	12	16	30	1		1	16	12	30	30
C	6	5 Rec., Merchandise	900	7.5	R	1	8	20	12	21	30	1		1	21	12	30	30
A	7	2 Rec., Pharmacy	360	3	R	1	3	20	12	21	30	1		1	21	12	30	30
A	8	3 Rec., Show Window	1550	13	C	1.25	16	20	12	16	30	1		1	16	12	30	30
B	9	9 Style L, 1 E	927	7.7	C	1.25	10	20	12	16	30	1		1	16	12	30	30
B	10	15 Style D, Pharmacy	1110	9.3	C	1.25	12	20	12	16	30	1		1	16	12	30	30
C	11	Track Show Window	600	5	C	1.25	7	20	12	16	30	1		1	16	12	30	30
C	12	4 Rec., S. Basement	720	6	R	1	6	20	12	21	30	1		1	21	12	30	30
C	13	Track Show Window	600	5	C	1.25	7	20	12	16	30	1		1	16	12	30	30
A	14	3 Rec., N. Basement	540	4.5	R	1	5	20	12	21	30	1		1	21	12	30	30
B	15	Sign	1200	10	C	1.25	13	20	12	16	30	1		1	16	12	30	30
B	16	Evaporator		3.2		1												
C	16	Compressor	8604	20	H	2.25	52	50	10	32	38	0.9	5	0.8	44	8	55	40
A	16	Condenser		3.2		1												
C	17	1 Receptacle, Roof	1500	13	N	1	13	20	12	16	38	0.9	5	0.8	22	12	30	22

TABLE A-18
Drugstore – Load Summary

Connection Summary:	VA Total
Connected Load Phase A	4503
Connected Load Phase B	5343
Connected Load Phase C	4272
Balanced Loads	8604
Connection Total	22,722
Loads Summary:	**VA Total**
Continuous	8838
Noncontinuous + Receptacle	5280
Highest Motor	8604
Other Motors	0
Loads Summary Total	22,722

TABLE A-19
Drugstore – Feeder Selection

Load Summary, 208Y/120 Volt, 3 Wire	Calculated	OCPD
Continuous Load	8838	11,048
Noncontinuous + Receptacle Load	5280	5280
Highest Motor Load	8604	10,755
Other Motor Load	0	0
Growth	5680	7100
Calculated Load & OCPD Load	28,402	34,183
OCPD Selection	**Input**	**Output**
OCPD Load Volt-Amperes & Amperes	34,183	95
OCPD Rating		100
Minimum Conductor Size		1 AWG
Minimum Ampacity		91
Phase Conductor Selection	**Input**	**Output**
Ambient Temperature & Correction Factor	28°C	1
Current-Carrying Conductors & Adjustment Factor	4	0.8
Derating Factor		0.8
Minimum Allowable Ampacity		114
Conductor Size		1 AWG
Conductor Type, Allowable Ampacity	THHN	150
Ampacity		120
Voltage Drop, 0.9 pf, per 100 ft	2.3	1.1%
Neutral Conductor Selection	**Input**	**Output**
OCPD Load Volt-Amperes	34,183	
Balanced Load Volt-Amperes	−8604	
Nonlinear Load Volt-Amperes	4962	
Total Load Volt-Amperes & Amperes	30,541	85
Minimum Neutral Size		4 AWG
Minimum Allowable Ampacity		106
Neutral Conductor Type & Size	THHN	3 AWG
Allowable Ampacity & Ampacity	110	88
Raceway Size Determination	**Input**	**Output**
Feeder Conductors Size & Total Area	1 AWG	0.4686
Neutral Conductor Size & Area	3 AWG	0.0973
Grounding Conductor Size & Area		0
Total Conductor Area		0.5659
Raceway Type & Trade Size	RMC	1¼

TABLE A-20
Insurance Office – Loading Schedule

	Count	VA/Unit	*NEC*®	Connected	Calculated	Balanced	Nonlinear
General Lighting:							
NEC® 220.12	1374	3	4122				
Style E luminaire (fixture)	10	144		1440			1440
Style O luminaire (fixture)	16	143		2288			2288
Style P luminaire (fixture)	2	87		174			174
Style Q luminaire (fixture)	4	143		572			572
Totals:			4122	4474	4474		
Other Loads:							
Copier outlets	1	1500		1500			1500
Computer outlets	9	500		4500			4500
Roof receptacle	1	1500		1500			
Totals:				7500	7500		
Receptacle Outlets							
NEC® 220.14(I)	15	180		2700			
NEC® 220.14(L)	56	180		10,080			
			11,390	12,780	11,390		11,390
Motors and Appliances:	**Amperes**						
Air Conditioning:							
Compressor	20.2	360		7272		7272	
Evaporator	3.2	208		666		666	
Condenser	3.2	208		666		666	
Totals:				8604	8604		
TOTAL LOADS:					31,968	8604	21,864

TABLE A-21
Insurance Office – Panelboard Worksheet

Phase	Circuit Number	Load/Area Served	Calculated Volt-Amperes	Calculated Amperes	Load Type	Load Modifier	OCPD Selection Amperes	OCPD Rating	Minimum Conductor Size AWG	Minimum Ampacity	Ambient Temp. °C	Correction Factor	Current-Carrying Conductors	Adjustment Factor	Minimum Allowable Ampacity	Conductor Size AWG	Allowable Ampacity	Ampacity
A	1	14 2 Lamp, Style O, Staff	1218	10	C	1.25	13	20	12	16	30	1		1	16	12	30	30
A	2	10 Rec. South Wall Staff	1800	15	R	1	15	20	12	21	30	1		1	21	12	30	30
B	3	14 1 Lamp, Style O; 2 P, Staff	958	8	C	1.25	10	20	12	16	30	1		1	16	12	30	30
B	4	10 Rec. South Wall Staff	1800	15	R	1	15	20	12	21	30	1		1	21	12	30	30
C	5	6 Style E, 2 O, Reception	1150	9.6	C	1.25	12	20	12	16	30	1		1	16	12	30	30
C	6	10 Rec. South Wall Staff	1800	15	R	1	15	20	12	21	30	1		1	21	12	30	30
A	7	4 Style Q, 4 E, Computer	1148	9.6	C	1.25	12	20	12	16	30	1		1	16	12	30	30
A	8	3 Rec. Computer Room	1500	13	C	1.25	16	20	12	16	30	1	6	0.8	20	12	30	24
B	9	13 Rec. East & South Walls	2340	20	R	1	20	20	12	21	30	1		1	21	12	30	30
B	10	3 Rec. Computer Room	1500	13	C	1.25	16	20	12	16	30	1	6	0.8	20	12	30	24
C	11	13 Rec. East & North Walls	2340	20	R	1	20	20	12	21	30	1		1	21	12	30	30
C	12	3 Rec. Computer Room	1500	13	C	1.25	16	20	12	16	30	1	6	0.8	20	12	30	24
A		Evaporator		3.2		1												
B	13	Compressor	8604	20	H	2.25	52	50	10	32	38	0.9	5	0.8	44	8	55	40
C		Condenser		3.2		1												
A	14	4 Floor Rec. Staff Office	720	6	R	1	6	20	12	21	30	1		1	21	12	30	30
A	15	Copy Machine	1500	13	N	1	13	20	12	16	30	1		1	16	12	30	30
B	16	5 Rec. West Wall Staff Office	900	7.5	R	1	8	20	12	21	30	1		1	21	12	30	30
B	17	Roof Receptacle	1500	13	N	1	13	20	12	16	38	0.9	5	0.8	22	12	30	21
C	18	6 Rec. Reception Room	1080	9	R	1	9	20	12	21	30	1		1	21	12	30	30

TABLE A-22
Insurance Office – Load Summary

Connection Summary:	VA Total
Phase A Loads	7886
Phase B Loads	8998
Phase C Loads	7870
Balanced Loads	8604
Connection Total	33,358
Loads Summary:	**VA Total**
Continuous	8974
Noncontinuous + Receptacle	15,780
Highest Motor	8604
Other Motors	0
Loads Summary Total	33,358

TABLE A-23
Insurance Office – Feeder Selection

Load Summary, 208Y/120 Volt, 3 Phase, 4 Wire	Calculated	OCPD
Continuous Load	8974	11,218
Noncontinuous + Receptacle Load	15,780	15,780
Highest Motor Load	8604	10,755
Other Motor Load	0	0
Growth	8340	10,425
Calculated Load & OCPD Load	41,698	48,178
OCPD Selection	**Input**	**Output**
OCPD Load Volt-Amperes & Amperes	48,178	134
OCPD Rating		150
Minimum Conductor Size		1/0 AWG
Minimum Ampacity		126
Phase Conductor Selection	**Input**	**Output**
Ambient Temperature & Correction Factor	28°C	1
Current-Carrying Conductors & Adjustment Factor	4	0.8
Derating Factor		0.8
Minimum Allowable Ampacity		158
Conductor Size		1/0 AWG
Conductor Type & Allowable Ampacity	THHN	170
Ampacity		136
Voltage Drop, 0.9 pf, per 100 ft		3.27
Neutral Conductor Selection	**Input**	**Output**
OCPD Load Volt-Amperes	41,698	
Balanced Load Volt-Amperes	−8604	
Nonlinear Load Volt-Amperes	21,864	
Total Load Volt-Amperes & Amperes	54,958	153
Minimum Neutral Size		2/0 AWG
Minimum Allowable Ampacity		191
Neutral Conductor Type & Size	THHN	2/0 AWG
Allowable Ampacity & Ampacity	195	156
Raceway Size Determination	**Input**	**Output**
Feeder Conductors Size & Total Area	1/0 AWG	0.5565
Neutral Conductor Size & Area	2/0 AWG	0.2223
Grounding Conductor Size & Area	8 AWG	0.0366
Total Conductor Area		0.8154
Raceway Type & Size	RMC	1½

TABLE A-24
Owner's – Loading Schedule

	Count	VA/Unit	*NEC*®	Connected	Calculated	Balanced	Nonlinear
General Lighting:							
Style E luminaire (fixture)	4	144		576			576
Style H luminaire (fixture)	4	87		348			348
Style J luminaire (fixture)	3	150		450			
Style K luminaire (fixture)	3	192		576			576
Style L luminaire (fixture)	6	87		522			522
Style N luminaire (fixture)	4	60		240			
Totals:				2712	2712		
Other Loads:							
Receptacle outlets	8	180		1440			
Telephone outlets	2	1250		2500			
Totals:				3940	3940		
Motors and Appliances:	**Amperes**						
Sump pump	9.4	120		1128			
Circulating pump	4.4	120		528			
Circulating pump	4.4	120		528			
Circulating pump	4.4	120		528			
Circulating pump	4.4	120		528			
Circulating pump	4.4	120		528			
Totals:				3768	3768		
TOTAL LOADS:					10,420	0	2022

TABLE A-25
Owner's Panelboard Worksheet

Phase	Circuit Number	Load/Area Served	Calculated Volt-Amperes	Calculated Amperes	Load Type	Load Modifier	OCPD Selection Amperes	OCPD Rating	Minimum Conductor Size AWG	Minimum Ampacity	Ambient Temp. °C	Correction Factor	Current-Carrying Conductors	Adjustment Factor	Minimum Allowable Ampacity	Conductor Size AWG	Allowable Ampacity	Ampacity
A	1	3 Style E, 4 N, 2nd Floor	672	5.6	C	1.25	7	20	12	16	38	0.9		1	18	12	30	27
A	2	3 Style K, Exterior Lighting	576	4.8	C	1.25	6	20	12	16	38	0.9		1	18	12	30	27
B	3	2 Style H, 2nd Floor Corridor	174	1.5	C	1.25	2	20	12	16	38	0.9		1	18	12	30	27
B	4	2 Exterior Receptacles	360	3	R	1	3	20	12	21	38	0.9		1	23	12	30	27
C	5	2 Recep. 2nd Floor Utility	360	3	N	1	3	20	12	16	38	0.9		1	18	12	30	27
C	6																	
A	7	3 Style J, 1 E, East & West	594	5	N	1	5	20	12	16	38	0.9		1	18	12	30	27
A	8	6 Style L, Utility Area	522	4.4	N	1	5	20	12	16	38	0.9		1	18	12	30	27
B	9	1 Recp. Telephone Panel	1250	10	C	1.25	14	20	12	16	38	0.9		1	18	12	30	27
B	10	4 Recp. Utility Area	720	6	R	1	6	20	12	21	38	0.9		1	23	12	30	27
C	11	1 Recp. Telephone Panel	1250	10	C	1.25	14	20	12	16	38	0.9		1	18	12	30	27
C	12	2 Style H, 2nd Floor Corridor	174	1.5	C	1.25	2	20	12	16	38	0.9		1	18	12	30	27
A	13	Drugstore Hot-Water Pump	528	4.4	M	1.25	6	20	12	16	38	0.9		1	18	12	30	27
A	14	Bakery Hot-Water Pump	528	4.4	M	1.25	6	20	12	16	38	0.9		1	18	12	30	27
B	15	Insurance Office Hot Water	528	4.4	M	1.25	6	20	12	16	38	0.9		1	18	12	30	27
B	16	Beauty Salon Hot-Water Pump	528	4.4	M	1.25	6	20	12	16	38	0.9		1	18	12	30	27
C	17	Doctor's Office Hot Water	528	4.4	M	1.25	6	20	12	16	38	0.9		1	18	12	30	27
C	18	Sump Pump	1128	9.4	M	1.25	12	20	12	16	38	0.9		1	18	12	30	27
A	19																	
A	20																	

366   Appendix

TABLE A-26
Owner's – Feeder Selection

Load Summary, 208Y/120 Volt, 3 Phase, 4 Wire	Calculated	OCPD
Continuous Load	4096	5120
Noncontinuous + Receptacle Load	2556	2556
Highest Motor Load	1128	1410
Other Motor Load	2640	2640
Growth	2605	3256
Calculated Load & OCPD Load	13,025	14,982
OCPD Selection	**Input**	**Output**
Selection Amperes	14,982	42
OCPD Rating		45
Minimum Conductor Size		6 AWG
Minimum Ampacity		41
Phase Conductor Selection	**Input**	**Output**
Ambient Temperature & Correction Factor	38°C	0.91
Current-Carrying Conductors & Adjustment Factor	3	1
Derating Factor		0.91
Minimum Allowable Ampacity		45
Conductor Size		6 AWG
Conductor Type & Allowable Ampacity	THHN	75
Ampacity		68
Voltage Drop, 0.9 pf, 3 ph, per 100 ft		2.94
Neutral Conductor Selection	**Input**	**Output**
Total Load	14,982	
Balanced Load	0	
Nonlinear Load	2022	
Total Load Volt-Amperes & Amperes	17,004	47
Minimum Neutral Size		6 AWG
Minimum Allowable Ampacity		68
Neutral Conductor Type & Size	THHN	6 AWG
Allowable Ampacity & Ampacity	75	68
Raceway Size Determination	**Input**	**Output**
Feeder Conductors Size & Area	6 AWG	0.1521
Neutral Conductor Size & Area	6 AWG	0.0507
Grounding Conductor Size & Area		0
Total Conductor Area		0.2028
Raceway Type & Size	RMC	¾

TABLE A-27
Owner's – Panelboard Summary

Connection Summary:	VA Total
Connected Load Phase A	3420
Connected Load Phase B	3560
Connected Load Phase C	3440
Balanced Loads	0
Connection Total	10,420
Loads Summary:	**VA Total**
Continuous	4096
Noncontinuous + Receptacle	2556
Highest Motor	1128
Other Motors	2640
Loads Summary Total	10,420

TABLE A-28
Useful Formulas

TO FIND	SINGLE PHASE	THREE PHASE	DIRECT CURRENT
AMPERES when kilovolt-amperes are known	$\dfrac{\text{kVA x }1000}{E}$	$\dfrac{\text{kVA x }1000}{E \times 1.73}$	not applicable
AMPERES when horsepower is known	$\dfrac{\text{hp} \times 746}{E \times \% \text{ eff.} \times \text{pf}}$	$\dfrac{\text{hp} \times 746}{E \times 1.73 \times \% \text{ eff.} \times \text{pf}}$	$\dfrac{\text{hp} \times 746}{E \times \% \text{ eff.}}$
AMPERES when kilowatts are known	$\dfrac{\text{kw} \times 1000}{E \times \text{pf}}$	$\dfrac{\text{kW} \times 1000}{E \times 1.73 \times \text{pf}}$	$\dfrac{\text{kW} \times 1000}{E}$
KILOWATTS	$\dfrac{I \times E \times \text{pf}}{1000}$	$\dfrac{I \times E \times 1.73 \times \text{pf}}{1000}$	$\dfrac{I \times E}{1000}$
KILOVOLT-AMPERES	$\dfrac{I \times E}{1000}$	$\dfrac{I \times E \times 1.73}{1000}$	not applicable
HORSEPOWER	$\dfrac{I \times E \times \% \text{ eff.} \times \text{pf}}{746}$	$\dfrac{I \times E \times 1.73 \times \% \text{ eff.} \times \text{pf}}{746}$	$\dfrac{I \times E \times \% \text{ eff.}}{746}$
WATTS	$E \times I \times \text{pf}$	$E \times I \times 1.73 \times \text{pf}$	$E \times I$

I = amperes	E = volts	kW = kilowatts	kVA = kilovolt-amperes
hp = horsepower		% eff. = percent efficiency	pf = power factor

METRIC SYSTEM OF MEASUREMENT

The following table provides useful conversions of English customary terms to SI terms, and SI terms to English customary terms. The metric system is known as the International System of Units (SI), taken from the French *Le Système International d'Unites.* Whenever the slant line / is found, say it as "per." Be practical when using the values in the following table. Present-day use of calculators and computers provides many "places" beyond the decimal point. You must decide how accurate your results must be when performing a calculation. When converting centimeters to feet, for example, one could be reasonably accurate by rounding the value of 0.03281 to 0.033. When a manufacturer describes a product's measurement using metrics, but does not physically change the product, it is referred to as a "soft conversion." When a manufacturer actually changes the physical measurements of the product to standard metric size or a rational whole number of metric units, it is referred to as a "hard conversion." When rounding off numbers, be sure to round off in such a manner that the final result does not violate a "maximum" or "minimum" value, such as might be required by the *National Electrical Code.*® The rule for discarding digits is that if the digits to be discarded begin with a 5 or higher, increase the last digit retained by one; for example: 6.3745, if rounded to three digits, would become 6.37. If 6.3745 were rounded to four digits, it would become 6.375. The following information has been developed from the latest U.S. Department of Commerce and the National Institute of Standards and Technology publications covering the metric system.

Multiply This Unit(s)	By This Factor	To Obtain This Unit(s)
acre	4 046.9	square meters (m²)
acre	43 560	square feet (ft²)
ampere hour	3 600	coulombs (C)
angstrom	0.1	nanometers (nm)
atmosphere	101.325	kilopascals (kPa)
atmosphere	33.9	feet of water (at 4°C)
atmosphere	29.92	inches of mercury (at 0°C)
atmosphere	0.76	meters of mercury (at 0°C)
atmosphere	0.007 348	tons per square inch
atmosphere	1.058	tons per square foot
atmosphere	1.0333	kilograms per square centimeter
atmosphere	10 333	kilograms per square meter
atmosphere	14.7	pounds per square inch
bar	100	kilopascals (kPa)
barrel (oil, 42 U.S. gallons)	0.158 987 3	cubic meters (m³)
barrel (oil, 42 U.S. gallons)	158.987 3	liters (L)
board foot	0.002 359 737	cubic meters (m³)
bushel	0.035 239 07	cubic meters (m³)
Btu	778.16	feet-pounds
Btu	252	grams-calories
Btu	0.000 393 1	horsepower-hours
Btu	1 054.8	joules (J)
Btu	1.055 056	kilojoules (kJ)
Btu	0.000 293 1	kilowatt-hours (kWH)
Btu per hour	0.000 393 1	horsepower (hp)
Btu per hour	0.293 071 1	watts (W)
Btu per degree Fahrenheit	1.899 108	kilojoules per kelvin (kJ/K)
Btu per pound	2.326	kilojoules per kilogram (kJ/kg)
Btu per second	1.055 056	kilowatts (kW)
calorie	4.184	joules (J)
calorie, gram	0.003 968 3	Btus
candela per foot squared (cd/ft²)	10.763 9	candelas per meter squared (cd/m²)

Note: The former term *candlepower* has been replaced with the term *candela*.

candela per meter squared (cd/m²)	0.092 903	candelas per foot squared (cd/ft²)
candela per meter squared (cd/m²)	0.291 864	footlambert*
candela per square inch (cd/in²)	1550.003	candelas per square meter (cd/m²)

Celsius = (Fahrenheit − 32) × 5⁄9
Celsius = (Fahrenheit − 32) × 0.555555
Celsius = (0.556 × Fahrenheit) − 17.8

Note: The term *centigrade* was officially discontinued in 1948 and was replaced by the term *Celsius*. The term *centigrade* may still be found in some publications.

centimeter (cm)	0.032 81	feet (ft)
centimeter (cm)	0.393 7	inches (in.)
centimeter (cm)	0.01	meters (m)
centimeter (cm)	10	millimeters (mm)
centimeter (cm)	393.7	mils
centimeter (cm)	0.010 94	yards
circular mil	0.000 005 067	square centimeters (cm²)
circular mil	0.785 4	square mils (mil²)
circular mil	0.000 000 785 4	square inches (in²)
cubic centimeter (cm³)	0.061 02	cubic inches (in³)

Multiply This Unit(s)	By This Factor	To Obtain This Unit(s)
cubic foot per second	0.028 316 85	cubic meters per second (m³/s)
cubic foot per minute	0.000 471 947	cubic meters per second (m³/s)
cubic foot per minute	0.471 947	liters per second (L/s)
cubic inch (in.³)	16.39	cubic centimeters (cm³)
cubic inch (in.³)	0.000 578 7	cubic feet (ft³)
cubic meter (m³)	35.31	cubic feet (ft³)
cubic yard per minute	12.742 58	liters per second (L/s)
cup (c)	0.236 56	liters (L)
decimeter	0.1	meters (m)
dekameter	10	meters (m)
Fahrenheit = (⁹⁄₅ Celsius) + 32		
Fahrenheit = (Celsius × 1.8) + 32		
fathom	1.828 804	meters (m)
fathom	6.0	feet
foot	30.48	centimeters (cm)
foot	12	inches
foot	0.000 304 8	kilometers (km)
foot	0.304 8	meters (m)
foot	0.000 189 4	miles (statute)
foot	304.8	millimeters (mm)
foot	12 000	mils
foot	0.333 33	yard
foot, cubic (ft³)	0.028 316 85	cubic meters (m³)
foot, cubic (ft³)	28.316 85	liters (L)
foot, board	0.002 359 737	cubic meters (m³)
cubic feet per second (ft³/s)	0.028 316 85	cubic meters per second (m³/s)
cubic feet per minute (ft³/min)	0.000 471 947	cubic meters per second (m³/s)
cubic feet per minute (ft³/min)	0.471 947	liters per second (L/s)
foot, square (ft²)	0.092 903	square meters (m²)
footcandle	10.763 91	lux (lx)
footlambert*	3.426 259	candelas per square meter (cd/m²)
foot of water	2.988 98	kilopascals (kPa)
foot pound	0.001 286	Btus
foot pound-force	1 055.06	joules (J)
foot pound-force per second	1.355 818	joules (J)
foot pound-force per second	1.355 818	watts (W)
foot per second	0.304 8	meters per second (m/s)
foot per second squared	0.304 8	meters per second squared (m/s²)
gallon (U.S. liquid)	3.785 412	liters (L)
gallons per day	3.785 412	liters per day (L/d)
gallons per hour	1.051 50	milliliters per second (mL/s)
gallons per minute	0.063 090 2	liters per second (L/s)
gauss	6.452	lines per square inch
gauss	0.1	millitesla (mT)
gauss	0.000 000 064 52	webers per square inch
grain	64.798 91	milligrams (mg)
gram (g) (a little more than the weight of a paper clip)	0.035 274	ounce (avoirdupois)
gram (g)	0.002 204 6	pound (avoirdupois)
gram per meter (g/m)	3.547 99	pounds per mile (lb/mile)
grams per square meter (g/m²)	0.003 277 06	ounces per square foot (oz/ft²)

Multiply This Unit(s)	By This Factor	To Obtain This Unit(s)
grams per square meter (g/m²)	0.029 494	ounces per square yard (oz/yd²)
gravity (standard acceleration)	9.806 65	meters per second squared (m/s²)
quart (U.S. liquid)	0.946 352 9	liters (L)
horsepower (550 ft·lbf/s)	0.745 7	kilowatts (kW)
horsepower	745.7	watts (W)
horsepower hours	2.684 520	megajoules (mJ)
inch per second squared (in/s²)	0.025 4	meters per second squared (m/s²)
inch	2.54	centimeters (cm)
inch	0.254	decimeters (dm)
inch	0.025 4	meters (m)
inch	25.4	millimeters (mm)
inch	1 000	mils
inch	0.027 78	yards
inch, cubic (in³)	16 387.1	cubic millimeters (mm³)
inch, cubic (in³)	16.387 06	cubic centimeters (cm³)
inch, cubic (in³)	645.16	square millimeters (mm²)
inches of mercury	3.386 38	kilopascals (kPa)
inches of mercury	0.033 42	atmospheres
inches of mercury	1.133	feet of water
inches of water	0.248 84	kilopascals (kPa)
inches of water	0.073 55	inches of mercury
joule (J)	0.737 562	foot pound-force (ft·lbf)
kilocandela per meter squared (kcd/m²)	0.314 159	lambert*
kilogram (kg)	2.204 62	pounds (avoirdupois)
kilogram (kg)	35.274	ounces (avoirdupois)
kilogram per meter (kg/m)	0.671 969	pounds per foot (lb/ft)
kilogram per square meter (kg/m²)	0.204 816	pounds per square foot (lb/ft²)
kilogram meter squared (kg·m²)	23.730 4	pounds foot squared (lb·ft²)
kilogram meter squared (kg·m²)	3 417.17	pounds inch squared (lb·in²)
kilogram per cubic meter (kg/m³)	0.062 428	pounds per cubic foot (lb/ft³)
kilogram per cubic meter (kg/m³)	1.685 56	pounds per cubic yard (lb/yd³)
kilogram per second (kg/s)	2.204 62	pounds per second (lb/s)
kilojoule (kJ)	0.947 817	Btus
kilometer (km)	1 000	meters (m)
kilometer (km)	0.621 371	miles (statute)
kilometer (km)	1 000 000	millimeters (mm)
kilometer (km)	1 093.6	yards
kilometer per hour (kg/hr)	0.621 371	miles per hour (mph)
kilometer, square (km²)	0.386 101	square miles (mile²)
kilopound-force per square inch	6.894 757	megapascals (MPa)
kilowatts (kW)	56.921	Btus per minute
kilowatts (kW)	1.341 02	horsepower (hp)
kilowatts (kW)	1 000	watts (W)
kilowatt-hour (kWh)	3 413	Btus
kilowatt-hour (kWh)	3.6	megajoules (MJ)
knots	1.852	kilometers per hour (km/h)
lamberts*	3 183.099	candelas per square meter (cd/m²)
lamberts*	3.183 01	kilocandelas per square meter (kcd/m²)
liter (L)	0.035 314 7	cubic feet (ft³)
liter (L)	0.264 172	gallons (U.S. liquid)
liter (L)	2.113	pints (U.S. liquid)

Multiply This Unit(s)	By This Factor	To Obtain This Unit(s)
liter (L)	1.056 69	quarts (U.S. liquid)
liter per second (L/s)	2.118 88	cubic feet per minute (ft³/min)
liter per second (L/s)	15.850 3	gallons per minute (gal/min)
liter per second (L/s)	951.022	gallons per hour (gal/hr)
lumen per square foot (lm/ft²)	10.763 9	lux (lx); plural: luces
lumen per square foot (lm/ft²)	1.0	footcandles
lumen per square foot (lm/ft²)	10.763	lumens per square meter (lm/m²)
lumen per square meter (lm/m²)	1.0	lux (lx)
lux (lx)	0.092 903	lumens per square foot (footcandle)
maxwell	10	nanowebers (nWb)
megajoule (MJ)	0.277 778	kilowatt hours (kWh)
meter (m)	100	centimeters (cm)
meter (m)	0.546 81	fathoms
meter (m)	3.280 9	feet
meter (m)	39.37	inches
meter (m)	0.001	kilometers (km)
meter (m)	0.000 621 4	miles (statute)
meter (m)	1 000	millimeters (mm)
meter (m)	1.093 61	yards
meter, cubic (m³)	1.307 95	cubic yards (yd³)
meter, cubic (m³)	35.314 7	cubic feet (ft³)
meter, cubic (m³)	423.776	board feet
meter per second (m/s)	3.280 84	feet per second (ft/s)
meter per second (m/s)	2.236 94	miles per hour (mph)
meter, square (m²)	1.195 99	square yards (yd²)
meter, square (m²)	10.763 9	square feet (ft²)
mho per centimeter (mho/cm)	100	siemens per meter (S/m)

Note: The older term "mho" has been replaced with "siemens." The term "mho" may still be found in some publications.

micro inch	0.025 4	micrometers (µm)
mil	25.4	micrometers (µm)
mil	0.025 4	millimeters (mm)
mil	2.540×10^{-3}	centimeters (cm)
mil	8.333×10^{-5}	feet
mil	0.001	inches
mil	2.540×10^{-8}	kilometers
mil	2.778×10^{-5}	yards
miles per hour	1.609 344	kilometers per hour (km/h)
miles per hour	0.447 04	meters per second (m/s)
miles per gallon	0.425 143 7	kilometers per liter (km/L)
miles	1.609 344	kilometers (km)
miles	5 280	feet
miles	1 609	meters (m)
miles	1 760	yards
miles (nautical)	1.852	kilometers (km)
miles squared	2.590 000	kilometers squared (km²)
millibar	0.1	kilopascals (kPa)
milliliter (mL)	0.061 023 7	cubic inches (in.³)
milliliter (mL)	0.033 814	fluid ounces (U.S.)
millimeter (about the thickness of a dime)	0.1	centimeters (cm)

Multiply This Unit(s)	By This Factor	To Obtain This Unit(s)
millimeter (mm)	0.003 280 8	feet
millimeter (mm)	0.039 370 1	inches
millimeter (mm)	0.001	meters (m)
millimeter (mm)	39.37	mils
millimeter (mm)	0.001 094	yards
millimeter, square (mm^2)	0.001 550	square inches (in.^{2})
millimeter, cubic (mm^3)	0.000 061 023 7	cubic inches (in.^{3})
millimeter of mercury	0.133 322 4	kilopascals (kPa)
ohm	0.000 001	megohms
ohm	1 000 000	micro ohms
ohm circular mil per foot	1.662 426	nano ohms meter (n$\Omega\cdot$m)
oersted	79.577 47	amperes per meter (A/m)
ounce (avoirdupois)	28.349 52	grams (g)
ounce (avoirdupois)	0.062 5	pounds (avoirdupois)
ounce, fluid	29.573 53	milliliters (mL)
ounce (troy)	31.103 48	grams (g)
ounce per foot, square (oz/ft^2)	305.152	grams per meter squared (g/m^2)
ounces per gallon (U.S. liquid)	7.489 152	grams per liter (g/L)
ounce per yard, square (oz/yd^2)	33.905 7	grams per meter squared (g/m^2)
pica	4.217 5	millimeters (mm)
pint (U.S. liquid)	0.473 176 5	liters (L)
pint (U.S. liquid)	473.177	milliliters (mL)
pound (avoirdupois)	453.592	grams (g)
pound (avoirdupois)	0.453 592	kilograms (kg)
pound (avoirdupois)	16	ounces (avoirdupois)
poundal	0.138 255	newtons (N)
pound foot (lb·ft)	0.138 255	kilograms meter (kg·m)
pound foot per second	0.138 255	kilograms meter per second (kg·m/s)
pound foot, square (lb·ft^2)	0.042 140 1	kilograms meter squared (kg·m^2)
pound-force	4.448 222	newtons (N)
pound-force foot	1.355 818	newton meters (N·m)
pound-force inch	0.112 984 8	newton meters (N·m)
pound-force per square inch	6.894 757	kilopascals (kPa)
pound-force per square foot	0.047 880 26	kilopascals (kPa)
pound per cubic foot (lb/ft^3)	16.018 46	kilograms per cubic meter (kg/m^3)
pound per foot (lb/ft)	1.488 16	kilograms per meter (kg/m)
pound per foot, square (lb/ft^2)	4.882 43	kilograms per meter squared (kg/m^2)
pound per foot, cubic (lb/ft^3)	16.018 5	kilograms per meter cubed (kg/m^3)
pound per gallon (U.S. liquid)	119.826 4	grams per liter (g/L)
pound per second (lb/s)	0.453 592	kilograms per second (kg/s)
pound inch, square (lb·in.^{2})	292.640	kilograms millimeter squared (kg·mm^2)
pound per mile	0.281 849	grams per meter (g/m)
pound square foot (lb·ft^2)	0.042 140 11	kilograms square meter (kg/m^2)
pound per cubic yard (lb/yd^3)	0.593 276	kilograms per cubic meter (kg/m^3)
quart (U.S. liquid)	946.353	milliliters (mL)
square centimeter (cm^2)	197 300	circular mils
square centimeter (cm^2)	0.001 076	square feet (ft^2)
square centimeter (cm^2)	0.155	square inches (in^2)

Multiply This Unit(s)	By This Factor	To Obtain This Unit(s)
square centimeter (cm²)	0.000 1	square meters (m²)
square centimeter (cm²)	0.000 119 6	square yards (yd²)
square foot (ft²)	144	square inches (in.²)
square foot (ft²)	0.092 903 04	square meters (m²)
square foot (ft²)	0.111 1	square yards (yd²)
square inch (in.²)	1 273 000	circular mils
square inch (in.²)	6.451 6	square centimeters (cm²)
square inch (in.²)	0.006 944	square feet (ft²)
square inch (in.²)	645.16	square millimeters (mm²)
square inch (in.²)	1 000 000	square mils
square meter (m²)	10.764	square feet (ft²)
square meter (m²)	1 550	square inches (in.²)
square meter (m²)	0.000 000 386 1	square miles
square meter (m²)	1.196	square yards (yd²)
square mil	1.273	circular mils
square mil	0.000 001	square inches (in.²)
square mile	2.589 988	square kilometers (km²)
square millimeter (mm²)	1 973	circular mils
square yard (yd²)	0.836 127 4	square meters (m²)
tablespoon (tbsp)	14.786 75	milliliters (mL)
teaspoon (tsp)	4.928 916 7	milliliters (mL)
therm	105.480 4	megajoules (MJ)
ton (long) (2,240 lb)	1 016.047	kilograms (kg)
ton (long) (2,240 lb)	1.016 047	metric tons (t)
ton, metric	2 204.62	pounds (avoirdupois)
ton, metric	1.102 31	tons, short (2,000 lb)
ton, refrigeration	12 000	Btus per hour
ton, refrigeration	4.716 095 9	horsepower-hours
ton, refrigeration	3.516 85	kilowatts (kW)
ton (short) (2,000 lb)	907.185	kilograms (kg)
ton (short) (2,000 lb)	0.907 185	metric tons (t)
ton per cubic meter (t/m³)	0.842 778	tons per cubic yard (ton/yd³)
ton per cubic yard (ton/yd³)	1.186 55	tons per cubic meter (ton/m³)
torr	133.322 4	pascals (Pa)
watt (W)	3.412 14	Btus per hour (Btu/hr)
watt (W)	0.001 341	horsepower
watt (W)	0.001	kilowatts (kW)
watt-hour (Wh)	3.413	Btus
watt-hour (Wh)	0.001 341	horsepower-hours
watt-hour (Wh)	0.001	kilowatt-hours (kW/hr)
yard	91.44	centimeters (cm)
yard	3	feet
yard	36	inches
yard	0.000 914 4	kilometers (km)
yard	0.914 4	meters (m)
yard	914.4	millimeters
yard, cubic (yd³)	0.764 55	cubic meters (m³)
yard, square (yd²)	0.836 127	square meter (m²)

*These terms are no longer used, but may still be found in some publications.

GLOSSARY

ELECTRICAL TERMS COMMONLY USED IN THE APPLICATION OF THE *NEC*®

The following terms are used extensively in the application of the *National Electrical Code.*® When added to the terms related to electric phenomena they constitute a "culture literacy" for the electrical construction industry. Many of these definitions are quotes or paraphrases of definitions given by the *NEC*®; these have been marked with an *.

Accent Lighting: Directional lighting to emphasize a particular object or to draw attention to a part of the field in view.

Accessible (wiring methods)*: Capable of being removed or exposed without damaging the building structure or finish, or not permanently closed in by the structure or finish of the building.

Accessible (equipment)*: Admitting close approach: not guarded by locked doors, elevation, or other effective means.

Accessible, Readily*: Capable of being reached quickly for operation, renewal, or inspections, without requiring those to whom ready access is requisite to climb over or remove obstacles or to resort to portable ladders, chairs, and so on.

Addendum: Modification (change) made to the construction documents (plans and specifications) during the bidding period.

Adjustment Factor: A multiplier that is used to reduce the ampacity of a conductor when there are more than factors for more than three current-carrying conductors in a raceway or cable with load diversity. See *NEC*® *Table 310.15(B)(2)*.

AL/CU: Terminal marking on switches and receptacles rated 30-ampere and greater suitable for use with aluminum, copper, and copper-clad aluminum conductors. If not marked, suitable for copper conductors only.

Allowable Ampacity: The values given in *NEC*® *Table 310.16, Table 310.17, Table 310.18,* and *Table 310.19*.

Ambient Lighting: Lighting throughout an area that produces general illumination.

American National Standards Institute (ANSI): An organization that identifies industrial and public requirements for national consensus standards and coordinates and manages their development, resolves national standards problems, and ensures effective participation in international standardization. ANSI does not itself develop standards. Rather, it facilitates development by establishing consensus among qualified groups. ANSI ensures that the guiding principles—consensus, due process, and openness—are followed.

Ampacity*: The current in amperes that a conductor can carry continuously under the conditions of use without exceeding its temperature rating.

Ampere: The measurement of intensity of rate of flow of electrons in an electric circuit. An ampere is the amount of current through a resistance of 1 ohm under a pressure of 1 volt.

Appliance*: Utilization equipment, generally other than industrial, normally built in standardized sizes or types, that is installed or connected as a unit to perform one or more functions such as clothes washing, air conditioning, food mixing, deep frying, and so forth.

Approved*: Acceptable to the authority having jurisdiction.

Arc-Fault Circuit Interrupter (AFCI)*: A device intended to provide protection from the effects of arc faults by recognizing characteristics unique to arcing and by functioning to de-energize a circuit when an arcing fault is detected.

As-Built Plans: When required, a modified set of working drawings prepared for a construction project that includes all variances from the original working drawings that occurred during the project construction.

Authority Having Jurisdiction (AHJ): The organization, office, or individual responsible for approving equipment, materials, an installation, or a procedure. An AHJ is usually a governmental body that has legal jurisdiction over electrical installations. The AHJ has the responsibility for making interpretations of the rules, for deciding upon the approval of equipment and materials, and for granting the special permission where required. Where a specific interpretation or deviation from the *NEC®* is given, get it in writing. See *90.4* of the *NEC.®*

Average Rated Life (lamp): How long it takes to burn out the lamp. For example, a 60-watt lamp is rated 1000 hours. The 1000-hour rating is based on the point in time when 50 percent of a test batch of lamps burn out and 50 percent are still burning.

Backlight: Illumination from behind a subject directed substantially parallel to a vertical plane through the optical axis of the camera.

Ballast: Used for energizing fluorescent lamps. It is constructed of a laminated core and coil windings or solid-state electronic components.

Ballast factor: The percentage of rated light output that can be achieved with a specific ballast.

Bare Lamp: A light source with no shielding.

Bonding*: The permanent joining of metallic parts to form an electrically conductive path that will ensure electrical continuity and the capacity to conduct safely any current likely to be imposed.

Bonding Jumper*: A reliable conductor to ensure the required electrical conductivity between metal parts required to be electrically connected.

Bonding Jumper, Circuit*: The connection between portions of a conductor in a circuit to maintain the required ampacity of the circuit.

Bonding Jumper, Equipment*: The connection between two or more portions of the equipment grounding conductor.

Bonding Jumper, Main*: The connection between the grounded circuit conductor and the equipment grounding conductor at the service.

▶**Bonding Jumper, System:** The connection between the grounded circuit conductor and the equipment grounding conductor at a separately derived system. ◀

Branch-Circuit*: The circuit conductors between the final overcurrent device protecting the circuit and the outlet(s).

Branch-Circuit, Appliance*: A branch-circuit supplying energy to one or more outlets to which appliances are to be connected; such circuits are to have no permanently connected luminaires (lighting fixtures) not a part of an appliance.

Branch-Circuit, General Purpose*: A branch-circuit that supplies two or more receptacles or outlets for lighting and appliances.

Branch-Circuit, Individual*: A branch-circuit that supplies only one utilization equipment.

Branch-Circuit, Multiwire*: *A branch-circuit consisting of two or more ungrounded conductors that have a voltage between them, and a grounded conductor that has equal voltage between it and each ungrounded conductor of the circuit and that is connected to the neutral or grounded conductor of the system.*

Branch-Circuit Selection Current*: Branch-circuit selection current is the value in amperes to be used instead of the rated-load current in determining the ratings of motor branch-circuit conductors, disconnecting means, controllers, and branch-circuit short-circuit and ground-fault protective devices wherever the running overload protective device permits a sustained current greater than the specified percentage of the rated-load current. The value of branch-circuit selection current will always be equal to or greater than the marked rated-load current.

Brightness: In common usage, the term *brightness* usually refers to the strength of sensation that results from viewing surfaces or spaces from which light comes to the eye.

Candela: The international unit (SI) of luminous intensity. Formerly referred to as "candle," as in "candlepower."

Candlepower: A measure of intensity mathematically related to lumens.

Carrier: A wave having at least one characteristic that may be varied from a known reference value by modulation.

Carrier Current: The current association with a carrier wave.

Circuit Breaker*: A device designed to open and close a circuit by nonautomatic means and to open the circuit automatically on a predetermined overcurrent without damage to itself when properly applied within its rating.

CC: A marking on wire connectors and soldering lugs indicating they are suitable for use with copper-clad aluminum conductors only.

CC/CU: A marking on wire connectors and soldering lugs indicating they are suitable for use with copper or copper-clad aluminum conductors only.

CU: A marking on wire connectors and soldering lugs indicating they are suitable for use with copper conductors only.

CO/ALR: Terminal marking on switches and receptacles rated 15 and 20 amperes that are suitable for use with aluminum, copper, and copper-clad aluminum conductors. If not marked, they are suitable for copper or copper-clad conductors only.

▶**Coordination (Selective):** Localization of an overcurrent condition to restrict outages to the equipment affected, accomplished by the choice of overcurrent-protective devices or settings.◀

Conductor*:

Bare: A conductor having no covering or electrical insulation whatsoever. (See "Conductor, Covered.")

Covered: A conductor encased within material of composition or thickness that is not recognized by this *Code* as electrical insulation. (See "Conductor, Bare.")

Insulated: A conductor encased within material of composition and thickness that is recognized by this *Code* as electrical insulation.

Continuous Load*: A load where the maximum current is expected to continue for 3 hours or more.

Contract: An agreement between two or more parties, especially one that is written and enforceable by law.

Contract Modifications: After the agreement has been signed, any additions, deletions, or modifications of the work to be done are accomplished by change order, supplemental instruction, and field order. They can be issued at any time during the contract period.

Contractor: A properly licensed individual or company that agrees to furnish labor, materials, equipment, and associated services to perform the work as specified for a specified price.

Correction Factor: A multiplier that is used to reduce the ampacity of a conductor when the ambient temperature is less or greater than 86°F (30°C). These multipliers are found at the bottom of *NEC® Table 310.16*. Sometimes referred to as a derating factor.

CSA: This is the Canadian Standards Association that develops safety and performance standards in Canada for electrical products similar to but not always identical to those of UL (Underwriters Laboratories) in the United States.

Current: The flow of electrons through an electrical circuit, measured in amperes.

Derating Factor: See "Adjustment Factor" and "Correction Factor."

Details: Plans, elevations, or sections that provide more specific information about a portion of a project component or element than smaller scale drawings.

Diagrams: Nonscaled views showing arrangements of special system components and connections not possible to clearly show in scaled views. A *schematic diagram* shows circuit components and their electrical connections without regard to actual physical locations. A *wiring diagram* shows circuit components and the actual electrical connections.

Dimmer: A switch with components that permits variable control of lighting intensity. Some dimmers have electronic components; others have core and coil (transformer) components.

Direct Glare: Glare resulting from high luminances or insufficiently shielded light sources in the field of view. Usually associated with bright areas, such as luminaires (fixtures), ceilings, and windows that are outside the visual task of region being viewed.

Disconnecting Means*: A device or group of devices, or other means by which the conductors of a circuit can be disconnected from their source of supply.

Downlight: A small, direct lighting unit that guides the light downward. It can be recessed, surface mounted, or suspended.

Drawings: Graphic representations of the work to be done. They show the relationship of the materials to each other, including sizes, shapes, locations, and connections. The drawings may include schematic diagrams showing such things as mechanical and electrical systems. They may also include schedules of structural elements, equipment, finishes, and other similar items.

Dry Niche Luminaire (fixture)*:   A luminaire (fixture) intended for installation in the wall of a pool or fountain in a niche that is sealed against the entry of pool water.

Dwelling Unit*:   ▶One or more rooms for the use of one or more persons as a housekeeping unit with space for eating, living, cooking, and sleeping, and permanent provisions for sanitation.◀

Efficacy: The total amount of light energy (lumens) emitted from a light source divided by the total lamp and ballast power (watts) input, expressed in lumens per watt.

Elevations: Views of vertical planes, showing components in their vertical relationship, viewed perpendicularly from a selected vertical plane.

Feeder*:   All circuit conductors between the service equipment or the source of a separately derived system and the final branch-circuit overcurrent device.

Fill Light: Supplementary illumination to reduce shadow or contract range.

Fine Print Note (FPN)*:   Explanatory material is in the form of fine print notes (FPN). Fine print notes are informational only and are not enforceable as a requirement by the *Code*.

Fire Rating: The classification indicating in time (hours) the ability of a structure or component to withstand fire conditions.

Flame Detector: A radiant energy-sensing fire detector that detects the radiant energy emitted by a flame.

Fluorescent Lamp: A lamp in which electric discharge of ultraviolet energy excites a fluorescing coating (phosphor) and transforms some of that energy to visible light.

Footcandle: The unit used to measure how much total light is reaching a surface. One lumen falling on 1 square foot of surface produces an illumination of 1 footcandle.

Fully Rated System: Panelboards are marked with their short-circuit rating in RMS symmetrical amperes. In a fully rated system, the panelboard short-circuit current rating will be equal to the lowest interrupting rating of any branch-circuit breaker or fuse installed. All devices installed shall have an interrupting rating greater than or equal to the specified available fault current. (See "Series-Rated System.")

Fuse: An overcurrent protective device with a fusible link that operates and opens the circuit on an overcurrent condition.

General Conditions: A written portion of the contract documents set forth by the owner stipulating the contractor's minimum acceptable performance requirements, including the rights, responsibilities, and relationships of the parties involved in the performance of the contract. General conditions are usually included in the book of specifications but are sometimes found in the architectural drawings.

General Lighting: Lighting designed to provide a substantially uniform level of illumination throughout an area, exclusive of any provision for special lighting.

Glare: The sensation produced by luminance within the visual field that is sufficiently greater than the luminance to which the eyes are adapted to cause annoyance, discomfort, or loss in visual performance and visibility.

Ground*:   A conducting connection, whether intentional or accidental, between an electrical circuit or equipment and the earth, or to some conducting body that serves in place of the earth.

Grounded*:   Connected to the earth or to some conducting body that serves in place of the earth.

Grounded Conductor*:   A system or circuit conductor that is intentionally grounded.

▶**Ground Fault*:**   An unintentional, electrically conducting connection between an ungrounded or grounded conductor of an electrical circuit and the normally non-current carrying conductors, metallic enclosures, metallic raceways, metallic equipment, or earth.◀

Ground-Fault Circuit-Interrupter (GFCI)*:   A device intended for the protection of personnel that functions to de-energize a circuit or portion thereof within an established period of time when a current to ground exceeds the values established for a Class A device.

▶**Ground-Fault Current Path, Effective*:**   An intentionally constructed, permanent, low-impedance electrically conductive path designed and intended to carry current under ground-fault conditions from the point of a ground fault on a wiring system to the electrical supply source and that facilitates the operation of the overcurrent protective device or ground fault detectors on high-impedance grounded systems.◀

Ground-Fault Protection of Equipment (GFPE)*: A system intended to provide protection of equipment from damaging line-to-ground fault currents by operating a disconnecting means to open all ungrounded conductors of the faulted circuit. This protection is provided at current levels less than those required to protect conductors from damage through the operation of a circuit overcurrent device.

Grounding Conductor*: A conductor used to connect equipment or the grounded circuit of a wiring system to a grounding electrode or electrodes.

Grounding Conductor, Equipment*: The conductor used to connect the non-current-carrying metal parts of equipment, raceways, and other enclosures to the system grounded conductor, the grounding electrode conductor, or both, at the service equipment or at the source of a separately derived system.

▶**Grounding Electrode:** A device that establishes an electrical connection to the earth.◀

▶**Grounding Electrode Conductor*:** The conductor used to connect the grounding electrode(s) to the equipment grounding conductor, to the grounded conductor, or to both, at the service, at each building or structure where supplied by a feeder(s) or branch circuit(s), or at the source of a separately derived system.◀

HACR: A circuit breaker that has been tested and found suitable for use on heating, air-conditioning, and refrigeration equipment. The circuit breaker is marked with the letters *HACR*. Equipment that is suitable for use with HACR circuit breakers will be marked for use with HACR-type circuit breakers.

Halogen Lamp: An incandescent lamp containing a halogen gas that recycles tungsten back onto the tungsten filament surface. Without the halogen gas, the tungsten would normally be deposited onto the bulb wall.

Heat Alarm: A single- or multiple-station alarm responsive to heat.

Hermetic Refrigerant Motor-Compressor: A combination consisting of a compressor and motor, both of which are enclosed in the same housing, with no external shaft or shaft seals; the motor operates in the refrigerant.

High-Intensity Discharge Lamp (HID): A general term for a mercury, metal halide, or high-pressure sodium lamp.

High-Pressure Sodium Lamp: A high-intensity discharge light source in which the light is primarily produced by the radiation from sodium vapor.

Horsepower (tools): Horsepower is a measurement of motor torque multiplied by speed. Horsepower also refers to the rate of work (power) an electric motor is capable of delivering. (See "Torque," "Horsepower, Rated," and "Horsepower, Maximum Developed.")

Horsepower, Maximum Developed: If a motor is required to work harder than its idle speed, the motor becomes overloaded and must develop extra horsepower. The most horsepower that can be drawn from a motor to handle this extra effort is referred to as its maximum developed horsepower. (See "Horsepower," "Torque," and "Horsepower, Rated.")

Horsepower, Rated: Rated horsepower is a motor's running torque at its rated running speed. The motor can be run continuously at its rated horsepower without overheating. If the motor is required to give an extra spurt of effort while running, the motor becomes overloaded and develops extra horsepower to compensate. The most horsepower that can be drawn from a motor to handle this extra effort is its maximum developed horsepower. (See "Torque," "Horsepower," and "Horsepower, Maximum Developed.")

Household Fire Alarm System: A system of devices that produces an alarm signal in the household for the purpose of notifying the occupants of the presence of a fire so that they will evacuate the premises.

Identified*: (as applied to equipment) Recognizable as suitable for the specific purpose, function, use, environment, application, and so on, where described in a particular *Code* requirement.

Identified* (conductor): The identified conductor is the insulated grounded conductor. For sizes 6 AWG or smaller, the insulated grounded conductor shall be identified by a continuous white or gray outer finish or by three continuous white stripes on other than green insulation along its entire length. For sizes larger than 6 AWG, the insulated grounded conductor shall be identified either by a continuous white or gray outer finish or by three continuous white stripes on other than green insulation along its entire length, or at the time of installation by a distinctive white marking at its terminations.

Identified* (terminal): The identification of terminals to which a grounded conductor is to be connected shall be substantially white in color. The identification of other terminals shall be of a readily distinguishable different color.

IEC: The International Electrotechnical Commission is a worldwide standards organization. These standards differ from those of Underwriters Laboratories. Some electrical equipment might conform to a specific IEC standard, but might not conform to the UL standard for the same item.

Illuminance: The amount of light energy (lumens) distributed over a specific area expressed as foot-candles (lumens/square foot) or luxces (lumens/ square meter).

Illumination: The act of illuminating or the state of being illuminated.

Immersion Detection Circuit Interrupter (IDCI): A device integral with grooming appliances that will shut off the appliance when it is dropped in water.

In Sight: Where this *Code* specifies that one equipment shall be "in sight from," "within sight from," or "within sight," and so on, of another equipment, the specified equipment is to be visible and not more than 50 ft (15 m) from the other.

Incandescent Filament Lamp: A lamp that provides light when a filament is heated to incandescence by an electric current. Incandescent lamps are the oldest form of electric lighting technology.

Indirect Lighting: Lighting by luminaire (fixtures) that distribute 90 to 100 percent of the emitted light upward.

Inductive Load: A load that is made up of coiled or wound wire that creates a magnetic field when energized. Transformers, core and coil ballasts, motors, and solenoids are examples of inductive loads.

Instant Start: A circuit used to start specially designed fluorescent lamps without the aid of a starter. To strike the arc instantly, the circuit utilizes higher open-circuit voltage than is required for the same length preheat lamp.

International Association of Electrical Inspectors (IAEI): A not-for-profit educational organization cooperating in the formulation and uniform application of standards for the safe installation and use of electricity, and collecting and disseminating information relative thereto. The IAEI is made up of electrical inspectors, electrical contractors, electrical apprentices, manufacturers, and governmental agencies.

Interrupting Rating*: The highest current at rated voltage that a device is intended to interrupt under standard test conditions.

Isolated Ground Receptacle: A grounding-type device in which the equipment ground contact and terminal are electrically isolated from the receptacle mounting means.

Kilowatt (kW): One thousand watts equals 1 kilowatt.

Kilowatt-hour (kWh): One thousand watts of power in 1 hour. One 100-watt lamp burning for 10 hours generates 1 kilowatt-hour. Two 500-watt electric heaters operating for 1 hour generates 1 kilowatt-hour.

Labeled*: Equipment or materials to which has been attached a label, symbol, or other identifying mark of an organization that is acceptable to the authority having jurisdiction and concerned with product evaluation, which maintains periodic inspection of production of labeled equipment or materials and by whose labeling the manufacturer indicates compliance with appropriate standards or performance in a specified manner.

Lamp: A generic term for a man-made source of light.

Light: The term generally applied to the visible energy from a source. Light is usually measured in lumens (a unit of luminous flux in terms of candelas), candelas (a unit of luminous intensity), or candlepower (luminous intensity expressed in candelas). When light strikes a surface, it is either absorbed, reflected, or transmitted.

Lighting Outlet*: An outlet intended for the direct connection of a lampholder, a luminaire (lighting fixture), or a pendant cord terminating in a lampholder.

▶**Listed*:** Equipment, materials, or services included in a list published by an organization that is acceptable to the authority having jurisdiction and concerned with evaluation of products or services, that maintains periodic inspection of production of listed equipment or materials, or periodic evaluation of services, and whose listing states that either the equipment, material, or services meets appropriate designated standards or has been tested and found suitable for a specified purpose. ◄

Load: The electric power used by devices connected to an electrical system. Loads can be figured in amperes, volt-amperes, kilovolt-amperes, or kilowatts. Loads can be intermittent, continuous intermittent, periodic, short-time, or varying. See the definition of "Duty" in the *NEC.*®

Load Center: A common name for residential panelboards. A load center may not be as deep as a panelboard, and generally does not contain relays or other accessories as are available for panelboards. Circuit breakers are "plug-in" as opposed to "bolt-in" types used in panelboards. Manufacturers' catalogs will show both load centers and panelboards. The UL standards do not differentiate.

Locked Rotor Current: The steady-state current taken from the line with the rotor locked and with rated voltage and frequency applied to the motor.

Low-Pressure Sodium Lamp: A discharge lamp in which light is produced from sodium gas operating at a partial pressure.

Lumens: The SI unit of luminous flux. The units of light energy emitted from the light source.

►**Luminaire (fixture)*:** A complete lighting unit consisting of a lamp or lamps together with the parts designed to distribute the light, to position and protect the lamps and ballast (where applicable), and to connect the lamps to the power supply. ◄ Prior to the *National Electrical Code*® adopting the international definition of *luminaire*, the commonly used term in the United States was and in most instances still is *lighting fixture*. It will take years for the electrical industry to totally change and feel comfortable with the term *luminaire*.

Luminaire (fixture) Efficiency: The total lumen output of the luminaire (fixture) divided by the rated lumens of the lamps inside the luminaire (fixture).

Lux: The SI (International System) unit of illumination. One lumen uniformly distributed over an area of 1 square meter.

Maximum Continuous Current (MCC): A value that is determined by the manufacturer of

hermetic refrigerant motor compressors under high load (high refrigerant pressure) conditions. This value is established in the *NEC*® as being no greater than 156 percent of the marked rated amperes or the branch-circuit selection current.

Maximum Overcurrent Protection (MOP): A term used with equipment that has a hermetic motor compressor(s). MOP is the maximum ampere rating for the equipment's branch-circuit overcurrent protective device. The nameplate will specify the ampere rating and type (fuses or circuit breaker) of overcurrent protective device to use.

Mercury Lamp: A high-intensity discharge light source in which radiation from the mercury vapor produces visible light.

Metal Halide Lamp: A high-intensity discharge light source in which the light is produced by the radiation from mercury together with halides of metals such as sodium and scandium.

Minimum Circuit Ampacity (MCA): A term used with equipment that has a hermetic motor compressor(s). MCA is the minimum ampere rating value used to determine the proper size conductors, disconnecting means, and controllers. MCA is determined by the manufacturer, and is marked on the nameplate of the equipment.

National Electrical Code (*NEC*®)**:** The electrical code published by the National Fire Protection Association. This *Code* provides for practical safeguarding of persons and property from hazards arising from the use of electricity. It does not become law until adopted by federal, state, and local laws and regulations. The *NEC*® is not intended as a design specification nor an instruction manual for untrained persons.

National Electrical Manufacturers Association (NEMA): NEMA is a trade organization made up of many manufacturers of electrical equipment. They develop and promote standards for electrical equipment.

National Fire Protection Association (NFPA): Located in Quincy, MA. The NFPA is an international standards making organization dedicated to the protection of people from the ravages of fire and electric shock. The NFPA is responsible for developing and writing the *National Electrical Code,*® *The Sprinkler Code,*®

The Life Safety Code,® *The National Fire Alarm Code,*® and over 295 other codes, standards, and recommended practices. The NFPA may be contacted at (800) 344-3555 or on their Web site at http://www.nfpa.org.

Nationally Recognized Testing Laboratory (NRTL):   The term used to define a testing laboratory that has been recognized by OSHA; for example, Underwriters Laboratories.

Neutral: The neutral conductor is the grounded conductor in a circuit consisting of three or more conductors from the same source. (There is no neutral conductor in a two-wire circuit, although electricians many times refer to the white grounded conductor as the neutral.) The voltages from the ungrounded conductors to the grounded conductor are of equal magnitude. See the definition of "Branch-Circuit, Multiwire" in the *NEC.*®

▶**Neutral Conductor*:**   A conductor, other than a grounding conductor, that is connected to the common point of a wye connection in a polyphase system or the point of a symmetrical system that is normally at zero voltage.◀

In residential wiring, the neutral conductor is the grounded conductor in a branch-circuit or feeder consisting of three conductors. The ungrounded conductors are connected to opposite phases in the panelboard. The voltages from the ungrounded conductors to the grounded conductor are of equal magnitude. There is no neutral conductor in a two-wire circuit, although electricians many times refer to the white grounded conductor as the neutral.

See the definition of "Multiwire Branch-Circuit" in the *NEC.*®

No Niche Luminaire (fixture)*:   A luminaire (fixture) intended for installation above or below the water without a niche.

Noncoincidental Loads: Loads that are not likely to be on at the same time. Heating and cooling loads would not operate at the same time. See *NEC*® 220.21.

Notations: Words on working drawings providing specific information.

Occupational Safety and Health Administration (OSHA):   OSHA is a division of the U.S. Department of Labor. OSHA develops, administers, and enforces regulations relating to safety in the workplace. Electrical regulations are cov-

ered in *Part 1910, Subpart S.* To simplify the regulations, *Part 1920, Subpart S* contains only the most common performance requirements of the *NEC.*® The *NEC*® must still be referred to, in conjunction with OSHA regulations.

Ohm: A unit of measure for electric resistance. An ohm is the amount of resistance that will allow 1 ampere to flow under a pressure of 1 volt.

Outlet*:   A point on the wiring system at which current is taken to supply utilization equipment.

Overcurrent*:   Any current in excess of the rated current of equipment or the ampacity of a conductor. It may result from overload, short circuit, or ground-fault.

Overcurrent Device: Also referred to as an overcurrent protection device. A form of protection that operates when current exceeds a predetermined value. Common forms of overcurrent devices are circuit breakers, fuses, and thermal overload elements found in motor controllers.

Overload*:   Operation of equipment in excess of normal, full-load rating, or of a conductor in excess of rated ampacity that, when it persists for a sufficient length of time, would cause damage or dangerous overheating. The current is contained in its normal path. A fault, such as a short circuit or ground fault, is not an overload.

Panelboard*:   A single panel or group of panel units designed for assembly in the form of a single panel; including buses, automatic overcurrent devices, and equipped with or without switches for the control of light, heat, or power circuits; designed to be placed in a cabinet or cutout box placed in or against a wall or partition and accessible only from the front.

Performance Bond: (1) A written form of security from a surety (bonding) company to the owner, on behalf of an acceptable prime or main contractor or subcontractor, guaranteeing payment to the owner in the event the contractor fails to perform all labor, materials, equipment, or services in accordance with the contract. (2) The surety companies generally reserve the right to have the original prime or main or subcontractor remedy any claims before paying on the bond or hiring other contractors.

Photometry: The pattern and amount of light that is emitted from a luminaire (fixture), normally represented as a cross-section through the fixture distribution pattern.

Plans: See "Working Drawings."

Power Supply: A source of electrical operating power including the circuits and terminations connecting it to the dependent system components.

Preheat Fluorescent Lamp Circuit: A circuit used on fluorescent lamps wherein the electrodes are heated or warmed to a glow stage by an auxiliary switch or starter before the lamps are lighted.

Proposal: A written offer from a bidder to the owner, preferably on a prescribed proposal form, to perform the work and to furnish all labor, materials, equipment, and/or service for the prices and terms quoted by the bidder.

Punch List (Inspection List):   A list of items of work requiring immediate corrective or completion action by the contractor prepared by the owner or an authorized representative.

▶**Qualified Person*:**   One who has skills and knowledge related to the construction and operation of the electrical equipment and installations and has received safety training on the hazards involved.◀

Raceway*:   An enclosed channel of metal or nonmetallic materials designed expressly for holding wires, cables, or busbars, with additional functions as permitted in this *Code*. Raceways include, but are not limited to, rigid metal conduit, rigid nonmetallic conduit, intermediate metal conduit, liquidtight flexible conduit, flexible metallic tubing, flexible metal conduit, electrical nonmetallic tubing, electrical metallic tubing, underfloor raceways, cellular concrete floor raceways, cellular metal floor raceways, surface raceways, wireways, and busways.

Rapid-Start Fluorescent Lamp Circuit: A circuit designed to start lamps by continuously heating or preheating the electrodes. Lamps must be designed for this type of circuit. This is the modern version of the "trigger start" system. In a rapid-start two-lamp circuit, one end of each lamp is connected to a separate starting winding.

Rated-Load Current: The rated-load current for a hermetic refrigerant motor-compressor is the current resulting when the motor-compressor is operated at the rated load, rated voltage, and rated frequency of the equipment it serves.

Rate of Rise Detector: A device that responds when the temperature rises at a rate exceeding a predetermined value.

Receptacle*:   A receptacle is a contact device installed at the outlet for the connection of a single contact device. A single receptacle is a single contact device with no other contact device on the same yoke. A multiple receptacle is two or more contact devices on the same yoke.

Receptacle Outlet*:   An outlet where one or more receptacles are installed.

Resistive Load: An electric load that produces heat. Some examples of resistive loads are the elements in an electric range, ceiling heat cables, and electric baseboard heaters.

Root-Mean-Square (RMS):   The square root of the average of the square of the instantaneous values of current or voltage. It is the effective or heating value of sinusoidal wave form. For example, assume the RMS value of voltage line to neutral is 120 volts. During each electrical cycle, the voltage rises from zero to a peak value ($120 \times 1.4142 = 169.7$ volts), back through zero to a negative peak value, then back to zero. The RMS value is 0.707 of the peak value ($169.7 \times 0.707 = 120$ volts). The root-mean-square value is what an electrician reads on an ammeter or voltmeter.

Schedules: Tables or charts that include data about materials, products, and equipment.

Scope: A written range of view or action; outlook; hence, room for the exercise of faculties or function; capacity for achievement; all in connection with a designated project.

Sections: Views of vertical cuts through and perpendicular to components, showing their detailed arrangement.

Series-Rated System: Panelboards are marked with their short-circuit rating in RMS symmetrical amperes. A series-rated panelboard will be determined by the main circuit breaker or fuse, and branch-circuit breaker combination tested in accordance to UL *Standard 489*. The series-rating will be less than or equal to the interrupting rating of the main overcurrent device, and greater than the interrupting rating of the branch-circuit overcurrent devices. (See "Fully Rated System.")

Service*:   The conductors and equipment for delivering energy from the serving utility to the wiring system of the premises served.

Service Conductors*: The conductors from the service point to the service disconnecting means.

Service Drop*: The overhead service conductors from the last pole or other aerial support to and including the splices, if any, connecting to the service-entrance conductors at the building or other structure.

Service Equipment*: The necessary equipment, usually consisting of a circuit breaker or switch and fuses, and their accessories, located near the point of entrance of supply conductors to a building or other structure, or an otherwise defined area, and intended to constitute the main control and means of cutoff of the supply.

Service Lateral*: The underground service conductors between the street main, including any risers at a pole or other structure or from transformers, and the first point of connection to the service-entrance conductors in a terminal box or meter or other enclosure with adequate space, inside or outside the building wall. Where there is no terminal box, meter, or other enclosure with adequate space, the point of connection shall be considered to be the point of entrance of the service conductors into the building.

Service Point*: A service point is the point of connection between the facilities of the serving utility and the premises wiring.

Shall: Indicates a mandatory requirement.

Short Circuit: A connection between any two or more conductors of an electrical system in such a way as to significantly reduce the impedance of the circuit. The current flow is outside of its intended path, thus the term *short circuit*. A short circuit is also referred to as a fault.

Should: Indicates a recommendation or that which is advised but not required.

Site: The place where a structure or group of structures was or is to be located (as in *construction site*).

Smoke Alarm: A single or multiple station alarm responsive to smoke.

Smoke Detector: A device that detects visible or invisible particles of combustion.

Sone: A sone is a unit of loudness equal to the loudness of a sound of 1 kilohertz at 40 decibels above the threshold of hearing of a given listener.

Specifications: Text setting forth details such as description, size, quality, performance, and workmanship, etc. Specifications that pertain to all of the construction trades involved might be subdivided into "General Conditions" and "Supplemental General Conditions." Further subdividing the specifications might be specific requirements for the various contractors such as electrical, plumbing, heating, masonry, and so forth. Typically, the electrical specifications are found in Division 16.

Split-Circuit Receptacle: A receptacle that can be connected to two branch-circuits. These receptacles may also be used so that one receptacle is live at all times, and the other receptacle is controlled by a switch. The terminals on these receptacles usually have breakaway tabs so the receptacle can be used either as a split-circuit receptacle or as a standard receptacle.

Standard Network Interface (SNI): A device usually installed by the telephone company at the demarcation point where their service leaves off and the customers service takes over. This is similar to the service point for electrical systems.

Starter: A device used in conjunction with a ballast for the purpose of starting an electric discharge lamp.

Structure*: That which is built or constructed.

Subcontractor: A qualified contractor subordinate to the prime or main contractor.

Surface-Mounted Luminaire (fixture): A luminaire (fixture) mounted directly on the ceiling.

Surge Suppressor: A device that limits peak voltage to a predetermined value when voltage spikes or surges appear on the connected line.

Suspended (pendant) Luminaire (fixture): A luminaire (fixture) hung from a ceiling by supports.

Switches*:

General-Use Switch: A switch intended for use in general distribution and branch-circuits. It is rated in amperes, and it is capable of interrupting its rated current at its rated voltage.

Motor-Circuit Switch: A switch, rated in horsepower, capable of interrupting the maximum operating overload current of a motor of the same horsepower rating as the switch at the rated voltage.

Symbols: A graphic representation that stands for or represents another thing. A symbol is a simple way to show such things as lighting outlets, switches, and receptacles on an electrical working drawing. The American Institute of Architects has developed a comprehensive set of symbols that represents just about everything used by all building trades. When an item cannot be shown using a symbol, then a more detailed explanation using a notation or inclusion in the specifications is necessary.

Task Lighting: Lighting directed to a specific surface or area that provides illumination for visual tasks.

Terminal: A screw or a quick-connect device where a conductor(s) is intended to be connected.

Thermocouple: A pair of dissimilar conductors so joined at two points that an electromotive force is developed by the thermoelectric effects when the junctions are at different temperatures.

Thermopile: More than one thermocouple connected together. The connections may be series, parallel, or both.

Torque: Torque is a measurement of rotation or turning force. Torque is measured in ounce inches (oz. in.), ounce feet (oz. ft.), and pound feet (lb. ft.). (See "Horsepower," "Horsepower, Rated," and "Horsepower, Maximum Developed.")

Transient Voltage Surge Suppressors (TVSS): A device that clamps transient voltages by absorbing the major portion of the energy (joules) created by the surge, allowing only a small, safe amount of energy to enter the actual connected load.

Troffer: A recessed lighting unit, usually long and installed with the opening flush with the ceiling. The term is derived from "trough" and "coffer."

UL: Underwriters Laboratories is an independent not-for-profit organization that develops standards, and tests electrical equipment to these standards.

UL-Listed: Indicates that an item has been tested and approved to the standards established by UL for that particular item. The UL Listing Mark may appear in various forms, such as the letters *UL* in a circle. If the product is too small for the marking to be applied to the product, the marking must appear on the smallest unit container in which the product is packaged.

UL-Recognized: Refers to a product that is incomplete in construction features or limited in performance capabilities. A "Recognized" prod-

uct is intended to be used as a component part of equipment that has been "listed." A "Recognized" product must not be used by itself. A UL product may contain a number of components that have been "Recognized."

Ungrounded Conductor: The conductor of an electrical system that is not intentionally connected to ground. This conductor is referred to as the "hot" or "live" conductor.

Volt: The difference of electric potential between two points of a conductor carrying a constant current of 1 ampere, when the power dissipated between these points is equal to 1 watt. A voltage of 1 volt can push 1 ampere through a resistance of 1 ohm.

Voltage (of a circuit)*: The greatest root-mean-square (effective) difference of potential between any two conductors of the circuit concerned.

Voltage (nominal)*: A nominal value assigned to a circuit or system for the purpose of conveniently designating its voltage class (e.g., 120/240 volts, 480Y/277 volts, 600 volts).

   The actual voltage at which a circuit operates can vary from the nominal within a range that permits satisfactory operation of equipment.

Voltage to Ground*: For grounded circuits, the voltage between the given conductor and that point or conductor of the circuit that is grounded; for ungrounded circuits, the greatest voltage between the given conductor and any other conductor of the circuit.

Voltage Drop: Also referred to as IR drop. Voltage drop is most commonly associated with conductors. A conductor has resistance. When current is flowing through the conductor, a voltage drop will be experienced across the conductor. Voltage drop across a conductor can be calculated using Ohm's law: $E = I \times R$.

Volt-ampere: A unit of power determined by multiplying the voltage and current in a circuit. A 120-volt circuit carrying 1 ampere is 120 volt-amperes.

Watertight: Constructed so that moisture will not enter the enclosure under specified test conditions.

Watt: A measure of true power. A watt is the power required to do work at the rate of 1 joule per second. Wattage is determined by multiplying voltage times amperes times the power factor of the circuit: $W = E \times I \times PF$.

Weatherproof*: Constructed or protected so that exposure to the weather will not interfere with successful operation. Rainproof, raintight, or watertight equipment can fulfill the requirements for weatherproof where varying weather conditions other than wetness, such as snow, ice, dust, or temperature extremes, are not a factor.

Wet Niche Luminaire (fixture)*: A luminaire (fixture) intended for installation in a forming shell mounted in a pool.

Working Drawings: Views of horizontal planes, showing components in their horizontal relationship. A set of construction drawings. Also called plans.

Work Order: A written order, signed by the owner or a representative, of a contractual status requiring performance by the contractor without negotiation of any sort.

Zoning: Restrictions of areas or regions of land within specific geographical areas based on permitted building size, character, and uses as established by governing urban authorities.

WEB SITES

The following list of Internet World Wide Web sites has been included for your convenience. These are current as of time of printing. Web sites are a "moving target" with continual changes. We do our utmost to keep these Web sites current. If you are aware of Web sites that should be added, deleted, or that are different than shown, please let us know. Most manufacturers will provide catalogs and technical literature upon request either electronically or by mail. You will be amazed at the amount of electrical and electronic equipment information that is available on these Web sites.

You can search the Web using a search engine such as http://www.google.com, http://www.dogpile.com, http://www.yahoo.com, or http://www.askjeeves.com. Just type in a key word (for example: fuses) and you will be led to many Web sites relating to fuses. It's easy!

Web sites marked with an asterisk * are closely related to the "home automation" market.

ACCUBID	http://www.accubid.com
Acme Electric Corporation	http://www.acmepowerdist.com
ActiveHome*	http://www.X10-beta.com/activehome/
Active Power	http://www.activepower.com
Adalet	http://www.adalet.com
Advance Transformer Co.	http://www.advancetransformer.com
Advanced Cable Ties	http://www.advancedcableties.com
AEMC Instruments	http://www.aemc.com
AFC Cable Systems	http://www.afcweb.com
Air-Conditioning & Refrigeration Association	http://www.ari.org
Alcan Cable	http://www.cable.alcan.com
Alflex	http://www.alflex.com
Allen-Bradley	http://www.automation.rockwell.com
Allen-Bradley	http://www.ab.com
Allied Moulded Products	http://www.alliedmoulded.com
Allied Support Products	http://www.alliedsupport.com
Allied Tube & Conduit	http://www.alliedtube.com
Alpha Wire	http://www.alphawire.com
ALTO Lamps	http://www.altolamp.com
The Aluminum Association, Inc.	http://www.aluminum.org
American Council for an Energy Efficient Economy	http://www.aceee.org
American Gas Association	http://www.aga.org
American Heart Association	http://www.americanheart.org
American Institute of Architects	http://www.aia.org
American Insulated Wire Corp.	http://www.aiwc.com
American Lighting Association	http://www.americanlightingassoc.com
American National Standards Institute	http://www.ansi.org
American Pipe & Plastics, Inc.	http://www.ampipe.com
American Power Conversion	http://www.apcc.com
American Public Power Association	http://www.appanet.org

American Society of Heating, Refrigerating, and Air-Conditioning Engineers	http://www.ashrae.org
American Society for Testing and Materials	http://www.astm.org
American Society of Mechanical Engineers	http://www.asme.org
American Society of Safety Engineers	http://www.asse.org
AMP*	http://www.amp.com
Amprobe	http://www.amprobe.com
Angelo	http://www.angelobrothers.com
Anixter (data & telecommunications cabling)*	http://www.anixter.com
Appleton Electric Co.	http://www.appletonelec.com
ArcWear	http://www.arcwear.com
Arlington Industries, Inc.	http://www.aifittings.com
Arrow Fasteners	http://www.arrowfastener.com
Arrow Hart Wiring Devices	http://www.arrowhart.com
Assisted Living: The World of Assisted Technology	http://www.abledata.com
Associated Builders & Contractors	http://www.abc.org
Association of Cabling Professions	http://www.wireville.com
Association of Home Appliance Manufacturers	http://www.aham.org
Automated Home Technologies*	http://www.autohometech.com
Automatic Switch Co. (ASCO)	http://www.asco.com
AVO Training Institute	http://www.avotraining.com
Baldor Motors & Drives	http://www.baldor.com
R. W. Beckett Corporation (oil burner technical info.)	http://www.beckettcorp.com
Belden Wire & Cable Co.	http://www.belden.com
Bell, Hubbell Inc.	http://www.hubbell-bell.com
Bender	http://www.bender.org
Best Power	http://www.bestpower.com
BICC General	http://www.generalcable.com
BICSI (Building Industry Consulting Services International, a telecommunications assoc.)*	http://www.bicsi.org
BidStreet	http://www.BidStreetUSA.com
Black and Decker	http://www.blackanddecker.com
The Blue Book (construction)	http://www.thebluebook.com
Bodine Company	http://www.bodine.com
BOCA	http://www.bocai.org
boltswitch, inc.	http://www.boltswitch.com
Brady Worldwide, Inc.	http://www.whbrady.com
Bridgeport Fittings Inc.	http://www.bptfittings.com
Broan	http://www.broan.com
Bryant – Hubbell Inc.	http://www.hubbell-bryant.com
Building Industry Consulting Service International, Inc.	http://www.bisci.org
BuildPoint	http://www.buildpoint.com
Burndy Electrical Products	http://www.fciconnect.com
Bussmann, Cooper Industries	http://www.bussmann.com
Cable Telecommunications Engineers*	http://www.scte.org
Caddy	http://www.erico.com
Cadweld	http://www.erico.com

Canadian Standards Association	http://www.csa-international.org
Cantex	http://www.cantexinc.com
Carlon, A Lamson & Sessions Co.	http://www.carlon.com
Carol Cable	http://www.generalcable.com
Casablanca Fan Company	http://www.casablancafanco.com
Caterpillar	http://www.cat.com
Caterpillar	http://www.cat-engines.com
CEBus Industry Council*	http://www.cebus.org
CEE News magazine	http://www.ceenews.com
Center for Disease Control (CDC)	http://www.cdc.gov
Centralite*	http://www.centralite.com
CEPro*	http://www.ce-pro.com
Channellock Inc	http://www.channellock.com
ChannelPlus*	http://www.multiplextechnology.com/channelplus/
Chromalox	http://www.chromalox.com
Circon Systems Corporation*	http://www.circon.com
Clifford of Vermont*	http://www.cliffordvt.com
Coleman Cable	http://www.colemancable.com
Columbia Lighting	http://www.columbia-ltg.com
CommScope* (coax cable mfg.)	http://www.commscope.com
Communications Systems, Inc.	http://www.commsystems.com
Computer and Telecommunications Page*	http://www.cmpcmm.com/cc/
ConnectHome*	http://www.connecthome.com
Construction Specifications Institute	http://www.csinet.org
Construction Terms/Glossary	http://www.constructionplace.com/glossary.html
Consumer Electronic Association (CEA)*	http://www.ce.org
Consumer Product Safety Commission (CPSC)	http://www.cpsc.gov
Continental Automated Building Association (CABA)*	http://www.caba.org
Contrast Lighting (recessed)	http://www.contrastlighting.com
Copper (technical information)	http://energy.copper.org
Copper Development Association	http://www.copper.org
Cooper Industries	http://www.cooperindustries.com
Here you will find links to all Cooper Divisions for electrical products, lighting, wiring devices, power and hand tools.	
Craftsman Tools	http://www.craftsman.com
Crouse-Hinds	http://www.crouse-hinds.com
Cummins Power Generation*	http://www.cumminspower.com
Custom Electronic Design & Installation Association*	http://www.cedia.net
Custom Electronic Professional*	http://www.ce-pro.com
Cutler-Hammer	http://www.cutler-hammer.com
DABMAR	http://www.dabmar.com
Daman Tools	http://www.damantools.com
Danaher Power Solutions	http://www.danaherpoersolutions.com
Daniel Woodhead Company	http://www.danielwoodhead.com
Day-Brite Lighting	http://www.daybrite.com
Delmar Publishers	http://www.delmarlearning.com

Delmar (electrical)	http://www.delmarelectric.com
Department of Energy	http://www.energy.gov
Dimension Express (to obtain roughing-in dimensions from all appliance manufacturers)	http://www.dexpress.com
F. W. Dodge	http://www.fwdodge.com
DOMOSYS Corporation*	http://www.domosys.com
Douglas Lighting Control	http://www.douglaslightingcontrol.com
Dremel	http://www.dremel.com
Dual-Lite	http://www.dual-lite.com
Dynacom Corporation	http://www.dynacomcorp.com
Eaton (links to all divisions)	http://www.eatonelectrical.com
Edison Electric Institute	http://www.eei.org
EC Online	http://www.econline.com
EC & M magazine	http://www.ecmweb.com
Edwards Signaling and Security Products	http://www.edwards-signals.com
ELAN Home Systems*	http://www.elanhomesystems.com
Electric-Find	http://www.electric-find.com
Electric Smarts	http://www.electricsmarts.com
Electri-flex	http://www.electriflex.com
Electrical Apparatus Service Association	http://www.easa.com
Electrical Contracting & Engineering News	http://www.ecenmag.com
Electrical Contractor magazine	http://www.ecmag.com
The Electrical Contractor Network	http://www.electrical-contractor.net
Electrical Designers Reference	http://www.edreference.com
Electrical Distributor magazine	http://www.tedmag.com
Electrical Reliability Services, Inc. (formerly Electro-Test)	http://www.electro-test.com
Electrical Safety	http://www.electrical-safety.com
Electrician.com	http://www.electrician.com
Electro-Test Inc.	http://www.electro-test.com
Electronic Design Technology*	http://www.edt.biz
Electronic House*	http://www.electronichouse.com
Electronic Industry Alliance*	http://www.eia.org
Emerson	http://www.emersonfans.com
Encore Wire Limited	http://www.encorewire.com
Energy Efficiency & Renewable Energy Network (U.S. Dept. of Energy)	http://www.eere.energy.gov
Energy Star	http://www.energystar.gov
Environmental Protection Agency	http://www.epa.gov
Erico	http://www.erico.com
Essex Group, Inc.*	http://www.essexgroup.com
Estimation Inc.	http://www.estimation.com
ETL Semko	http://www.etlsemko.com
Exide Technologies	http://www.exide.com
Factory Mutual (FM Global)	http://www.fmglobal.com
Faraday	http://www.faradayfirealarms.com
FCI/Burndy (telecom connectors)	http://www.fciconnect.com
Federal Communications Commission*	http://www.fcc.gov
Federal Pacific	http://www.federalpacific.com
Federal Signal Corp.	http://www.federalsignal.com

Ferraz-Shawmut	http://www.ferrazshawmut.com
Fiber Optic Association, Inc.*	http://www.thefoa.org
Fibertray	http://www.ditel.net
First Alert	http://www.firstalert.com
Flex-Core	http://www.flex-core.com
Fluke Corporation	http://www.fluke.com
Fluke Networks*	http://www.flukenetworks.com
Fox Meter Inc.	http://www.arcfaulttester.com
Fox Meter Inc.	http://www.foxmeter.com
Gardner Bender	http://www.gardnerbender.com
GE Appliances	http://www.geappliances.com
GE Electrical Distribution & Control	http://www.ge.com/edc
GE Industrial Systems	http://www.geindustrial.com
GE Lighting	http://www.gelighting.com
GE Total Lighting Control	See Horton Control
General Cable	http://www.generalcable.com
Generators:	
Coleman Powermate	http://www.colemanpowermate.com
Cummins Power Generation	http://www.cumminspowergeneration.com
Generac Power Systems	http://www.generac.com
Gen-X	http://www.genxnow.com
Gillette Generators, Inc.	http://www.gillettegenerators.com
Kohler Power Systems	http://www.kohlergenerators.com
Onan Corporation	http://www.onan.com
Reliance Electric	http://www.reliance.com
Genesis Cable Systems*	http://www.genesiscable.com
Global Engineering Documents	http://www.global.ihs.com
Grayline, Inc. (tubing)	http://www.graylineinc.com
Greenlee Inc.	http://www.greenlee.com
Greyfox Systems*	http://www.greyfox.com
Guth Lighting	http://www.guth.com
The Halex Company	http://www.halexco.com
Hand Tools Institute	http://www.hti.org.
Handicapped Suggestions	http://www.abledata.com
The World of Assisted Technology	
Hatch Transformers, Inc	http://www.hatchtransformers.com
Heavy Duty	http://www.milwaukeetool.com
Heyco Products, Inc.	http://www.heyco.com
Hilti Corporation	http://www.hilti.com
Hinkley Lighting	http://www.hinkleylighting.com
HIOKI (measuring instruments)	http://www.hioki.co.jp/
Hitachi Cable Manchester, Inc.*	http://www.hcm.hitachi.com
Hoffmann	http://www.hoffmanonline.com
Holophane	http://www.holophane.com
Holt, Mike Enterprises	http://www.mikeholt.com
Home Automation and Networking Association*	http://www.homeautomation.org
Home Automation, Inc.*	http://www.homeauto.com
Home Automation Links*	http://www.asihome.com
Home Automation Systems*	http://www.smarthome.com
Home Controls, Inc.*	http://www.homecontrols.com

Home Director, Inc.*	http://www.homedirector.net
HomeStar*	http://www.homestar.net
Homestore*	http://www.homestore.com
Home Touch*	http://www.home-touch.com
Home Toys*	http://www.hometoys.com
Honeywell	http://www.honeywell.com
Honeywell*	http://www.honeywell.com/yourhome
Housing Urban Development	http://www.hud.gov.
Houston Wire & Cable Company	http://www.houwire.com
Hubbell Incorporated (links to all divisions)	http://www.hubbell.com
Hubbell Incorporated	http://www.hubbell-premise.com
Hubbell Lighting Incorporated	http://www.hubbell-ltg.com
Hubbell Wiring Devices	http://www.hubbell-wiring.com
Hunt Control Systems, Inc.	http://www.huntdimming.com
Hunter Fans	http://www.hunterfan.com
Husky	http://www.mphusky.com
ICBO	http://www.icbo.org
Ideal Industries, Inc.	http://www.idealindustries.com
Illuminating Engineering Society of North America	http://www.iesna.org
ILSCO	http://www.ilsco.com
Independent Electrical Contractors	http://www.ieci.org
Institute of Electrical and Electronic Engineers, Inc.	http://www.ieee.org
Instrumentation, Systems, and Automation Society	http://www.isa.org
Insulated tools (certified insulated products)	http://www.insulatedtools.com
Intel's Home Network	http://www.intel.com/anypoint
Intermatic. Inc.	http://www.intermatic.com
International Association of Electrical Inspectors (IAEI)	http://www.iaei.org
International Association of Plumbing & Mechanical Officials	http://www.iapmo.org
International Brotherhood of Electrical Workers	http://www.ibew.org
International Code Council (ICC)	http://www.iccsafe.org
International Dark-Sky Association	http://www.darksky.org
International Electrotechnical Commission (IEC)	http://www.iec.ch/
International Organization for Standardization (ISO)	http://www.iso.org
Intertec Testing Services (ETL)	http://www.etlsemko.com
Jacuzzi	http://www.jacuzzi.com
Jefferson Electric	http://www.jeffersonelectric.com
Johnson Controls, Inc.	http://www.johnsoncontrols.com
Joslyn	http://www.joslynmfg.com
Juno Lighting, Inc.	http://www.junolighting.com
KatoLight	http://www.katolight.com
Kenall Lighting	http://www.kenall.com
Kichler Lighting	http://www.kichler.com
Kidde (home safety)	http://www.kiddeus.com
King Safety Products	http://www.kingsafety.com
King Wire Inc.	http://www.kingwire.com
Klein Tools, Inc.	http://www.kleintools.com
Kohler Generators	http://www.kohler.com
Leviton Mfg. Co., Inc (links to all divisions)	http://www.leviton.com

Lew Electric	http://www.lewelectric.com
Liebert	http://www.liebert.com
Lighting	http://www.lightsearch.com
Lighting Design Forum	http://www.qualitylight.com
Lighting Research Center	http://www.lrc.rpi.edu
Lightning Protection Institute	http://www.lightning.org
Lightolier	http://www.lightolier.com
Lightolier Controls*	http://www.lolcontrols.com
Litetouch*	http://www.litetouch.com
Lithonia Lighting	http://www.lithonia.com
Littelfuse, Inc.	http://www.littelfuse.com
LonWorks *	http://www.lonworks.echelon.com
Lucent Technologies*	http://www.lucent.com
Lumisistemas, S.A. de C.V	http://www.lumisistemas.com
Lutron Electronics Co., Inc.	http://www.lutron.com
Lutron Home Theater*	http://www.ultimatehometheater.com
MagneTek (links to all divisions)	http://www.magnetek.com
Makita Tools	http://www.makita.com
Manhattan/CDT	http://www.manhattancdt.com
McGill	http://www.mcgillelectrical.com
Megger	http://www.megger.com/us/index.asp
MET Laboratories, Inc.	http://www.metlabs.com
MI Cable Company	http://www.micable.com
Midwest Electric Products, Inc.	http://www.midwestelectric.com
Milbank Manufacturing Co.	http://www.milbankmfg.com
Milwaukee Electric Tool	http://www.milwaukeetool.com
Minerallac	http://www.minerallac.com
Molex Inc.	http://www.molex.com
Motor Control	http://www.motorcontrol.com
Motorola Lighting Inc.	http://www.motorola.com
National Association of Electrical Distributors	http://www.naed.org
National Association of Home Builders	http://www.nahb.org
National Association of Manufacturers	http://www.nam.org
National Association of Radio and Telecommunications Engineers, Inc.	http://www.narte.com
National Conference of States on Building Codes and Standards	http://www.ncsbcs.org
National Electrical Contractors Association (NECA)	http://www.necanet.org
National Electrical Manufacturers Association (NEMA)	http://www.nema.org
National Electrical Safety Foundation	http://www.nesf.org
National Fire Protection Association	http://www.nfpa.org
National Fire Protection Association National Electrical Code® information	http://www.nfpa.org
National Institute for Occupational Safety & Health	http://www.cdc.gov/niosh
National Institute of Standards and Technology (NIST)	http://www.nist.gov
National Joint Apprenticeship and Training Committee (NJATC)	http://www.njatc.org
National Lighting Bureau	http://www.nlb.org

National Resource for Global Standards	http://www.nssn.org
National Safety Council	http://www.nsc.org
National Spa and Pool Institute	http://www.nspi.org
National Systems Contractors Association*	http://www.nsca.org
NECDirect	http://www.necdirect.org
Newton's Electrician	http://www.electrician.com
Nora Lighting	http://www.noralighting.com
NuTone	http://www.nutone.com
Occupation Health & Safety	http://www.ohsonline.com
Occupational Safety & Health Administration	http://www.osha.gov
Okonite Co.	http://www.okonite.com
Olflex Cable	http://www.olflex.com
Onan Corporation	http://www.onan.com
ONEAC Corporation	http://www.oneac.com
OnQ Technologies*	http://www.onqtech.com
Optical Cable Corporation*	http://www.occfiber.com
Ortronics Inc.	http://www.ortronics.com
Osram Sylvania Products, Inc.	http://www.sylvania.com
Overhead door information: Door & Access System Manufacturers Association	http://www.dasma.com
OZ /Gedney	http://www.o-zgedney.com
Panasonic Lighting	http://www.panasonic.com/lighting
Panasonic Technologies	http://www.panasonic.com
Panduit Corporation	http://www.panduit.com
Pass and Seymour, Legrand	http://www.passandseymour.com
Pelican Rope Works	http://www.pelicanrope.com
Penn-Union Corp.	http://www.penn-union.com
Phase-A-Matic, Inc.	http://www.phase-a-matic.com
Philips	http://www.philips.com
Philips Lamps	http://www.ALTOlamp.com
Philips Lighting Company	http://www.philipslighting.com
Philtek Power Corporation	http://www.philtek.com
Phoenix Contact	http://www.phoenixcon.com
Pigeon Hole, The	http://www.thepigeonhole.com
Plant Engineering On-Line	http://www.plantengineering.com
Popular Home Automation*	http://www.pophome.com
Porter Cable	http://www.porter-cable.com
Power Quality Monitoring	http://www.pqmonitoring.com
Power-Sonic Corporation	http://www.power-sonic.com
Premise Wiring Products	http://www.unicomlink.com
Prescolite	http://www.prescolite.com
Pringle Electrical Mfg. Co.	http://www.pringle-elec.com
Progress Lighting	http://www.progresslighting.com
Radix Wire Company	http://www.radix-wire.com
RCA*	http://www.rca.com
RACO	http://www.racoinc.com
Rayovac Corporation	http://www.rayovac.com
Refrigeration Service Engineering Society	http://www.rses.org
Regal Fittings	http://www.regalfittings.com

Reliable Power Meters	http://www.reliablemeters.com
Remke Industries, Inc.	http://www.remke.com
Rhodes, M. H. (timers)	http://www.mhrhodes.com
Rigid Tool Company	http://www.ridgid.com
Robertson Transformer Co.	http://www.robertsontransformer.com
Robicon	http://www.robicon.com
Robroy Industries – Conduit Div.	http://www.robroy.com
Romex Cable	http://www.biccgeneral.com
RotoZip Tool Corp.	http://www.rotozip.com
Russelectric	http://www.russelectric.com
Safety Technology International, Inc.	http://www.sti-usa.com
S&C Electric Company	http://www.sandc.com
SBCCI	http://www.sbcci.org
Sea Gull Lighting	http://www.seagulllighting.com
Seatek, Inc.	http://www.SeatekCo.com
SECURITY*	
Caddx*	http://www.caddx.com
Digital Monitoring Products*	http://www.dmpnet.com
Home Automation*	http://www.homeauto.com
Honeywell Security*	http://www.security.honeywell.com/sce
NAPCO Security Systems*	http://www.napcosecurity.com
Siemens (main site)	http://www.siemens.com
Siemens Energy & Automation	http://www.sea.siemens.com
The Siemon Company*	http://www.siemon.com
Sensor Switch Inc.	http://www.sensorswitchinc.com
Shat-R-Shield	http://www.shat-r-shield.com
Silent Knight	http://www.silentknight.com
Simplex Grinnell	http://www.simplexgrinnell.com
Simpson Electric Co.	http://www.simpsonelectric.com
Skil Tools	http://www.skil.com
Smart Home*	http://www.smarthome.com
Smart Home Pro*	http://www.smarthomepro.com
Smart House, Inc.*	http://www.smart-house.com
Smoke & Fire Extinguisher Signs	http://www.smokesign.com
Sola/Hevi Duty	http://www.sola-hevi-duty.com
Southwire Co.	http://www.mysouthwire.com
SP Products, Inc.	http://www.spproducts.com
A.W. Sperry Instruments, Inc.	http://www.awsperry.com
Stanley Tools	http://www.stanleyworks.com
Starfield Controls*	http://www.starfieldcontrols.com
Steel Tube Institute	http://www.steeltubeinstitute.org
Straight Wire*	http://www.straightwire.com
Square D, Groupe Schneider	http://www.squared.com
Square D (Homeline)	http://www.squared.com/retail
Square D, Power Logic	http://www.powerlogic.com
Sunbelt Transformers	http://www.sunbeltusa.com
Superior Electric	http://www.superiorelectric.com
Superior Essex Group	http://www.superioressex.com
Suttle	http://www.suttleonline.com

Sylvania	http://www.sylvania.com
Sylvania Lighting International	http://www.sli-lighting.com
Technical Consumer Products Inc.	http://www.springlamp.com
Telecommunications Industry Association (TIA)*	http://www.tiaonline.org
Terms /Construction /Glossary	http://www.constructionplace.com/glossary.html
Thermostats, Communicating*	http://www.homeauto.com
This Old House	http://www.thisoldhouse.org
Thomas & Betts Corporation	http://www.tnb.com
Thomas Lighting	http://www.thomaslighting.com
Thomas Register	http://www.thomasregister.com
3M Electrical Products	http://www.3m.com/elpd
Tork	http://www.tork.com
Touch-Plate Lighting Controls	http://www.touchplate.com
Trade Service Corporation	http://www.tradeservice.com
Trade Service Systems	http://www.tradepower.com
Triplett Corporation	http://www.triplett.com
TYCO (links to all divisions)	http://www.tyco.com
Tyton Hellermann	http://www.tytonhellermann.com
UEI Test Equipment	http://www.ueitest.com
Underwriters Laboratories	http://www.ul.com
UL Standards Department	http://ulstandardsinfonet.ul.com
Unistrut (Tyco)	http://www.unistrut.com
Unity Manufacturing	http://www.unitymfg.com
Universal Lighting Technologies	http://www.universalballast.com
USA Wire and Cable, Inc.	http://www.usawire-cable.com
U.S. Department of Energy (DOE)	http://www.doe.gov
U.S. Department of Labor	http://www.dol.gov
U.S. Department of Labor, Employment & Training Administration	http://www.doleta.gov
U.S. Environment Protection Agency (EPA)	http://www.epa.gov
Ustec* (structured wiring)	http://www.ustecnet.com
Vantage Automation & Lighting Control*	http://www.vantagecontrols.com
Watt Stopper	http://www.wattstopper.com
Waukesha Electric Systems	http://www.waukeshaelectric.com
Werner Ladder Company	http://www.wernerladder.com
Westinghouse Lighting	http://www.westinghouselighting.com
Wheatland Tube Company	http://www.wheatland.com
Wireless Industry*	http://www.wireless.about.com
Wireless Thermostats*	http://www.wirelessthermostats.com
Wiremold	http://www.wiremold.com
Woodhead Industries	http://www.danielwoodhead.com
X-10*	http://www.x10.com
X-10 Pro Automation Products*	http://www.x10pro.com
Zenith Controls, Inc.*	http://www.geindustrialsystems.com
Zircon	http://www.zircon.com

CODE REFERENCE INDEX

Note: Page numbers in **bold type** reference tables and figures.

INDEX

Note: Page numbers in **bold text** reference non-text material, such as figures and tables.